Wireless Personal Communications Systems

Wireless Personal Communications Systems

David J. Goodman

ADDISON-WESLEY

An Imprint of Addison Wesley Longman, Inc.

Reading, Massachusetts • Harlow, England • Menlo Park, California
Berkeley, California • Don Mills, Ontario • Sydney
Bonn • Amsterdam • Tokyo • Mexico City

Many of the designations used by manufacturers and sellers to distinguish their products are claimed as trademarks. Where those designations appear in this book and Addison Wesley Longman, Inc., was aware of a trademark claim, the designations have been printed with initial capital letters.

The author and publisher have taken care in preparation of this book, but make no expressed or implied warranty of any kind and assume no responsibility for errors or omissions. No liability is assumed for incidental or consequential damages in connection with or arising out of the use of the information or programs contained herein.

The publisher offers discounts on this book when ordered in quantity for special sales. For more information, please contact:

Corporate & Professional Publishing Group
Addison Wesley Longman, Inc.
One Jacob Way
Reading, Massachusetts 01867

Library of Congress Cataloging-in-Publication Data

Goodman, David J., 1939–
     Wireless personal communications systems / David J. Goodman.
         p.   cm. — (Addison-Wesley wireless communications series)
     Includes index.
     ISBN 0-201-63470-8
     1. Personal communication service systems.   I. Title.
     II. Series.
     TK5103.485.G66   1997
     621.3845—dc21   97–23431
                                                                CIP

Text printed on recycled and acid-free paper.

ISBN 0-201-63470-8
 1 2 3 4 5 6 7 8 9—MA—0100999897
First printing, August 1997

Readers are encouraged to contact David Goodman with questions or comments at http://winwww.rutgers.edu/wpcs

I recently read the introduction to **Mrs. Beeton's Book of Cookery and Household Management** *(published in 1861 by S.O. Beeton) in which the author states, "I must frankly own that had I known, beforehand, that this book would have cost me the labours which it has, I should never have been courageous enough to commence it." I can't quite bring myself to say that about this book. However, the people closest to me would say that, if they had known what this effort would do to my personality, they would have persuaded me not to begin. Thank you Lizzie, Mom, Leila, Lissy, Laura, Meghan, Jackie, Michael for your love, your patience, and your gentle, amused indulgence.*

CONTENTS

2   PRINCIPLES OF PERSONAL COMMUNICATIONS

LIST OF FIGURES

Chapter Six

Chapter Seven

Chapter Eight

Chapter Nine

LIST OF TABLES

PREFACE

I have written this book to introduce professionals and students with diverse backgrounds to the technology of wireless personal communications systems. I expect that many readers will be new to the subject and that others, who are familiar with specific systems or technologies, will read the book in order to broaden their knowledge. Although the book describes complex information systems, I have written it in simple prose to help people quickly get the information they need. This approach distinguishes this book from two large bodies of literature on personal communications systems: textbooks and papers that rely on mathematical exposition, and standards documents that provide all the information necessary to build communications equipment.

A major aim of the book is to provide readers with insight into "the big picture" of how personal communications systems work. Fortified with this insight, readers will be in good shape to absorb the detailed knowledge they need to meet their objectives. The book begins with descriptions of two telephone calls—one using conventional wired telephones and the other using two cellular phones. Observing the considerably higher complexity of the cellular call, readers will begin to understand the challenges addressed by personal communications technology. The details of the nine systems presented in this book comprise a variety of answers to these challenges.

The existence of so many competing techniques for addressing the same set of challenges is an intriguing aspect of personal communications. Almost all other areas of information technology converge quickly into one or two standards. Videotape formats and personal computer operating systems are two examples. By contrast, this book documents a large and growing number of standards for personal communications. Chapter 2 offers one explanation for this diversity in the form of a long list of criteria for judging the merits of personal communications systems. Many of these criteria are inconsistent in the sense that doing a good job with respect to some of them inevitably compromises others. Each system reflects a different set of priorities adopted by system designers faced with a complex list of inconsistent design goals.

Chapter 2 describes—in general terms—the operations performed by all personal communications systems. Then it presents a framework, used in the remainder of the book, for studying specific systems. This framework helps readers understand how individual technologies, such as radio modulation and database management, contribute to system operation. It also encourages readers to compare and contrast the techniques adopted in specific systems. The emphasis throughout the book is on a systems approach, which is particularly important in personal communications. More than in other areas of information technology, it is essential, when developing or adjusting one part of a system, to understand how that part relates to the rest of the system. In personal communications, doing a good job on one component of a system is no guarantee that the overall effect will be positive.

Chapters 3 through 8 use the framework established in Chapter 2 to study nine systems that deliver Personal Communications Services to a rapidly growing population of more than 100 million subscribers in all parts of the world. Although the technology of personal communications is in a state of flux, many of the new developments will be absorbed into the systems presented in this book. As a consequence these systems will, for the foreseeable future, serve the majority of the world's personal communications subscribers. The material presented in this book covers the fundamentals of system operation. I hope it will endure as a valuable reference to professionals performing a broad range of tasks.

In writing a book to introduce personal communications to these professionals, I have anticipated a readership with widely diverse backgrounds. To ease readers into the field, Chapter 9 contains twelve brief tutorials, each on a subject that plays an important role in personal communications. Each tutorial is a glimpse of a deep technical subject, addressed in its own specialized books and courses. Many readers will already be acquainted with some of the tutorial topics but will require an introduction to others to understand how they influence practical systems. Chapter 9 will make it possible for readers unfamiliar with specific technical subjects to read this book without making detours to the library to fill in the gaps.

By placing the tutorials at the end of the book rather than at the beginning, I have broken ranks with traditional technical education and with other books on personal communications. The more common approach is to start with individual pieces and eventually to explain how they come together in a system. This book begins with a cellular telephone call. Although an everyday event for tens of millions of people, each phone call coordinates a large of array of technologies. Knowledge of how each specialty makes the phone call possible will help people gain the insights they seek when they plunge into the details.

I have been teaching wireless communications for several years to senior undergraduates and first-year graduate students in three departments at Rutgers University: Electrical and Computer Engineering, Computer Science, and Communications. In a one-semester course, I weave the systems material in Chapters 1 through 8 with the technical material of Chapter 9. Homework and exams include simple computational exercises that reveal relationships among important phenomena and questions that call on students to exercise their judgment and analytic skills. For example, the first homework question is, "Why does a cellular telephone have a SEND button, while none is necessary on a conventional phone?" In responding to this and other questions, my students often give me new insights, both into the subject and into their approach to their work.

To stimulate readers of this book to think critically about the material presented, I conclude each chapter with ten review exercises. Some of them can help readers determine whether they have grasped key points of a chapter. Others will help them form their own judgments about the systems described.

As an early academic enthusiast of wireless personal communications, I have the pleasure of seeing many former students contributing in a variety of ways to this dynamic area of information technology. I hope that when you read this book you will sense the exciting potential of personal communications and join us in inventing its future.

ACKNOWLEDGMENTS

When I thought about acknowledging the help I received writing this book, I was overwhelmed, and humbled, by the size of my task. I began working in this field in 1978, and I have taught courses in it since 1989. I owe my basic knowledge and point of view to extraordinary professors at Imperial College in the 1960s and visionary mentors at Bell Laboratories in the 1970s. My work on this book has been enriched by research collaborators, academic colleagues, reviewers, and hundreds of students.

Although it is impractical to acknowledge everyone by name, I would like to thank the following reviewers for their advice to me and to Addison Wesley Longman: John Burruss, Ken Davidson, Patrick Doyle, Michael Garyantes, Mario Gerla, Mohammad Hossain, Bijan Jabbari, Charles Jackson, Kari Kalliojarvi, Richard Kane, Randy Katz, Allen Levesque, Scott Marcus, Craig Matthias, James Norris, Raymond Pickholtz, Riku Pirhonen, John Radpour, Theodore Rappaport, Fred Seelig, Matthew Taylor, Alan Triggs, Tonis Vaga, Rajiv Vijayan, Andrew Viterbi, Branimir Vojcic, T. A. Wilkinson, Zong Wu, and Valerie Zelenty.

I also had help from the following (present and former) WINLAB people: Elizabeth Cleerdin, Dick Frenkiel, Sudheer Grandhi, Jack Holtzman, Sarath Kumar, Sanjiv Nanda, Andy Ogielski, Greg Pollini, and Roy Yates.

Philomena Genatempo, my secretary, not only contributed a lot to this book, but also kept the other parts of my job going while I was absorbed in writing it. Noreen DeCarlo, the word-processing whiz, was invaluable in getting us out of a few jams with the manuscript.

CHAPTER ONE

What Is Personal Communications?

1.1 The Person in Personal Communications

Personal communications begins with a person. In the late 1990s, personal communications systems represent a major step forward in the dramatic story that began ten years earlier with the arrival of cellular and cordless telephones. When people spend billions of dollars to buy and use these new kinds of telephones, they demonstrate a strong desire to control their information. Mobility is at the heart of personal communications. People transmit and receive information wherever they are and whenever they choose, even when they are moving. They want to produce and acquire information in formats they choose—including sounds, text, still pictures, moving pictures, keyboard operations, mouse movements, and pen strokes. The promise of personal communications is to make all kinds of information available anywhere, anytime, at low cost to a large mobile population.

Although this promise has widespread appeal, practical systems that deliver it remain several years in the future. At the moment we have a collection of services and products, each capable of performing some of the functions of a complete personal communications system. We require new technology to merge today's separate systems into the integrated information delivery system of the twenty-first century. This book describes the late-1990s communications systems that are the stepping-stones on the route to personal communications. The subject of this book is the technology of personal communications. The book takes a systems view. It identifies the essential functions that are performed by personal communications systems, and examines several specific systems to find out how each one performs these functions. In contrast to other areas of information technology, present-day personal communications systems conform to a wide range of competing standards. People working or studying in this field will find it useful to explore the reasons why there are many different technical solutions to the same problems and to identify the relative strengths and weaknesses of the alternatives. This book presents a framework for addressing these issues. The first two chapters take a general view of the subject. The subsequent chapters describe specific communications systems embodied in published standards. By using the same general framework to examine each specific system, readers will be in a strong position to recognize similarities and differences and to identify the merits and weaknesses of each system.

The present chapter proceeds, in Section 1.2, with a definition of a personal communications system and describes the user's perception of a personal communications service. The popular acronym for personal communications is *PCS*. Section 1.3 offers six possibilities for the last word of this abbreviation. Together, the six *S* words encompass the person in personal communications as well as the equipment, networks, government regulations, and industry arrangements that make personal communications possible. To introduce the technical challenges of personal communications, Section 1.4 describes two phone calls that take place in the same office building. The first one is a conventional call between two wired phones. The second call connects two cellular phones. It requires a large number of system operations that are absent in the conventional call. These operations respond to the inherent technical challenges of personal communications. Section 1.5 introduces these challenges and indicates some of the technologies devised to overcome them. Section 1.6 describes the evolution of personal communications systems since their origins around 1980. Finally, Section 1.7 briefly introduces the systems presented in the remainder of this book, and Section 1.8 describes four other categories of wireless communications systems.

1.2 Essential Ingredients

With mobility at the heart of personal communications, wireless information transfer is essential. Therefore, we adopt the following definition:

> *A personal communications system provides people with wireless access to information services.*

This is a deliberately broad definition that applies to a wide variety of existing and future systems, including residential cordless telephones, cellular networks, mobile data networks, and certain land mobile radio systems. Of course, it also includes Personal Communications Services and Personal Communications Networks, which are the official designations of systems operating in specific geographical areas and frequency bands, established by government regulators. In fact, the main distinguishing characteristic of the official PCSs and PCNs is their treatment by authorities. In terms of technologies and services offered, many of them are virtually identical to cellular telephone systems. An important, exciting aspect of personal communications systems is their dynamic nature. Most are growing rapidly in use and are also evolving to incorporate new technology in order to meet the rising expectations of their users. The contents of this chapter and the next apply to personal communications in general. In subsequent chapters, we encounter a variety of practical systems with various official designations including cellular, cordless, PCS, and PCN.

Figure 1.1 lists, from the point of view of the human user, the key attributes of advanced personal communications. Although they all appear in some form in the practical systems of 1997, new technology will be needed to assemble them in the unified personal communications systems of the next century.

Personal information machine (PIM) is the name I give to the information device carried by the person. Like a telephone, a PIM will have a microphone, an earphone, and a keypad. It will also have a display screen. It will be comfortable to carry. PIMs will come in a variety of forms to accommodate the diverse needs and budgets of their owners. Some will be wearable, like watches. Others will be pocket-sized, while the most elaborate ones will incorporate powerful laptop computers.

A *personal address* replaces a person's conventional telephone numbers. Each conventional telephone number is associated with a specific location, such as the person's residence, office, or vehicle. A personal address, by contrast, remains with the person as he or she changes location.

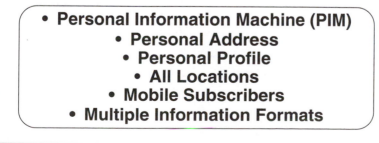

- **Personal Information Machine (PIM)**
- **Personal Address**
- **Personal Profile**
- **All Locations**
- **Mobile Subscribers**
- **Multiple Information Formats**

Figure 1.1 Characteristics of personal communications.

To control his or her personal information services, each subscriber will have a *personal profile*. One component of the personal profile is a directory with the names and personal addresses of people frequently called. The personal profile also contains details of services selected by the subscriber, which may include calling party identification, voice mail, and selective call forwarding. Advanced profiles will automatically examine arriving information and process it according to the subscriber's preferences. In the case of telephone calls, for example, the profile will tell the system to direct some calls to PIM, others to voice mail, and forward others to another individual. These directions will depend on such factors as the source of the phone call, the time of day, and the subscriber's location.

Personal communications systems will use the address and profile of each subscriber to deliver information where and when the subscriber chooses. Personal information services will be available in *all locations*. Subscribers will maintain communications as they *change location*. Personal information services will accept and deliver information in the *formats* selected by the people who use them.

This is the vision of personal communications. The present reality is more modest. We have a collection of products and services that have expanded rapidly in popularity since their origins around 1980. They include cellular telephones, cordless telephones, pagers, and portable computers. In the late 1990s, Personal Communications Services and Personal Communications Networks are making their appearance. They are continuations of the trends set in motion by the products and services introduced in the 1980s. Personal communications, in the 1990s, begins to unite the separate products of the 1980s into a general-purpose PIM.

1.3 The S in PCS

In the United States, the shorthand term for personal communications is *PCS* [FCC, 1990b]. This terminology was adopted by the Federal Communications Commission in 1990 to refer to Personal Communications Services. The nature of these services remains to be determined by industry and by consumers. For the moment, PCS means different things to different people. Figure 1.2 lists some of these meanings.[1] Each one plays an important role in delivering to people the advantages of personal

---

[1] Mr. Jacques Bursztejn of Alcatel introduced me to this approach to describing aspects of personal communications.

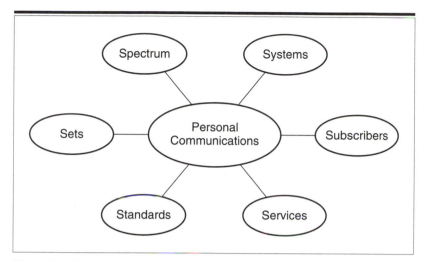

Figure 1.2 Dimensions of personal communications.

communications. Together the six dimensions of personal communications generate a high volume of activity in the technical, business, and government communities.

Services and *applications* go hand in hand. The two words are often used as synonyms. Strictly speaking, services are the technical means of delivering applications to people. For example, applications of voice services include telephone conversations, emergency calls, access to recorded announcements, and voice mail. Data services deliver electronic mail, access to the World Wide Web, and a wide range of specialized business and consumer applications. Cellular telephones primarily offer voice services. Emerging PCS networks feature an expanding array of non-voice services. Many of them reflect the ability of personal communications technologies to bring to a mobile population a wide range of existing applications. We can also expect, as the technology matures, to see the emergence of new applications tailored to the needs of people on the move.

Sets are personal communications consumer products, the PIMs carried by people. I expect PIMs in the twenty-first century to take the place of today's telephones, which are the result of a hundred years of gradual evolution. Today's telephone, like its earliest ancestor, has one medium, sounds, for delivering information to its users. Users introduce information in the form of sounds and keystrokes or dial movements. Basic PIMs will add text and graphics presented on a visual display. Other versions will be capable of accepting pen inputs, and in some cases sending and receiving moving pictures. No doubt there will be many types of information terminals corresponding to the diverse needs, budgets, and preferences of a mass population of personal communications subscribers. In contrast with most of today's telephones, each PIM will be portable and configured to meet the needs of its owner.

With no tethers to the remainder of the communications system, each PIM will communicate by means of unguided radio or infrared radiation. The signals emitted by personal communications equipment have to conform to government *spectrum* management regulations. These regulations have a strong influence on the nature of personal communications services and their costs.

Standards are essential to telecommunications. Information transfer requires compatibility of transmitters and receivers. Beyond this basic requirement, the utility and economy of communications products expand rapidly with the number of compatible products. Thus there is a strong commercial incentive for widespread adherence to a small number of standards. Compact discs provide an outstanding example. One reason for their rapid adoption by the public is the fact that, from the beginning, there has been only one standard for recording and playback. All of the discs in the world are compatible with all players. This has not been the case with personal communications products. From the earliest days of cellular communications, there have been a large number of incompatible standards. This situation persists and there is little hope of convergence before the middle of the first decade of the twenty-first century. This is due to the instability of personal communications technology. The technology is new and advancing rapidly in the face of a high consumer demand for its products. This differs from the history of other information industries. Usually science, technology, and standards are in place before consumers discover applications. Personal computing is an example.

Systems are the principal subject of this book. A communications system is a collection of hardware and software assembled to perform certain tasks. Some of the thorniest problems in information technology are related to "systems integration." This is especially true of personal communications systems, in which the components of a system all influence one another. Readers of this book will acquire an understanding of these interactions, and will learn that high-quality components are necessary but not sufficient for successful systems. They will understand how the components need to interact in order to achieve high-quality communications.

Subscribers take us back to the basic purpose of personal communications: to deliver information services to people in a natural, convenient manner. This implies that people will have more control over their information than they are now accustomed to. When personal communications services mature, they will replace telephony as the dominant mode of interactive communications for the majority of the world's population.

Only in the next century will personal communications realize its full promise. To make this happen, we need technology that is not available today. One purpose of this book is to explain the technological challenges raised by personal communications and to describe the systems already in place that address these challenges. Questions for students and teachers at the beginning of a study of personal communications technology include: "What is special to this subject?" and "What knowledge is necessary to understand personal communications beyond acquaintance with established technologies such as telephony and computer communications?" To address these questions, it is useful to examine two telephone calls. One is a call from a conventional telephone in an office building to another one in the same building. The other call, also confined to a single building, uses two cellular phones.

1.4 Two Telephone Calls

Helen phones Mike, her colleague in the same office building. She lifts the handset of the telephone on her desk, hears a dial tone, and presses four buttons. The telephone on Mike's desk rings. Helen hears a ringing sound in her phone. Mike lifts the phone and says "Hello, this is Mike." Helen responds, and they speak to each other.

Susan, in the office next to Mike's, calls Bill, also in the same building. Susan uses her portable cellular phone. She presses some buttons on the phone until she sees Bill's name on the phone's screen. Then she presses a green SEND button and hears a ringing tone. Bill's cellular phone rings. He takes it from his pocket, presses a button on the phone, and says "Hello, this is Bill." Susan responds and they speak to each other.

Observing Helen, Mike, Susan, and Bill, we see that although the two phone calls differ in a few details, they are essentially similar. If, on the other hand, we examine the equipment set into operation by the two calls, we find profound differences. Helen's call to Mike uses technology that has progressed steadily for more than a hundred years. The technology that makes the cellular call possible is less than 20 years old and is changing rapidly. This new technology responds to the fundamental challenges of wireless personal communications. Readers of this book will gain an understanding of these challenges and the communications systems deployed to overcome them. We begin by describing, in general terms, how the two phone calls work.

1.4.1 Conventional Call

By lifting the handset on her telephone, Helen connects the two wires that go from the telephone to a switching machine. In telephone jargon we say that Helen takes the phone "off hook." This causes current to flow at the switching machine. The switch reacts by placing a dial tone on the wires connected to Helen's telephone. When Helen hears the dial tone and keys in Mike's telephone number, the switch interprets the beep transmitted each time that Helen presses a key on the telephone. Information stored in the switching machine's computer associates Helen's keystrokes with the pair of wires connected to Mike's phone. The switch makes Mike's phone ring by placing an electrical "alert" signal on these wires. It also places an audible ringing tone on Helen's pair of wires. When the switch learns that Mike has lifted his handset, it discontinues the ringing signals and connects Mike's pair of wires to Helen's.

We illustrate this call with two diagrams. Figure 1.3 displays three system elements: two telephones and a switch. It shows the interfaces between the telephones and the switch. It also shows the connection between the switch and the Public Switched Telephone Network (PSTN), which links the world's telephones. Figure 1.4 shows a sequence of communications moving between the switching machine and the two telephones. In this diagram, time advances from top to bottom. We will encounter diagrams similar to Figures 1.3 and 1.4 throughout this book. Figure 1.3 is an example of a network architecture. It indicates system elements and possible connections between the elements. We can examine network architectures at different levels of detail. For example, we

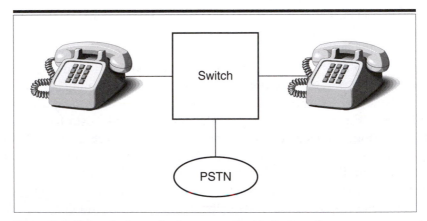

Figure 1.3 Network elements of a conventional phone call.

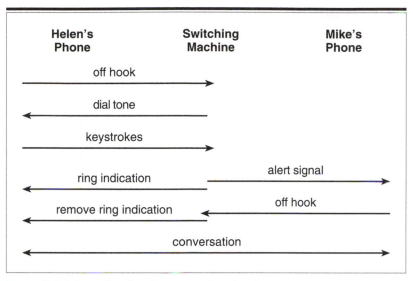

Figure 1.4 Information flow for conventional call.

could indicate, as separate system elements, components of the telephones or components of the switching machine and show the connections between these components.

Figure 1.4 is an information flow diagram. It shows the information that has to move among system elements in order to accomplish a specific communications task.

After we describe Susan's cellular phone call we will examine the reasons why wireless communications systems require technology not present in conventional systems. To anticipate this analysis, note two properties of the communication system in Figures 1.3 and 1.4:

- A pair of wires connects each telephone to the switching machine.

- Each pair of wires has its own telephone number (network address).

The network address of a telephone belongs to the pair of wires. When a person connects the telephone to another pair of wires, the network address changes.

Neither of these properties exists in the cellular system. The signals going to and from a cellular telephone travel through the air. The telephone retains its network address regardless of location.

1.4.2 Cellular Call

Channels, Cells

Now consider Susan's call to Bill. There are no wires connecting her cellular phone to the rest of the communications system. All of the information flowing to and from her telephone travels through the air. This communication takes place in *radio channels*. Like a television set, Susan's telephone can tune to any of a large number of channels. In contrast to the television set, which only receives information, a cellular telephone also transmits radio signals. These signals arrive at a *base station*. The base station is a stationary collection of equipment that communicates, by radio, with nearby cellular phones. The geographical area covered by a base station is a *cell*. A *cell site* is a synonym for base station. A cellular system consists typically of thousands of cells, each with its own base station. Together these cells cover a large region, such as a metropolitan area or an entire country.

The channels available to a cellular phone fall in two categories—*control channels* and *traffic channels*. Typically, control channels carry information, such as telephone numbers, that is necessary to coordinate system operation. Traffic channels carry user information, such as speech or facsimile signals. Conventional telephones do not use separate communications links for user information and system control information. The same pair of wires carries both types of information in the same frequency band. That is why we hear the signaling tones as we key in a telephone number. This is called *in-band signaling*. Cellular phones rely to a large extent on *out-of-band signaling*. With out-of-band signaling, network control information and user information travel separately.

Initialization

The story of Susan's call begins when she turns on her cellular phone. Lacking a pair of wires connected to the rest of the telephone system, Susan's telephone has to establish contact with a base station. It does so by scanning several control channels. Each control channel carries signals from one base station in the vicinity of the telephone. Different base stations in a neighborhood use different channels. The phone measures the strength of each signal it scans, and tunes to the channel with the strongest signal. This signal usually comes from the base station nearest to the telephone. When Susan's phone tunes to one channel, it decodes the signal in this channel and thereby acquires several items of information about the cell it occupies. One example of this information is how much

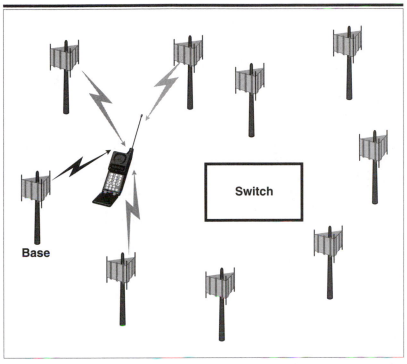

Figure 1.5 Susan's telephone scans a group of control channels and tunes to the strongest signal.

power Susan's phone should emit when it has to send information to the base station. Figure 1.5 illustrates this initialization process, which culminates in Susan's telephone tuning to the signal from the closest base station.

Service Request

When Susan brings Bill's name to the screen of her telephone and presses the SEND button, the telephone transmits a service request message to the base station. This message includes Bill's number, just as Helen had to transmit Mike's phone number to the switching machine when she made her call. In the case of Susan's cellular call, more information is needed. The service request message identifies Susan as the calling party. (This is unnecessary in the fixed call, because the switching machine computer associates Helen with the pair of wires carrying her service request. In our example, Helen's service request is the set of keystrokes for Mike's phone number.) The base station, on receiving Susan's request, relays it to

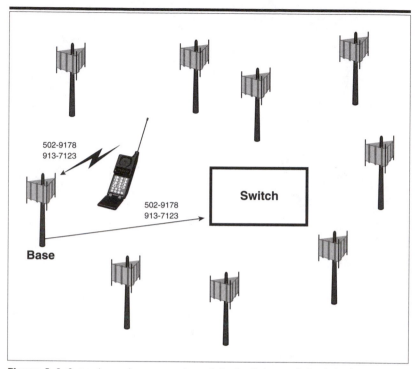

Figure 1.6 Susan's service request contains both her cellular telephone number and Bill's. The base station relays this request to the switch.

a *switch*. This is a telephone switching machine that has important responsibilities for all of the cellular communications in dozens of cells. To discharge its responsibilities, the switch communicates (by means of optical fibers, point-to-point microwave links, or telephone wires) with all of the base stations it controls. Figure 1.6 shows Susan's service request arriving at the switch.

Paging

The switch can establish connections to telephones throughout the world. It also manages the cellular phone service of a large number of subscribers. Each subscriber has a *home system* responsible for many details of his or her cellular service, including billing and individual service options. Voice mail and call forwarding are examples of popular service options. The switch in Figure 1.6 is part of the home system for both Susan and Bill. On recognizing Bill's phone number as one of its own, the switch attempts to establish communications with Bill. To do so, it sends identi-

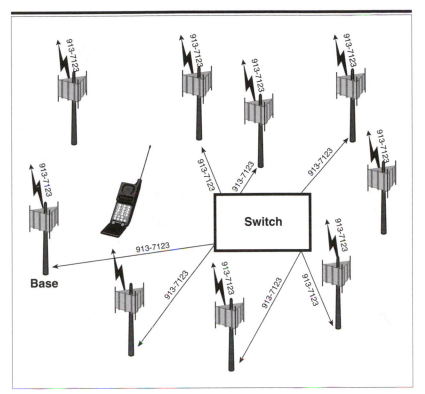

Figure 1.7 The switch, after receiving Susan's service request, commands several base stations to transmit paging messages containing Bill's number.

cal paging messages, containing Bill's phone number, to all of the base stations it controls. Each base station broadcasts Bill's phone number in a control channel. Figure 1.7 illustrates this paging process.

In our example, Bill is in the region served by the switch and he has turned on his cellular phone. Following an initialization procedure similar to Susan's shown in Figure 1.5, Bill's phone tunes to a control channel. Because Bill is in a different part of the building than Susan, Bill's control channel signal comes from a different base station than Susan's. Along with the other base stations connected to the switch, Bill's base station broadcasts his phone number in a paging message. When Bill's telephone detects its own phone number in the signal arriving on the control channel, it sends a response message to its current base station, which relays the message to the switch, as indicated in Figure 1.8. When the switch

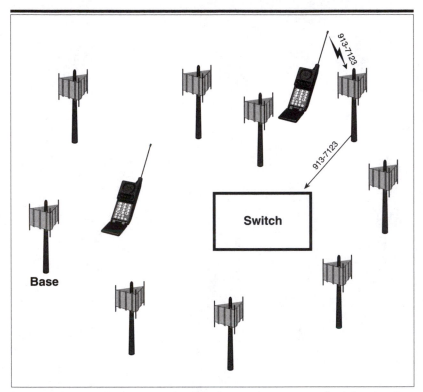

Figure 1.8 Bill's telephone responds to the paging message by informing the system of its location.

finds out which base station is in contact with Bill's phone, it can set up the call through that base station.

Channel Assignment

The next job of the switch is to assign a traffic channel to Bill's phone and another traffic channel to Susan's. These traffic channels will be dedicated to Susan and Bill's conversation. This is in contrast to the control channels used in Figures 1.5 through 1.8, which are shared by many telephones near their respective base stations. Figure 1.9 shows channel assignment messages relayed from the switch to Bill's and Susan's telephones by their base stations. When the switch receives confirmation that Bill's telephone has tuned to the assigned traffic channel, it sends an alert message to Bill's telephone. This message causes the phone to beep. When Bill responds by pushing a button on his phone, the beeping stops and the switch connects the traffic channel assigned to Bill to the traffic channel assigned to Susan, and their conversation begins. Figure 1.10 shows the connections in place at the beginning of the conversation.

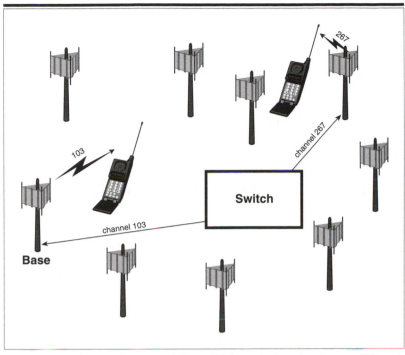

Figure 1.9 The switch commands Susan's phone to tune to channel 103 and Bill's phone to tune to channel 267.

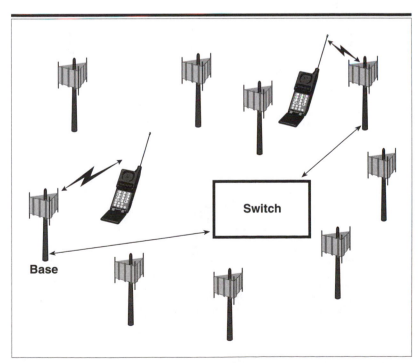

Figure 1.10 The cellular phone conversation is in progress with both voices following the paths shown.

Handoff

As complex as it is already, the story of the cellular phone call is not yet complete. We have to take into account the possibility of Susan and/or Bill changing location while the call is in progress. Imagine Bill on his way to lunch with Mike when the call begins. He carries the phone with him to Mike's car and continues the conversation with Susan as they drive to a restaurant. In this event, Bill is likely to move out of range of the base station he used initially.

When he moves out of range, the switch receives a report from the original base indicating that Bill's radio signal is becoming weaker. The switch then sends messages to several surrounding base stations asking for reports on the strength of the signal from Bill's phone. A base station near the restaurant reports a high signal strength. The switch then initiates a *handoff* procedure that transfers the call from the original base, near Bill's office, to the new base, near the restaurant. This will typically require Bill's phone to tune to a new traffic channel. All of this takes place within a matter of seconds. Susan and Bill are not aware of its happening as they carry on their conversation. Figure 1.11 shows the system configuration after the handoff takes place.

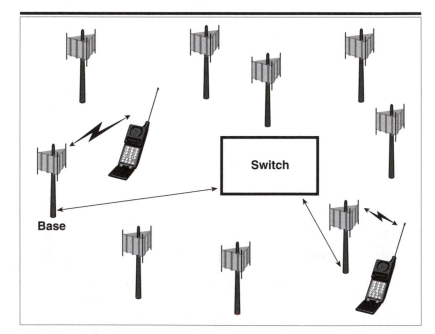

Figure 1.11 During the conversation, Bill moves to a new cell. The system rearranges itself to maintain the conversation.

Architecture and Information Flow

One reason for describing the two phone calls is to demonstrate that cellular systems are considerably more complex than conventional telephone systems. Another purpose is to provide an intuitive preview to the remainder of this book, which examines in detail the way that practical systems perform the functions described here. Corresponding to the architecture shown in Figure 1.3 for a fixed telephone system, we have Figure 1.12, which shows several elements of a cellular system: telephones, radio channels, base stations, a switch, and connections linking the base stations to the switch. It also shows the connection of the cellular switch to the worldwide Public Switched Telephone Network.

Figure 1.13 is similar to Figure 1.4: it summarizes the information that moves through the cellular system as the system sets up the call from Susan to Bill. We see that compared to setting up a conventional call, a communications system has to work considerably harder to set up a cellular phone call. We will also observe, as we examine specific systems, that the extra work does not stop at call setup. Cellular and other personal communications systems perform intricate network control operations, including those associated with handoff, to maintain a call. Even with no calls in progress, personal communications systems have work to do, keeping track of the locations of their mobile subscribers.

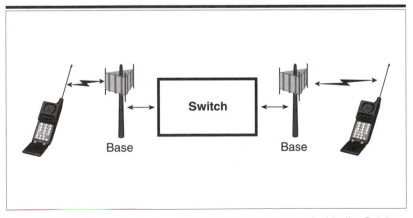

Figure 1.12 Network elements of a cellular system connected to the Public Switched Telephone Network.

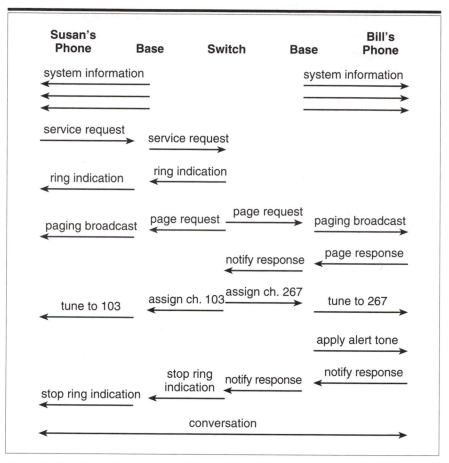

Figure 1.13 Information flow for setting up the cellular telephone call.

1.5 Technical Challenges

Before the first cellular system appeared on the scene, affluent countries boasted advanced telephone networks accessible nearly everywhere. These networks have grown in size and sophistication in the past decade, and they have been joined by the Internet, a collection of networks linking tens of millions of computers throughout the world. With this perspective, it is reasonable to ask why we need new technology—and new investments of tens of billions of dollars—to achieve the promise of personal communications. We find the answers to these questions in the differences between Susan's phone call and Helen's. To describe these differences we refer to three phenomena that play a major role in Susan's cellular phone call. None of them influences Helen's conventional call.

We refer to these phenomena as *mobility, ether,* and *energy.* A principal attraction of personal communications systems is their ability to provide information services to a *mobile* population. (There are alternatives to personal communications systems for delivering information to mobile subscribers. Call forwarding is one example.) The locations of mobile subscribers are not precisely known to the system before a call begins, and are subject to change while the call is in progress. To accommodate mobility, personal communications systems transmit and receive signals through the air. *Ether* denotes the properties of the open-air path between a cellular telephone and the rest of the communications network. *Energy* refers to the nature of the portable telephones preferred by three-quarters of cellular subscribers. These telephones carry their own lightweight power supplies. (In addition to portable phones, there are vehicle-mounted telephones, connected to the battery of a vehicle.)

None of these three phenomena was anticipated in the designs of the two global communications networks: the Public Switched Telephone Network and the Internet. Each of these networks serves terminals at fixed locations with wired connections to the rest of the network. The terminals obtain electrical power from the network, or from public power companies. The following paragraphs address some of the challenges imposed by mobility, ether, and energy, and refer to technology devised to meet the challenges.

1.5.1 Mobility

Mobility is the principal distinguishing characteristic of personal communications. With users changing their locations all the time, a personal communications network constantly has to rearrange itself.

In conventional networks, rearrangements are infrequent and complicated to achieve. For example, if Helen moves to a different office in her company, she can keep the same phone number. However, she has to give sufficient notice that she wants her phone number moved to a different location. This may require a week or longer. She can also maintain her computer service by making arrangements with the company's network administrator. This also requires advance notice. On the other hand, Susan and Bill carry their cellular phones with them and expect to place and receive phone calls wherever they are. They also expect the calls to continue as they move. In the example of Susan's phone call to Bill, Bill was in the area controlled by his home system, which simplified matters for the cellular telephone system. *Location management* techniques make it possible for subscribers to receive calls in other service areas. A *roaming subscriber* is a person in a service area that is not served by his or her home system. To accommodate changes in location while a call is in progress, we have seen (in the example of Susan's call to Bill) that cellular systems require *handoff* procedures.

1.5.2 Ether

The radio signals traveling between Susan's telephone and the remainder of the cellular system encounter other cellular transmissions. They are subject to fading and noise. Government regulations limit the communications bandwidth available to her cellular telephone company. Her signals are vulnerable to eavesdropping, and the cellular telephone company is vulnerable to unauthorized use of its facilities. The solutions to all of these problems are embodied in specialized technologies, as described in the following paragraphs.

Access

The wires between Helen's telephone and the switching machine carry signals dedicated to Helen's communications. Susan's signals travel through the air. There, they encounter noise and interference from a variety of sources, including signals going to and from other telephones served by Susan's cellular phone company. It follows that Susan's signals have to be transmitted in a way that will allow a receiver to extract Susan's information from the information associated with other conversations. To make this possible, each system has an *access technology* (see Section 9.1). The main categories of access technologies are frequency division multiple access (FDMA), time division multiple access (TDMA), and code division multiple access (CDMA). The access technology implemented in a practical system is one of the main distinguishing characteristics of the system. Two of the systems that we study in detail, NA-TDMA in Chapter 5, and CDMA in Chapter 6, derive their names from the access technologies they adopt.

Channel Impairments

Helen expects the wires connecting her phone to the switching machine to meet high standards. They should at all times be free of noise and distortion. If this is not the case, the telephone company will quickly rectify the situation. The quality of the signal path between Susan's cellular phone and its current base station changes all the time. It depends strongly on Susan's location. As she moves, the strength of the signals arriving at her phone and the base station changes. The exposure of these signals to noise and interference also changes as a result of Susan's movements and the behavior of other communications equipment in her area. Wireless communications systems deploy a large collection of *signal processing* techniques to overcome inevitable transmission impairments.

Bandwidth

If Helen wants to send and receive faxes while she has a phone call in progress, the telephone company will be pleased to sell her a separate line. Her computer has its own dedicated link to the company network, which displays important messages on her screen as soon as they arrive. A coaxial cable gives her access to 50 television channels. All of this costs money, which her company will spend if its business justifies it. By contrast, the communications capacity available to Susan's cellular phone company is limited by government regulation of radio spectrum. To increase the transmission capacity of their systems—either to accommodate more subscribers or to provide enhanced communications services—wireless operating companies have to rely on advanced technology. Approaches to conserving bandwidth include *channel reuse, signal compression,* and *efficient modulation and coding.*

Privacy and Security

To eavesdrop on Helen and Mike's conversation, someone has to make a special effort. That person has to gain access either to the company premises or to telephone company facilities and attach wiretapping equipment to the telephone system. The task is much simpler in the case of the cellular phone call. Susan and Bill's signals travel through the air and are accessible to people in a fairly wide area who have equipment that is not hard to obtain. To make eavesdropping harder, personal communications systems require *encryption* technologies.

Imagine someone in Helen's office making expensive phone calls without her permission. She or her company will have to pay for the calls. The telephone company holds its subscribers responsible for access to the telephone system. On the other hand, it is possible to charge calls to Susan's cellular phone bill without using her telephone. All that is necessary is to transmit signals, as in Figure 1.6, that appear to come from Susan's telephone. In this case it is the responsibility of the network to distinguish Susan's transmissions from unauthorized attempts to use the system. To do so, the network employs *authentication* techniques.

1.5.3 Energy

Helen's office phone is plugged into a wall outlet. If the electrical power in the office fails, the telephone can obtain sufficient power from the switching machine for Helen to make and receive calls. Her computer terminal relies on the public power supply. By contrast, Susan and Bill carry their own electrical power supplies, in the form of rechargeable batteries, inside their portable phones. Like the users of portable computers, they are sometimes acutely aware of time limits on their telephone use imposed by the size of the batteries. To extend battery life, cellular systems employ *power control* techniques and some phones are equipped for *discontinuous transmission*. In its brief history, cellular telephony has experienced a rapid increase in the importance of lightweight portable phones and other portable communications devices. As a consequence, energy conservation becomes an increasingly important consideration in the design of new systems.

1.6 Evolution of Personal Communications

The challenges of mobility, ether, and energy are formidable obstacles on the path to the unified personal communications system anticipated for the twenty-first century. As stepping-stones on that path, we have four sets of products and services, each fulfilling a fraction of the promise of personal communications. They appear in Figure 1.14 as cellular telephones, cordless telephones, mobile computing, and paging. They all came into existence in the 1970s and 1980s as separate products and services. With an expanding public appetite for personal communications, all of them attracted large, growing markets in the 1990s. As they mature, their differences become less distinct. This trend will continue as they merge into the twenty-first century personal communications systems anticipated in Section 1.2. Public awareness of this trend has been stimulated in the mid-1990s by the commercial appearance of communications products and services labeled "personal" by government regulators and network operators. For the most part, the services marketed as personal communications employ advanced cellular technology, either in cellular frequency bands or bands allocated for "personal communications" by regulators. Two practical systems (PHS and PACS, described in Sections 8.3 and 8.4) were designed from the outset to deliver personal communications services. The following paragraphs describe briefly each of the four stepping-stones to personal communications.

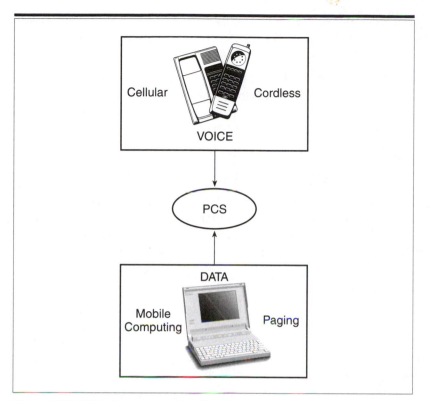

Figure 1.14 PCS will merge four families of wireless communications systems.

1.6.1 Cellular Networks

Of the four precursors of personal communications, cellular telephones have had the greatest commercial impact. Their technology is also the most complex. In the late 1990s, there are four major trends in cellular technology:

- With respect to radio transmission, second-generation digital technologies are appearing throughout the world next to existing first-generation analog systems.

- With respect to network infrastructure, the signaling and network control technologies of second-generation systems conform to published standards, in contrast to the proprietary signaling and control of first-generation systems.

- With respect to terminal equipment, there has been a strong shift in consumer preference from vehicle-mounted telephones to small portable units.

- With respect to services, cellular networks are expanding their scope to cover services originally delivered by paging networks and wireless data systems. Examples of these services include caller identification, transmission of text messages, and direct digital access to computer networks.

Although these trends are global, government policies in different parts of the world have stimulated the creation of a diverse set of technologies to satisfy the demand for cellular service. The original cellular telephones all relied on analog frequency modulation for voice transmission. Each country assigned its own spectrum bands to cellular service. Carrier frequencies were in the vicinity of 450 MHz, 850 MHz, and 900 MHz. A voice channel occupied 25 kHz or 30 kHz, depending on the transmission standard in place.

The largest market for a single transmission technique was the United States and Canada, where all analog cellular telephones and base stations conform to the AMPS (Advanced Mobile Phone System) standard [Electronic Industries Association, 1989]. It is paradoxical that subscribers in North America for a long time derived limited benefit from this continental conformity to a single transmission standard. This is due to the nature of cellular radio licenses, which are confined, for the most part, to small geographical areas. There are more than 700 service areas in the United States. Each one is served by two operating companies. As a result, the ability of an American subscriber to use his cellular telephone when he leaves his home service area depends on business arrangements between his operating company and the one that serves the area he is visiting.

Such business arrangements are necessary conditions for serving roaming subscribers, but they are not sufficient. This is because the scope of the AMPS specification is limited to the "air interface" between subscriber equipment and base station radios. Signaling systems that link base stations to switches and switches to other switches are proprietary in North America. Therefore, roaming between two service areas requires software and hardware to link the switching equipment in the two locations. To address this requirement and provide seamless service throughout the United States and Canada, the North American cellular industry adopted, in the 1990s, a signaling standard for communications between mobile switches [TIA, 1991]. This standard is the subject of Chapter 4.

In Europe, most countries formed national analog cellular networks from the outset. This gave their subscribers access to cellular service anywhere within their national boundaries. The original operating area of the NMT (Nordic Mobile Telephone) system covered Denmark, Finland, Norway, and Sweden. Equipment compatibility ended at national (or regional) frontiers, however. To have access to analog cellular service over a wide area of Western Europe, an individual had to carry five different cellular telephones.

By the mid-1980s, the popularity of cellular service in industrial countries threatened to overwhelm the capacity of the original spectrum bands assigned. The European Community responded to this challenge by assigning new spectrum, at 900 MHz, to cellular service and mandating a single continental compatibility standard GSM (Global System for Mobile Communications, Chapter 7) for future cellular services [Mouly and Pautet, 1990]. The GSM standard includes infrastructure signaling specifications as well as transmission specifications. Therefore, in addition to adding to the transmission capacity of cellular services, GSM addresses the problem of international roaming. GSM radio transmission is based on time division multiple access (TDMA). Figure 1.15 shows the European trend from a diverse set of first-generation cellular standards to a single second-generation system.

By contrast, North American regulators offered no new spectrum bands for cellular services. Instead, to meet growing demand, they modified the original cellular licensing arrangements in order to allow companies to introduce bandwidth-efficient transmission techniques [FCC, 1990a].

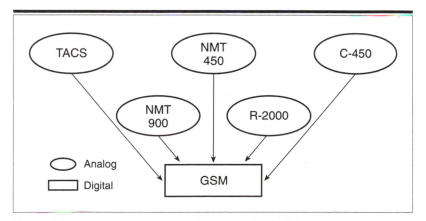

Figure 1.15 European move from diverse analog cellular systems to a single digital standard.

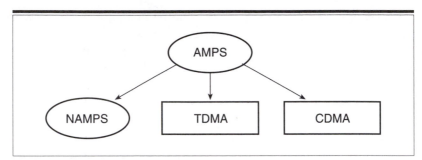

Figure 1.16 North American evolution from a single first-generation standard to a diverse set of second-generation systems.

There is no mandate for a national approach. Each company is free to use its licensed spectrum in the way it chooses. The industry's response to this flexibility has been to introduce three incompatible bandwidth-efficient transmission technologies. Two of them are digital and one is analog. All three are offered to subscribers within the context of *dual-mode systems.* In addition to one of the new bandwidth-efficient techniques, a dual-mode telephone is capable of operating with the original AMPS standard. One of the digital cellular standards in North America uses TDMA, as described in Chapter 5 [TIA, 1996d; TIA, 1992]. The other one uses code division multiple access (CDMA, described in Chapter 6) [TIA, 1993b]. The dual-mode analog standard is referred to as narrowband AMPS (NAMPS: Section 3.9.1) [TIA, 1993a]. Figure 1.16 illustrates the new diversity of technologies in North America in contrast to Figure 1.15, which displays the convergence in Europe.

Japan is a third source of cellular technology. In contrast to North American and European standards, which have been adopted in several countries, cellular standards developed in Japan are used only in their country of origin. At present there are four cellular systems in place in Japan [Kinoshita, 1991]. Two of them are analog in the 800 MHz band. One analog system has channels spaced at 25 kHz. The other uses narrower bandwidths. The two digital systems are based on TDMA. One operates in the 850 MHz band and the other at 1,500 MHz.

1.6.2 Cordless Telephones

In contrast to cellular systems, which are complete communications networks, residential cordless phones simply replace the telephone line cord with radio equipment that transmits signals between a telephone and the pair of telephone company wires within a residence. Each country establishes spectrum bands for cordless telephone operation. However, there is no need for compatibility standards governing residential cordless telephone operation. The only requirement is that each consumer's telephone be compatible with the base station connected to her telephone wires. The cordless telephones of the 1980s were all analog, operating in various frequency bands including 50 MHz (North America), 900 MHz (Europe), and 300 MHz (Japan). More recently, digital cordless phones operating at 900 MHz have appeared in North American shops. Residential cordless telephones are important stepping-stones to personal communications because they have attracted a mass market of consumers who have come to appreciate the convenience of using their phones where they choose within their homes. The main limitation of a cordless telephone is that it functions only within a limited distance from a single residential base station.

Two European standards for cordless telephone *systems* have been adopted to overcome this restriction and to make wireless communications available in a wide variety of indoor environments. The systems are CT2 [European Telecommunications Standards Institute, 1991a] (Cordless Telephone, second generation, Section 8.1) and DECT [European Telecommunications Standards Institute, 1991b] (Digital European Cordless Telecommunications, Section 8.2). There is also a Canadian standard referred to as CT2Plus [Radio Advisory Board, 1990], an enhanced version of CT2. These systems allow a single subscriber unit to use many different base stations such as residential base stations, base stations connected to business telephone systems (PBX), and *telepoint* base stations. Telepoint services have base stations in public areas, where they provide an alternative to wired public telephones. Cordless telephone base stations are also installed in commercial premises to provide wireless access to business telephone systems (Private Branch Exchanges, or PBX).

1.6.3 Mobile Computing

It can be argued that personal computing was the most significant development in information technology in the 1980s. By 1997, tens of millions of people had grown accustomed to having a computer on their desks, with each computer organized to meet the needs and preferences of its owner. Since the early 1990s, two powerful trends have propelled advances in personal computing. One is the popularity of portable computers: laptops, notebooks, and personal digital assistants. By spending billions of dollars on these personal computing devices, people indicate that they want their information to be part of themselves, rather than part of their office or their home. This was the same message they delivered

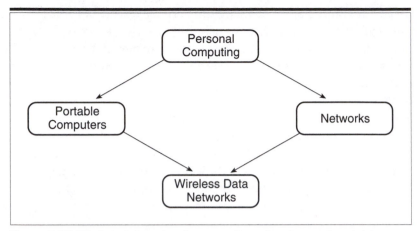

Figure 1.17 Wireless data networks merge the advantages of portable computers and networked personal computers.

when they expressed a strong preference for portable cellular phones over vehicle-mounted telephones.

The other major trend in personal computing is networking. It is not enough to have access to the information stored in a single computer. The computer has to communicate with other computers and a variety of information services. By investing heavily in modems and local area networks (LANs), consumers demonstrate their desire to be "connected." The simultaneous popularity of portable computing and networking poses a paradox, because portable computers are "disconnected" from the wires of conventional networks and from public power supplies. As illustrated in Figure 1.17, wireless data networks resolve this paradox. By making use of wireless data networks, the owners of portable computers retain the advantages of mobility and they remain connected to their important information services.

Wireless data technology is even less mature than cellular and cordless technologies. In 1997, there are several wireless local area networks on the market that use proprietary technologies. There are also two wireless local area network standards on the drawing boards [LaMaire et al., 1996]. In addition, there are three public networks in the United States accessible to mobile computing devices [Pahlavan and Levesque, 1995: Section 12.3]. Two of them are independent wireless data networks while the third uses cellular telephone channels. The main purpose of all three wide area data networks is to move short messages, such as electronic mail, to and from portable computers. The technology and price structures of wide area wireless data networks discourage long computing sessions.

A long-term goal of mobile computing is to deliver the widest possible range of information services to a mobile population [Badrinath and Imielinski, 1994]. This challenging ambition has stimulated a host of research efforts spanning diverse subject areas including multimedia broadband radio communications [Raychaudhuri, 1996], low-power information terminals [Sheng, Chandrakasan, and Brodersen, 1992], and communication protocols that function efficiently in the presence of impairments introduced by wireless communications.

1.6.4 Paging

Paging is the oldest of the personal communications precursors in Figure 1.14. It is also the simplest technically, and as a consequence the least expensive. Paging is a one-way service. All information travels from the network infrastructure to users. Another reason why paging is relatively simple is that the base stations have high power budgets leading to coverage areas hundreds or thousands of times greater than those of cellular, cordless, or wireless data base stations.

As paging technology progresses, information formats become increasingly diverse. As implied by their street name "beepers," the simplest pagers do no more than alert their users when someone wants to reach them. The person with the beeper then makes a phone call to learn why he or she was paged. Pagers introduced in the 1990s are more sophisticated. Some display the phone number of the person trying to reach the paging subscriber. Others receive text messages, including electronic mail. The most advanced pagers also receive voice mail in the form of prerecorded digital speech.

A major step forward in paging came with the award in the United States in 1995 of licenses for *narrowband Personal Communications Services*. License holders offer two-way paging services resembling some of the wireless data services described in the previous section.

With respect to technology and radio spectrum, paging is by far the least demanding of the four personal communications stepping-stones. In fact, all of the other systems included in Figure 1.14 perform paging operations to begin the process of delivering information to portable terminals. Many incorporate alphanumeric paging in the form of short message services.

1.6.5 Current Personal Communications

The earlier sections of this chapter refer to personal communications in the general sense, as defined in Section 1.2. In this section we focus on

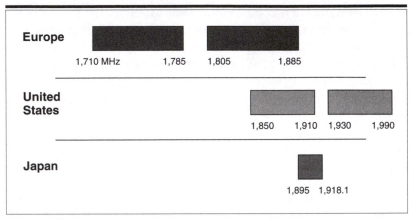

Figure 1.18 Frequency bands for personal communications.

systems specifically designated as personal communications systems by spectrum regulators in Europe, the United States, and Japan. All of these systems operate in frequency bands between 1,700 MHz and 2,000 MHz, as shown in Figure 1.18.

Europe

The term *personal communications* originated in 1989 in the United Kingdom to refer to spectrum assignments around 1,800 MHz as depicted at the top of Figure 1.18. The British government issued licenses in these bands for "Personal Communications Networks" with the aim of increasing the public's access to services that resembled cellular services in many ways [Department of Trade and Industry, 1989a]. In contrast to the original vehicular applications of cellular technology, however, Personal Communications Networks were designed from the outset to serve pedestrians carrying small portable telephones in urban areas. The British government prescribed GSM technology for this purpose. This decision was later adopted throughout Europe and embodied in a standard, referred to as DCS 1800, published by the European Telecommunications Standards Institute (ETSI). In 1997, there are Personal Communications Networks conforming to the DCS 1800 standard in operation in the United Kingdom, Germany, and Thailand.

North America

In the United States, the Federal Communications Commission issued licenses in 1995, 1996, and 1997 for "personal communications services."

Some of the licenses are in the 900 MHz band, for "narrowband services." These licenses cover limited bandwidths (at most 50 kHz), which are insufficient to support the advanced services described at the beginning of this chapter. The most prominent narrowband PCS service is two-way paging, described in Section 1.6.4.

Personal Communications Services, as described in Section 1.2, will be delivered by holders of wideband PCS licenses at the frequencies near 1,900 MHz, depicted in the center of Figure 1.18. These licenses cover bandwidths as large as 30 MHz (cellular companies operate with 25 MHz allocations). These licenses are not at all specific about the nature of the services to be provided in the PCS spectrum. License holders began offering services late in 1995. It is likely that there will be a wide diversity of service offerings. Like the cellular industry, PCS operators in the United States, which include the largest cellular companies, are certain to adopt a variety of technologies. Standards organizations, in 1994, adopted seven technical approaches to wideband PCS [Cook, 1994]:

- the North American TDMA cellular standard (see Chapter 5) [TIA, 1996d],

- the North American CDMA cellular standard (see Chapter 6) [TIA, 1993b],

- the GSM Pan-European cellular system operating in the 1,900 MHz PCS bands (Chapter 7) [Mouly and Paulet, 1992],

- the DECT European cordless system based on TDMA (see Section 8.2) [European Telecommunications Standards Institute, 1991b],

- a low-tier TDMA system designed for personal communications (see Section 8.4) [Noerpel, 1996], and

- two different CDMA techniques with signals that occupy 5 MHz bands.

The first three items are all extensions of cellular standards. They are referred to as *high-tier technologies* [Cox, 1995]. They operate with relatively high transmitter powers and are capable of serving terminals moving at vehicular speed. In 1997, each of the three high-tier technologies has been adopted by a substantial number of PCS license holders. PACS and DECT are examples of *low-tier technologies* with simpler equipment requirements. They operate with low transmitter powers and transmission techniques that cater to people moving at pedestrian speeds. PACS combines technical features of the Wide Area Communications System developed at Bell Communications Research [Bell, 1994] and the Japanese Personal Handyphone System. A few PCS license holders have announced plans to adopt PACS technology.

The United States has also set aside the band 1,910–1,930 MHz for "unlicensed PCS" services. In this band, people may operate any equipment that conforms to a *spectrum etiquette* approved by the Federal Communications Commission [Steer, 1994]. This etiquette protects each user to some extent from interference from other radio signals. However, it does not mandate equipment compatibility.

Japan

A system referred to as PHS (Personal Handyphone System—see Section 8.3) began operating in Japan in 1995 [Research and Development, 1993]. Like PACS and DECT, PHS is a low-tier TDMA technology, designed to serve people moving at pedestrian speeds with low-power terminals. PHS quickly attracted a large market of Japanese subscribers.

1.7 Systems Presented in This Book

The purpose of this book is to introduce professionals to the technologies of personal communications. Chapters 3 through 7 examine in detail the five high-tier technical standards that together will govern a majority of the world's cellular and PCS communications in the late 1990s and the early years of the next century. Chapter 8 describes, in less detail, four low-tier systems that are also in commercial operation in 1997. To set the stage for the analysis of specific systems, Chapter 2 addresses general issues pertaining to personal communications systems in order to provide a framework for studying specific systems. By following this framework, readers will gain insights into the design of each individual system. They will also be able to evaluate the relative strengths and weaknesses of different systems. The emphasis in each presentation is on the way system components interact to perform complex communications tasks. Readers will learn that, although the systems perform similar functions overall, they differ in the way they distribute tasks among system components, including terminals, base stations, and switches (see Figure 1.12).

Personal communications systems assemble a wide range of specialized technologies. To understand any given system fully, it is necessary to be acquainted with a large set of topics in information technology, which in themselves are subjects of specialized textbooks. Anticipating readers with diverse backgrounds and needs, Chapter 9 contains 12 brief technology tutorials. The aim of these tutorials is to help readers who are unfamiliar with specific technologies, such as speech coding and equalization, understand how the technologies fit into practical systems. The tutorials will give readers sufficient information to understand a system without the need to obtain specialized information from other sources. Eventually, many readers will plunge into detailed publications that will equip them with the tools and knowledge they need to meet their individual goals.

The only analog system covered in the book is AMPS (Chapter 3), which is used throughout North America, many other places in the Western Hemisphere, and a few Asian and Pacific countries. In the United States alone there are, in 1997, more than 40 million AMPS cellular telephones in use. In 1997, the North American cellular industry is accelerating its deployment of digital cellular technologies based on either TDMA radio signals (Chapter 5) or CDMA signals (Chapter 6). However, all of the digital terminals and base stations are parts of dual-mode systems capable of handling AMPS signals and one of the two digital standards. As a consequence, the number of AMPS cellular phones will continue to grow rapidly for the next several years and AMPS will continue to be an important practical technology for the foreseeable future. The situation is different in Europe where there is only one digital system, GSM (Chapter 7). GSM telephones have no analog capability, and it is anticipated that over the next several years, the number of analog phones in Europe will rapidly decrease. These projections led the industry to invest in enhancements to AMPS terminals and infrastructure, but not in other analog systems.

The GSM standard, described in Chapter 7, contains the details of a complete communications system, including interfaces among all system elements. By contrast, the AMPS standard and its digital descendants, covered in Chapters 3, 5, and 6, specify only the "air interface" between terminals and base stations. Originally, all of the other network interfaces were proprietary to each manufacturer of infrastructure equipment. This situation made it difficult to serve subscribers of one system who roam to geographical areas served by other systems. To facilitate service to roaming subscribers, the North American industry adopted standard interfaces between separate systems. These interfaces are embodied in Interim Standard 41, the subject of Chapter 4.

Chapter 8 contains summary descriptions of the four low-tier systems in commercial operation in 1997. CT2 (Cordless Telephone, Second Generation, Section 8.1) appears in a few public telepoint services in Europe and Asia. Its primary application is in business telephone systems. Business telephones are also the main application of DECT (Digital European Cordless Telecommunications, Section 8.2). PHS (Personal Handyphone System, Section 8.3) technology forms the basis of successful public networks in Japan, and PACS (Public Access Communications System, Section 8.4) has, in 1997, been adopted by a small number of PCS license holders in the United States.

1.8 Other Wireless Communications Systems

The systems described in Chapters 3 through 8 will, for the foreseeable future, meet a large majority of the world's demand for wireless personal communications. Meanwhile, in the late 1990s, many other wireless communications systems will emerge to serve special needs that are not met well by the existing systems. These systems include

- mobile communications satellites,

- wireless local area networks (LANs),

- wireless local loops, and

- wireless data networks.

Figure 1.19 indicates that satellites and wireless LANs occupy opposite corners of a plane that displays two properties of a system—the bit rate available to terminals and the coverage area. The systems described in this book are in intermediate positions. In this qualitative view, "coverage" can be interpreted to be either cell size or the service area of an entire system. A satellite system delivers low-bit-rate services (typically, 10 kb/s or less) over continental or global coverage areas, while a wireless LAN operates at Mb/s over a range measured in tens of meters. The high-tier systems have national coverage with bit rates on the order of 10–20 kb/s, while the low-tier systems typically cover metropolitan or suburban areas with a bit rate per terminal of 32 kb/s. Wireless local loops operate with fixed radio terminals. Without the challenge of mobility, they are able to achieve higher spectrum efficiency and other economies relative to communications systems that serve mobile subscribers. Wireless data systems serve mobile populations that have different service needs from the populations served by the other systems.

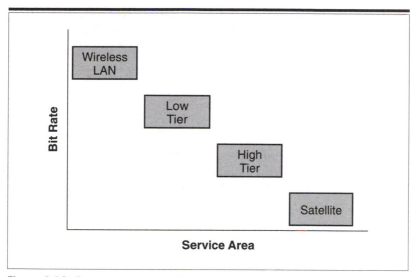

Figure 1.19 Coverage area and bit rate of four types of wireless communications systems.

1.8.1 Mobile Satellite Systems

A major trend in satellite communications in the 1990s has been toward smaller and smaller Earth terminals [Abrishamkar and Siveski, 1996]. This trend has produced a thriving direct broadcast satellite television industry, two-way communications between satellites and vehicles, two-way communications between satellites and ships, and the one-way Global Positioning System. With respect to personal communications, the main goal that can be served by satellites is ubiquitous coverage. Viewing a communications satellite as a base station, the cell dimensions are many orders of magnitude larger than those of terrestrial systems. Satellites can therefore provide communications services in areas where it is uneconomical to install the infrastructure of one of the systems described in this book.

Satellite systems that provide mobile communications services fall into categories distinguished by the orbits of the satellites. The first mobile satellites are in geosynchronous orbits (GEO), at a distance of 35,800 km above the equator. Geosynchronous satellites have the advantage of a simple network configuration. The cell size of one satellite is approximately one-third of the Earth's surface. On the other hand, due to high transmission path attenuation, geosynchronous satellites require high-power transmitters in both the satellite and the mobile terminal. Other disadvantages are long propagation path delays and poor radio coverage at high latitudes. To overcome these disadvantages, some of the mobile satellite systems planned for the late 1990s will operate in low Earth orbits (LEOs), on the order of 500 to 2,000 km above the Earth, and others will operate in medium Earth orbits (MEOs) at altitudes around 10,000 km. LEO and MEO satellites have smaller coverage areas than GEO satellites and they move with respect to the terminals they serve. Therefore, each system requires many satellites and network control that includes handoff from a satellite that moves out of range of a terminal to another satellite that moves into range. It is interesting to note that LEO and MEO satellite systems require handoff due to the mobility of base stations (in satellites) rather than the mobility of terminals.

The systems planned for introduction in 1998–2001 are diverse in their characteristics. Not only do they differ in orbit but also in the number of satellites per system (between 10 and 840), channel transmission rates (between 2,400 b/s and 2 Mb/s) and the sophistication of the communications tasks performed by each satellite. Some are "bent pipes," which simply receive signals from one place on Earth and relay them to another place. Others perform sophisticated switching operations.

1.8.2 Wireless Local Area Networks

Wireless LANs [LaMaire et al., 1996; Wickelgren, 1996] provide high throughput (Mb/s per second) communications between stationary or slowly moving terminals in small coverage areas (on the order of tens of meters in diameter). Although the wireless LANs developed in the early 1990s use proprietary transmission protocols, it is likely that products produced at the end of the decade will conform to two published standards: IEEE 802.11 [IEEE, 1996], and HIPERLAN [Wilkinson, Phipps, and Barton, 1995]. Both of these standards anticipate operation in unlicensed frequency bands, around 2.5 GHz, 5 GHz, and at higher frequencies. Another point of departure of wireless LANs from satellites and the systems studied in this book is that they allow terminals to communicate directly with one another, rather than through a network infrastructure containing base stations and switching equipment. Key issues in the design of wireless LANs derive from the distributed nature of the system architecture. Without the coordination of a base station, terminals contend for access to the same radio channel. Protocols are designed to promote fairness of access (equitable sharing of the channel among all terminals) and reliable operation even when some of the terminals are out of range of others (hidden-terminal problem).

1.8.3 Wireless Local Loops

The technical challenges of wireless local loops [Yu et al., 1997] are less stringent than those of communications systems that serve mobile terminals. Furthermore, operating companies have considerably less incentive to adopt systems that conform to published standards. As a consequence, the communications industry, in 1997, offers a wide variety of communications systems to serve as wireless local loops. Some are adaptations of the standard systems described in this book. Others are proprietary systems, designed from the outset for this application.

1.8.4 Wireless Data Networks

Wide area wireless data systems [Pahlavan and Levesque, 1995: Section 12.3] have been in operation since the early 1980s to serve specialized commercial needs including transaction processing, such as credit card verification; broadcast services, such as road traffic advisories; and interactive services, such as wireless electronic mail. These systems provide two-way, low-speed, packet-switched data communications. Two of the early systems were Ardis, deployed throughout the United States, and Mobitex, used in public data networks in the United States and several European countries. Ardis and Mobitex [Khan and Kilpatrick, 1995] operate in "specialized mobile radio" frequency bands between 800 MHz and 1,000 MHz. Channel bit rates range from 8 kb/s to 19.2 kb/s. The network architectures are similar to those of cellular systems. A newer system, introduced in the United States by Metricom, operates in an unlicensed frequency band at 900 MHz in the United States. The Metricom system relays packets through several radio transceivers between terminals and fixed-packet switches. The channel rate is 100 kb/s. In Europe, ETSI has adopted a standard for trans-European trunked radio (TETRA). It operates in 25 kHz channels between 380 MHz and 393 MHz with a transmission rate of 36 kb/s.

In addition to these specialized wireless data networks, most of the systems described in this book have adopted standards for packet data transmission, using the physical channels of the wireless personal communications system. These standards make it possible for service providers to offer wireless access to packet data networks, including the Internet. The first of these technologies to be deployed commercially is *cellular digital packet data* (CDPD), which uses AMPS (Chapter 3) radio channels. In GSM (Chapter 7), the packet data technology has the designation *generalized packet radio service* (GPRS).

Review Exercises

1. Why does a cellular phone have a SEND button? List as many reasons as you can.

2. One of the main purposes of a personal communications system is to serve a mobile population. Comment on the relationship of mobility to wireless communications. Give examples of wireless communications systems that are not mobile. Give examples of mobile telephone communications that are not wireless.

3. What is the difference between a personal telephone number and a conventional telephone number? What are some technical problems that have to be addressed in a system with personal telephone numbers?

4. What are some differences between wireless personal communications and conventional, wired telephone communications? List as many as you can.

5. Give some examples of control information transmitted to and from a conventional telephone. How does a wired telephone transmit and receive user information and control information?

6. Give some examples of control information that is transferred to and from a cellular phone but is not necessary in conventional telephones. Why do cellular systems have separate control channels and traffic channels?

7. Why are privacy and security more critical issues in the design of personal communications systems than in the design of conventional, wired systems?

8. Compare a residential cordless telephone and a cellular telephone with respect to the challenges of mobility, ether, and energy.

9. Why do cellular telephone systems require handoff procedures?

10. What are some advantages and disadvantages of using communications satellites to provide wireless personal communications?

CHAPTER TWO

Principles of Personal Communications

2.1 Presentations of Practical Systems

This chapter sets the stage for the presentations of practical systems in the remainder of the book. We take a broad view of personal communications, beginning with the perspective of the person using a system and progressing to the operations a system performs in order to meet the person's expectations. We then establish the framework adopted in the remainder of this book for examining practical systems.

We first describe the set of information services delivered to people by personal communications systems. We then examine system performance by introducing 21 figures of merit of individual systems. Together these figures of merit reflect the quality and utility of a system as perceived by its users and operators. They are also performance goals of system designers. As in most engineering tasks, the design and operation of practical systems involve compromises among conflicting goals. This book describes many alternative approaches to performing the technical operations required for personal communications. To understand why there are many approaches to performing similar tasks, readers will find it useful to refer to the priorities of systems designers who confront a long list of inconsistent objectives.

2.2 Information Services for People

The subject of this book is wireless access to information services. Although the long-term vision anticipates systems capable of delivering a wide range of services to a mobile population, each of today's systems has a limited information repertoire. The following paragraphs describe specific information services and comment on the ability of personal communications systems to deliver them.

2.2.1 Telephone Services

The principal information service delivered by personal communications systems is *voice telephony*. Obviously, this service makes it possible for a personal communications subscriber to have a conversation with somebody who is using another telephone. Beyond basic telephone service, many systems have special provisions for dealing with emergency calls from wireless terminals. In common with many fixed telephone systems, personal communications systems also offer a range of supplementary services. Examples include voice mail, call forwarding, calling number identification, call waiting, and conference calls.

2.2.2 Short Message Services

Some systems have a built-in capability to deliver *short message services* in addition to telephony. The simplest of these services are comparable to alphanumeric paging. They display on a screen of a personal information machine (PIM) on the order of 100 characters of text transmitted from a base station. An important advantage of this type of service over paging services is the ability of the terminal to acknowledge receipt of the message. Because a pager has no transmitter, the sender of a message cannot be certain that the message has been delivered, unless the paging subscriber uses another communications system to acknowledge receipt. The transmitter contained in every PIM also makes it possible for the personal communications subscriber to send short messages to other people. In addition to messages transmitted to and from individual terminals, *broadcast messages* comprise a third category of short message service. These are messages available to all terminals in a geographical area designated by the service provider. Examples of messages useful to a high population of subscribers are road traffic reports, winning lottery numbers, and stock market quotations.

2.2.3 Voiceband Data and Facsimile

Telephone systems are optimized for the transmission of human speech. To send digital data through a telephone network, a modem transforms the data into an audible signal that occupies the same frequency band (300–3,000 Hz) as telephone speech. Residential subscribers to on-line information services rely on data modems to send and receive information over telephone lines. Many people use modems for access to remote computers. Merchants use modems to authorize credit card transactions. Fax machines also contain modems that transform images into sounds and send them over telephone lines.

Like the telephone network, the radio transmission techniques of most personal communications systems are optimized for speech. Radio channels, however, differ significantly from conventional telephone lines. Most of the differences cause problems for modems and protocols designed for telephone lines. In analog radio systems, the principal problems are high interference levels, rapid variations in the strength of received signals, and circuit interruptions during handoffs.

When voiceband data signals appear in digital radio systems, they go through a complex set of transformations as indicated in Figure 2.1. A voiceband modem transforms digital data into waveforms that fit into the 3 kHz telephone bandwidth. Source coders designed for speech can have harmful effects on the 3 kHz voiceband data waveforms. In addition, digital errors in received radio signals further distort voiceband data signals and cause the remote modem to make errors in detecting the arriving data. As a consequence, it is usually not acceptable to send voiceband data signals designed for telephone lines through personal communications systems designed primarily for telephony. Personal communications systems that allow subscribers to communicate with fax machines and data modems require special adapters to move information reliably across radio links.

2.2.4 Direct Digital Access

Here the applications are similar to the ones for which people use voiceband modems, but the system operations are different. Subscriber terminals do not have modems that send and receive digital signals disguised as human voices. Instead, they gain access directly to the radio bit stream

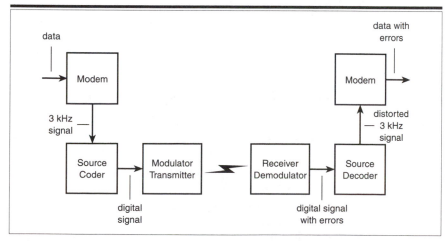

Figure 2.1 Voiceband data in a digital radio system.

of the digital personal communication system. However, compared with other digital terminals communicating with computers or information services, data terminals in personal communications systems require special techniques to deal with high error rates in radio links.

There are two basic approaches. With forward error correction, information bits flow in and out of the terminal at a constant rate. However, their error rates rise and fall as the strength of the arriving signal fluctuates. It is left to the end-to-end protocols of the data services to deal with the errors that are not suppressed by forward error correction. This approach is called *transparent data transmission*. The other approach is to introduce error detection to the radio link. When a decoder detects one or more errors, the personal communications system retransmits blocks of information where errors have been detected. In this case, the error rate of the radio link is always low. Rising and falling impairment levels on the channel cause the flow of information in and out of the terminal to slow down and speed up, respectively. Services that include error detection and retransmission are referred to as *nontransparent data services*. Table 2.1 is a summary of the main characteristics of transparent and nontransparent data services.

Typical rates for data transmission in telephone-oriented Personal Communications Services are 2,400 b/s to 9,600 b/s. Wide area wireless data services offer rates up to 19,200 b/s. Wireless local area networks operate at Mb/s rates.

2.2.5 Closed User Groups

There are many private mobile radio systems used by taxi companies, repair services, delivery services, public safety authorities, and other organizations. Within these systems, terminals are restricted to communicating with one another. One of the terminals may belong to a dispatcher who can broadcast to all other terminals simultaneously. In some systems,

Table 2.1 Characteristics of Digital Data Services

	Technique	Transmission Rate	Error Rate
Transparent	Forward error correction	Constant	Variable
Nontransparent	Error detection and retransmission	Variable	Constant

the other terminals can communicate only with the dispatcher and not with one another. It is possible for personal communications systems to offer this type of service to groups of subscribers. These are radio versions of the virtual private networks offered by public telephone companies.

2.2.6 Telemetry

Personal communications systems are capable of communicating status signals including security alarms, utility meter readings, and vending machine inventory from remote locations to administrative sites.

2.2.7 Wireless Local Loops

In telephone terminology, a *local loop* [Yu et al., 1997] contains the facilities that link a customer's premises with the local telephone company. Cellular and cordless telephones found their earliest applications in industrial countries where wired telephones already served nearly the entire population. However, in 1997, "universal service" exists in only a fraction of the world and there are many countries that aim to expand telephone service rapidly over the next several years. In many environments, radio links offer an economically attractive alternative to installing telephone wires in a vast number of customer premises. Personal communications systems can be deployed for this application. If the wireless subscriber line appears to the customer as a telephone jack at a fixed location, the technical challenges of personal communications described in Chapter 1 vanish (mobility) or are greatly attenuated (ether, energy). This leads to higher efficiency and lower costs. In the late 1990s, these economies are also attractive in countries with extensive wired telephone networks. Many of these countries are opening local telephone service to competition. To compete with long-standing telephone monopolies, some companies deploy wireless local loops to provide local telephone service.

2.2.8 Video and Other Broadband Services

The long-term goal of the Public Switched Telephone Network is to evolve to a set of broadband integrated services digital networks (B-ISDN). B-ISDN offers, in addition to voice, facsimile, and low-speed data services, access to information services with high transmission bit rates. The most demanding of these services involve transmission of moving pictures. One goal of personal communications is to deliver to a mobile population all of the information services available from fixed networks [Raychaudhuri, 1996]. Thus, broadband services will be attractive to users of future personal communications systems. These services are not, however, at present delivered by the systems described in this book.

2.3 Figures of Merit: Design Goals

This book examines several systems that provide people with access to information services. It is likely that all of them will be used for several years to come. An important question for somebody learning about personal communications is, "Why are there so many different approaches to performing the same task?" A major part of the answer lies in the large number of criteria for assessing the merits of a system. Many of the criteria are contradictory—to do a good job with respect to one means compromising with respect to another. Each system represents a different set of tradeoffs. To set the stage for presentations of individual systems, this section describes 21 measures of the quality of a personal communications system. Each measure can be viewed as a design goal of new systems or a figure of merit for existing systems.

Compromise is at the heart of engineering practice. The compromises of personal communications systems reflect tradeoffs between cost and performance that arise in the creation of most products and services. In personal communications we also find many conflicts among figures of merit that arise from the triple challenge of mobility, ether, and energy.

2.3.1 Summary of Figures of Merit

Consider the indicators of system quality encountered by subscribers as they select a system and participate in a communication session. Before subscribing to a system, a person considers the price and characteristics of the PIMs that operate with each system. He or she also considers the service price, the available communications services, the geographical coverage area, and the possibility of using the service when leaving the area served by the home system. In using the system, the person would welcome a comfortable, convenient user interface. He or she expects it to accept all service requests and to set up communications promptly. During a phone call or other communication session, the system should deliver information accurately and promptly. The quality of the signal should not deteriorate as a consequence of the motion of the subscriber, nor should the communication session be vulnerable to premature disconnection. The subscriber is also conscious of battery life, preferring a long usage interval before having to replace or recharge the terminal's battery.

The system operator benefits from low-cost equipment, economical operation of the communication links within the infrastructure, and efficient use of the radio spectrum allocated to the system. It is also an advantage early in the life of a new system to operate with the largest possible cells. This enables an operating company to initiate a service offering with relatively few base stations and gradually increase its investment as the number of subscribers (and revenues) grows. The operating company requires protection against unauthorized use of the network. The operator desires early access to new systems, and opportunities to incorporate further technical advances into existing systems.

Table 2.2 lists the 21 figures of merit implied by this description. The following paragraphs discuss them individually.

Table 2.2 Summary of Personal Communications Figures of Merit

Relevance	Figure of Merit	Measure
Subscribing to a system	Terminal price	money
	Terminal size and weight	cm^3 grams
	Service price	money/month money/minute money/message
	Range of services	
	Coverage area	km^2
	Roaming	
Using a system	User interface	
	Call blocking	percent
	Setup time	seconds
	Transmission quality	error rate signal-to-distortion ratio delay
	Privacy	
	Mobility	km/hr
	Call dropping	percent
	Battery life	hours
	Modes of operation	
Operating company	Infrastructure cost	money/cell
	Cell radius	km
	Spectrum efficiency	conversations per base station/MHz
	Network security	
	Early deployment	
	Adaptability	

2.3.2 Terminal Price

Each system has its own family of compatible PIMs. Radio transmission technology has a strong influence on the cost of manufacturing a personal information machine. Each system that we examine has its own air interface standard. The systems vary widely in complexity, with direct effects on terminal production costs. In addition, each system has its own way of assigning communications tasks to network elements. To the extent that a system makes more demands on its terminals, it raises the cost of the terminal. The PIMs that perform sophisticated information processing functions represent a marked departure from conventional telephones. Ordinary wired telephones are simple information processing devices connected to extremely complex switching equipment. Computer networks have opposite characteristics. The communications infrastructure is relatively simple, with most of the complexity appearing in the host terminals, which are sophisticated computers. With respect to complexity, personal communications sets take intermediate positions between simple telephones and complex computers.

2.3.3 Terminal Size and Weight

The size and weight of a PIM are important in personal communications services, with their emphasis on mobile subscribers carrying their own information terminals. In many terminals, the battery has the strongest influence on size and weight. Battery size is proportional to battery life (see Section 2.3.15), which, in turn, is influenced by many details of system design. Within the terminal itself, the complexity of the required information processing operations has the strongest influence on size, weight, and power consumption. For a given set of information processing functions, size and weight depend on the approach to circuit design in the terminal. Generally, customized integrated circuits, which are expensive to design and develop, have the potential of reducing terminal size, weight, and power consumption. An alternative approach is to assemble a terminal with standard components, including programmable microcontrollers and digital signal processors.

2.3.4 Service Price

A residential cordless telephone costs no more to use than an ordinary telephone. Cellular phone conversations incur monthly rental fees and substantial service charges every minute. The users of wireless data services pay for every unit of information transferred to or from a terminal.

These billing systems reflect the cost of providing services. Service price depends on the cost of infrastructure equipment and on the network architecture, which places demands on fixed communications links. The long-term goal of personal communications systems is to become the principal means of communications for most people. This goal implies more than a mass-market subscriber population, which is already in sight. It also implies that each subscriber will use personal communications networks for a high proportion of his or her communications. This requires service prices to be comparable to those of alternative, fixed systems. One approach to achieving this is to vary the price according to the location of the subscriber. With this approach, a person at home would pay the same fees as users of fixed networks. Traveling at high speeds, the costs would resemble those of cellular telephones. There would be intermediate costs (comparable to pay telephones) for people moving at pedestrian speeds in public places.

2.3.5 Range of Services

A personal communications system is useful to the extent that it meets the information needs of its users. The wider the range of information services it can deliver, the better the system. The systems described in this book begin with the principal aim of providing telephone service. Some of them also are inherently capable of delivering other services described in Section 2.2. Others add new service capabilities as the systems mature.

2.3.6 Coverage Area

Personal communications systems serve a mobile population. Enthusiasts speak of ubiquitous or even "universal" service, available everywhere. Although this is clearly a worthy goal, it is hard to achieve. Some systems are accessible in virtually every outdoor location in a nation or a group of nations. Others cover major population centers, and still others offer limited islands of coverage, such as within buildings or parts of buildings.

2.3.7 Roaming

Each personal communications operating company provides services in a specific geographical area. The area can be local or national. A subscriber who enters the service area of another operating company is a *roamer*. The roaming subscriber would like to obtain, conveniently, all of the communications services available to her in her home system. The possibility of doing so requires technology for coordinating the operations of the home and visited systems and business arrangements between the two service providers.

2.3.8 User Interface

The system should be easy to use, especially in view of the special requirements of a person in motion, sometimes driving a car. As elements of a *personal* communications system, the terminal and network services should be adaptable enough to meet the individual preferences of each subscriber.

2.3.9 Call Blocking

There should be a low probability that the system will deny a service request due to congestion. With limited bandwidth allocations and unpredictable user mobility, personal communications systems are particularly vulnerable to overload. A principal problem is finding all traffic channels at a cell occupied when a new service request arrives. Even with traffic channels available, the control channels used to set up new communications are vulnerable to overloads that result in a blocked call. The infrastructure consisting of network control computers and transmission links between base stations and switches can also experience congestion that causes the system to deny service requests.

2.3.10 Setup Time

Subscribers expect connections to be established in time intervals comparable with other networks. Congested control channels and switches, even when they do not cause rejection of service requests, can result in long delays. In addition, when a request arrives to set up a call or deliver other types of information to a mobile subscriber, the system has to search for the subscriber. The nature of the search procedure can influence the setup time.

2.3.11 Transmission Quality

The third column in Table 2.2 suggests that transmission quality is itself a multidimensional phenomenon. In general, a system should deliver information promptly and accurately, with the level of required promptness and accuracy dependent on the service provided. Each service has its own specific criteria of transmission quality. For high-quality telephony, the received voice should sound good. It should be free of distortion and background noise. It should not experience a long delay traveling from talker to listener. A received facsimile should look like the original image.

Text and digital data should be error free. For some applications, the text or data should arrive without long delay. Relative to other networks, service quality is an acute problem in personal communications because of transmission impairments due to interference, multipath propagation, and fading. Depending on the system, service quality can be impaired by brief interruptions when a terminal moves to a new cell.

2.3.12 Privacy

With signals traveling through the air, personal communications systems are vulnerable to eavesdropping. There are two privacy issues in personal communications systems. One issue is access to user information: the contents of conversations, electronic mail, facsimiles, and so on. The other issue is access to network control information. This information reveals the communications behavior of subscribers (whom they call, how long they speak, etc.) and the locations of subscribers (even when the subscribers do not use a system). The level of privacy in a personal communications system depends on the radio transmission technology and on data encryption techniques introduced to prevent eavesdropping on user information and network control information.

2.3.13 Mobility

Although mobility is at the heart of personal communications, systems differ in the degree of mobility they can support. At the desirable extreme, a system will maintain communications as its users move at high speed through a diverse set of geographical areas. This goal places demands on the radio transmission system, which has to adapt to transmission conditions that vary over a wide range in a short time interval when a terminal moves at high velocity. It also burdens the network control infrastructure by adding complexity to the tasks of tracking user location and performing handoffs. Some systems do not address this array of challenges fully, but instead cater mainly to users moving at low, pedestrian speeds, who are unlikely to move out of range of the base station where a call originated. With terminals anchored to fixed locations, wireless local loops (see Section 2.2.7) are at the low end of the mobility scale.

2.3.14 Call Dropping

There should be a low probability of interrupting an established phone call, facsimile transmission, data transaction, or other type of communication session. Vulnerability to call dropping is usually associated with mobility. A person with a call in progress could move into an area with poor signal coverage, or into an area where all channels are occupied by calls in progress. It is also possible that the movements of other users will fatally increase the interference experienced by a call and cause the call to be terminated by a system.

2.3.15 Battery Life

In contrast to conventional telephones and computers, PIMs carry their own power supplies in the form of disposable or rechargeable batteries. Clearly, it is a burden to a user to have to change or recharge the batteries frequently. Battery life is the time interval between these events. It has two components. *Talk time* is the number of hours that a terminal can participate in two-way communications. Talk time is shorter than *standby time*, which is the number of hours that a terminal is available to respond to a call request but is not actually engaged in a communication session. Except for sending brief status messages, the terminal does not use its transmitter in the standby condition. Battery life increases with battery size and weight. It decreases with the complexity of the communication tasks demanded by the system. A third factor that influences battery life is the location of the terminal. In dense urban areas, the terminal is near a base station and the transmitter requires relatively low power. This is also true of indoor and campus environments with base stations installed. In these situations, battery life is longer than it is when a terminal has to communicate with a distant base station, a likely situation in areas with low population density. Battery life is also influenced by various details of system design. For example, the digital systems described in Chapters 5 through 7 all allow terminals without a call in progress to enter a *sleep mode,* in which their receivers are turned off for the majority of the time in order to conserve battery power.

2.3.16 Modes of Operation

The value of a PIM to its owner is enhanced when the terminal can play a variety of roles, depending on the owner's location and requirements. For example, a telephone might function as a cordless phone when the subscriber is at home, as a cellular phone in a car, and as an office telephone when the person is at work. At home, calls would be routed through the subscriber line to the local telephone company. When the subscriber is traveling, calls would go through cellular base stations; when the subscriber is in the office, calls would be handled by the employer's private telephone system (PBX). These functions require special terminal capabilities and coordination of different systems—in this example, the cellular phone network, the local telephone system, and the office telephone system.

2.3.17 Infrastructure Cost

The network infrastructure consists of base stations, switches, databases, and the communications links connecting them. Base station costs, like those of terminals, depend on the radio transmission technology and on network architecture. However, different considerations apply at base stations, which handle several communications simultaneously, and terminals, which usually have to support no more than one communication at a time. For example, a time-division transmission technique, which allows the same radio equipment to serve many different channels simultaneously, produces economies at a base station but not at a terminal. Changing the distribution of network control tasks can add to the cost of terminals while creating base station economies, or vice versa. The topological organization of system elements affects the cost of the communications links among base stations, switching equipment, and external networks.

2.3.18 Cell Radius

Another important issue for infrastructure economics is the timing of required investments. It takes years for an operating company to build up a subscriber base that generates a profitable stream of revenues. The company would like its equipment expenses to flow gradually to meet rising demand. The alternative is to invest heavily at the outset, before there is any compensating revenue. The number of base stations necessary to meet radio coverage objectives (see Section 2.3.6) has a strong influence on start-up costs. This number is determined by the geographical range of a base station transmitter. The range depends on transmission technology.

2.3.19 Spectrum Efficiency

Personal communications system operators hold licenses to transmit radio signals within limited bands of frequencies. At any single location, these frequencies can carry only a small volume of communications. Personal communications systems achieve high capacity by means of frequency reuse (see Section 9.3). With frequency reuse, each channel carries different signals in different cells within a geographical service area. Thus, the total number of communications spread over the service area is considerably higher than the number that can be supported at a single location.

There are many ways to describe system capacity quantitatively. One measure is the total number of subscribers a system can serve within the constraints of certain quality objectives, such as binary error rate and call-blocking probability. This number depends on many factors, including:

- subscriber behavior (number of calls per unit time and call duration),

- the total bandwidth allocation,

- the radio technology employed, and

- the number of base stations in place.

To meet a growing demand for service, a system operator adds base stations, thus reducing cell size and increasing capacity. Capacity also increases proportionally with the size of a company's bandwidth allocation.

Various efficiency measures have been defined to isolate the effects of technology on capacity. Because overall system cost depends heavily on the number of base stations needed to serve a specific population, efficiency is often defined in terms of capacity per base station per MHz of assigned bandwidth. For telephony, this efficiency can be measured as the maximum number of simultaneous conversations per base station per MHz. A closely related measure is the number of Erlangs per base station per MHz. Erlang efficiency is proportional to the average number (which is a fraction of the maximum number) of conversations that a base station can handle while meeting a specified call-blocking objective (see Section 2.3.9). Services other than telephony have corresponding efficiency measures. Efficiency limits are imposed by interference from other signals in the same service area. If the number of simultaneous conversations is too high, the interference causes transmission quality to fall to unacceptable levels. Efficiency depends on radio technology and on network management operations referred to as *radio resources management*. As explained in Section 9.3, the two properties of a radio system with the strongest influence on efficiency are the bandwidth occupied by transmitted signals and the tolerance of the signals to interference.

2.3.20 Network Security

In conventional telephony, access to the network is the responsibility of each subscriber. The telephone company provides wires to the subscriber premises, and the subscriber has to take measures to prevent unauthorized individuals from generating large telephone bills. Computers rely on passwords to prevent unauthorized use of a system. Each user is responsible for protecting his or her own password. In personal communications, the system has to take far more responsibility for protection against unauthorized use of the network. Service requests travel through the air to reach the operating company, and the system requires techniques to determine whether these requests come from valid subscribers. Operating companies have found, to their expense, that first-generation cellular systems are particularly vulnerable to fraudulent use. Later systems incorporate more robust authentication technology to prevent unauthorized access to personal communications systems.

2.3.21 Early Deployment and Adaptability

All else being equal, the earlier the availability of a system, the better. However, with predictably rapid advances in information technology, all else is not equal. The impact of technical advances on personal communications is particularly strong. Some of the technologies deployed in 1997 were prohibitively expensive and/or complex in 1990. In 2002, there will be new opportunities to improve economy, quality, and efficiency. An important question for equipment vendors and system operators is when to introduce a new technology. Waiting too long can be costly. It deprives operating companies of revenue and consumers of the benefits of current technologies. It also adds to the difficulty of displacing a competitor with a strong market position, albeit one based on older technology. By anticipating future advances, systems can make the timing of their deployment a less critical issue for operating companies. Systems that anticipate developments can evolve and absorb new techniques as they become technically and economically feasible.

2.4 System Operations

The principal task of a personal communications system is to deliver to its subscribers information services, as described in Section 2.2. The most obvious aspect of performing this task is the transport of user information, such as a facsimile signal, from one part of the network to another. Beyond this basic operation, the system has to perform a large number of network control operations to meet the design goals described in Section 2.3. Each system performs these operations in its own way, by distributing specific functions among network elements. Table 2.3 lists six

Table 2.3 Network Operation Categories

Network Operation	Figure of Merit
User information transport	Transmission quality
	Spectrum efficiency
	Terminal price
	Infrastructure cost
Mobility management	Roaming
	Setup time
	Call blocking
	Infrastructure cost
Authentication and encryption	Network security
	Privacy
Call management	Available services
	Setup time
Radio resources management	Transmission quality
	Call blocking
	Call dropping
	Battery life
Operations, administration, and maintenance	Coverage area
	Transmission quality
	Call blocking
	Call dropping

categories of network operations and indicates up to four figures of merit that are strongly influenced by each category. They are introduced individually in the following paragraphs.

2.4.1 User Information Transport

The technical challenges of mobility, ether, and energy have a strong influence on the design of the *air interface* between base stations and terminals of a personal communications system. The technology for information transport across this interface is the most prominent feature of each practical system. The nature of the air interface reflects the tradeoffs among a large number of design goals including costs, coverage area, transmission quality, spectrum efficiency, and user mobility. The practical systems described in later chapters differ most in the details of their air interfaces. A major portion of this book is devoted to descriptions of these details.

2.4.2 Mobility Management

The operations in this category enable a system to deliver information to mobile subscribers. Mobility management procedures influence call setup time, probability of blocking, spectrum efficiency, and infrastructure costs. Each mobility management system represents a compromise between the effort required to track the locations of moving subscribers and the effort required to search for a subscriber when there is information to deliver. Two key procedures are *registration* and *paging*. A terminal registers its location by sending a message to the network. When it receives this message, the network learns which cell the terminal occupies. When the network has to send information to a terminal, it initiates a search procedure that involves broadcasting paging messages to cells in the vicinity of the cell where the terminal most recently registered.

If each terminal registers its location frequently, the network maintains accurate information on the locations of its terminals. This implies that when the network has to establish contact with a terminal, it can confine its search to a small number of cells in the vicinity of the cell that received the most recent registration message. Frequent registration can be costly, however. It consumes communications resources, including bandwidth on radio channels. It also requires frequent database updates. With less frequent registration, these costs diminish but there is more uncertainty about the terminal's location. Therefore, when the network has to deliver information (such as a call setup request) to the terminal, it has to search for the terminal over a relatively wide area by broadcasting paging messages to a large number of cells. Thus, infrequent registration adds to the network resources consumed in the paging process. Therefore, network operators have to establish mobility management procedures that efficiently balance the costs of registration and paging.

2.4.3 Authentication and Encryption

To address the design goal of network security, each system establishes procedures to verify that a terminal is authorized to use the system. These procedures are required when a terminal transmits a request to use the system, when it registers its location, and when it responds to a paging message. These procedures, to a degree, are analogous to the password controls imposed on computer users and the personal identity numbers entered into automatic teller machines. Compared with these applications, personal communications systems are far more vulnerable because authentication information travels through the air. As a consequence, the authentication procedures of personal communications systems are generally more elaborate than they are in other information systems.

Digital personal communications systems employ data encryption techniques to protect user information and sensitive network control information against eavesdropping. These encryption methods tend to be linked with the authentication operations performed by a system. Together, authentication procedures and encryption procedures are sometimes referred to as *network security operations* [Mouly and Pautet, 1992].

2.4.4 Call Management

The principal tasks in this category are call setup, call release, and delivery of special features. In the example of a telephone call, the personal communications network, often in cooperation with external networks, has to establish a series of communications links that connect the two telephones participating in the call. Similar operations are required for data transactions. At the end of the call, the network has to discontinue the connection and make the links available for other communications. Other tasks include call waiting notification, placing parties on hold, calling number identification, and call forwarding. To perform these tasks, personal communications systems, to a large extent, make use of procedures established for other networks. Thus, telephone services make extensive use of Signaling System 7 [Ramteke, 1994: Chapter 14] and ISDN [Ramteke, 1994: Chapter 15] protocols. Mobile computing applications also use established computer communications protocols, most prominently TCP (Transmission Control Protocol) [Halsall, 1992: Section 10.3] and IP (Internet Protocol) [Halsall, 1992: Section 9.6]. Because ISDN call management procedures and Internet protocols were originally intended for networks with stationary users, personal communications systems designers introduce additional procedures to accommodate mobility. A design goal for the new procedures is to minimize, or preclude entirely, the need to modify existing networks and consumer equipment.

2.4.5 Radio Resources Management

During a phone call or other communication session, radio resources management procedures are the main weapons deployed by a system to combat the triple challenge of mobility, ether, and energy. In general terms the task of radio resources management is to maintain high-quality communications between each terminal and a base station. To do this, the system has to ensure that the signals transmitted in any given communication session do not cause unnecessary interference to other communications. In

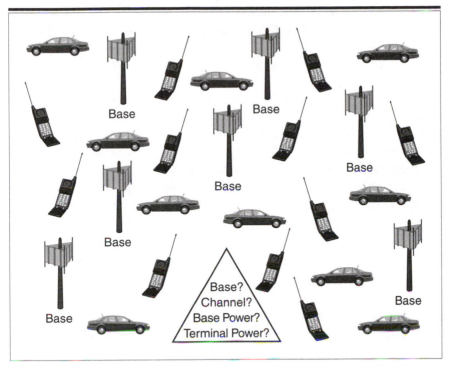

Figure 2.2 Radio resources management operations control four attributes of each communication session.

performing their tasks, radio resources management procedures dynamically manage four attributes of each communication, as illustrated in Figure 2.2, which shows a large number of portable and wireless terminals with communications sessions in progress. Each terminal can be at any location within a large service area, and its location can change as it communicates. There is also a set of stationary base stations at known locations. At any time, the system has to assign the following resources to each terminal:

- a base station,

- a physical channel (the nature of the channel depends on the access technology of a system, as described in Section 9.1),

- the power of the signal transmitted by the terminal, and

- the power of the signal transmitted to the terminal by the base station.

The assignments are subject to change as the terminals move. The procedures employed for assigning these resources influence several figures of merit. Those most strongly affected are call blocking, transmission quality, call dropping, mobility, battery life, spectrum efficiency, and infrastructure cost.

Although there is a concise statement of the radio resources management problem, the large number of figures of merit and the large number of interacting terminals make it difficult to evaluate prospective algorithms and to select effective ones. In effect, radio resources management is a very large combinatorial optimization problem with many interdependent objectives. To cope with this complexity, practical systems approach the big problem illustrated in Figure 2.2 by performing four separate tasks, shown in Figure 2.3.

Call Admission

Call admission procedures determine whether the system accepts a request to set up a new communication. Sometimes requests are denied because all channels are occupied at the base stations within range of the terminal. Even when there are channels available, the system sometimes rejects setup requests. In some cases, the system reserves one or more channels to satisfy potential handoff requests. This reflects the tradeoff between call-blocking probability and call dropping. Because interruption of a call

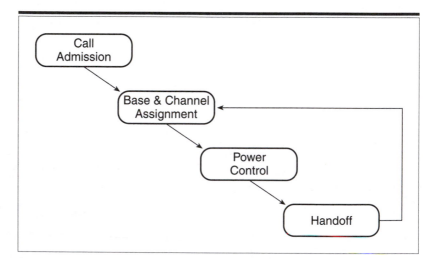

Figure 2.3 Four separate radio resources management tasks.

in progress is more annoying than failure to establish a call, the system accepts a higher-than-necessary call-blocking probability in order to reduce call dropping. Setup requests may also be rejected by a call admission procedure when the system determines that the new call would cause too much interference to calls in progress.

Base and Channel Assignment

When it accepts a call request, the system assigns the call to a base station and a channel. In most situations, the call goes through the base station nearest to the wireless terminal. An exception occurs when there is a high signal attenuation (due to a physical obstacle, for example) on the transmission path between the terminal and the nearest base station. If there is a more distant base station with a better signal path to the terminal, the system will assign the call to that base station. The call might also go through a more distant base station when no channels are available at the nearest base station.

Depending on the system, there are various ways that base stations are associated with channels. In cellular systems, the most common approach is *fixed channel allocation*, governed by a channel plan. This is the approach taken in cellular systems that use FDMA and TDMA, and is described in Section 9.3.2. Another approach, used in cordless telephone systems, is *dynamic channel allocation*. With this approach, all base stations have access to all channels. After selecting a base station to handle a new call, the system searches the entire set of channels for a channel that will allow the call to proceed with high transmission quality. The channel selection depends on which channels are already in use at the base station selected for the new call and the channels in use at surrounding base stations.

CDMA systems assign base stations and channels to calls in a different way. CDMA channels correspond to the binary codes assigned to individual calls. All codes can be used at all base stations, regardless of which other codes are already in use. As described in Section 9.8, CDMA systems maintain high transmission quality by limiting the number of calls in progress at a base station.

Power Control

Power control procedures determine the power levels for transmission to and from the terminal. If, during a call, the terminal moves closer to the base station it is using, the system may reduce the power levels in order to reduce interference to other calls. Conversely, as the terminal moves further from the base station, the system may increase the power levels in order to maintain transmission quality. Power control procedures can also promote long battery life by reducing the power radiated at a terminal to the minimum needed to meet transmission quality objectives.

Handoff

Eventually, *handoff criteria* adopted by the system may determine that the call would be better served by a different channel and/or base station. Handoff procedures then direct the terminal and infrastructure to move the call to the new channel and/or base station.

2.4.6 Operations, Administration, and Maintenance (OA&M)

The telephone industry uses the acronym *OA&M* to refer to procedures for monitoring the performance of communications systems, changing system configuration, and responding to malfunctions. In personal communications, these procedures have a strong influence on quality and economics. This is because personal communications systems are new and growing rapidly. Network operators serve new subscribers by adding equipment to existing base stations and installing new base stations and switches. Specific requirements for system expansion depend not only on the number of subscribers but also on their communication habits and their mobility. Because these characteristics of subscriber behavior are hard to predict, operators of personal communications systems rely heavily on network monitoring techniques and on planning strategies based on measurements and predictions. For the most part, these techniques are not addressed by the standards that govern other aspects of system operation. They are proprietary, competitive tools of each company, and as a consequence are outside the scope of this book.

2.5 A Framework for Studying Personal Communications Systems

The principal purpose of this book is to describe a diverse set of communications systems in a consistent manner. By adopting one approach to studying different systems, readers will be able to identify and understand the reasons for the differences and similarities between systems. The first three sections of this chapter examine systems from the outside. They describe system properties perceived by the users and operators of the systems. Section 2.4 looks inside a system in order to identify the operations a system performs to deliver services to users. This section presents our method for studying the collection of technologies assembled by each system in order to perform these operations.

Each system description begins with a brief summary of the system history. This background provides insights into the goals of the system designers. Although all 21 goals listed in Section 2.3 are relevant to any system, the designers of each system established their own priorities among conflicting goals. The introductory section of each chapter states the goals that had the highest priority in the system covered in the chapter; to understand how systems differ, it is useful to refer to the goals of the system designers.

The presentation of the details of each system begins with a description of system architecture consisting of network elements and interfaces between the elements. As previously stated, the most important interface in a personal communications system is the air interface, which is the set of techniques for transferring information between base stations and terminals. We examine the air interface of each system in three sections, corresponding to the lowest three layers of the Open Systems Interconnection (OSI) protocol summarized briefly in Section 9.10 [Halsall, 1992: Sections 1.4 and 1.5]. Many of the systems we describe refer explicitly to the OSI in their formal specifications. In other systems, the correspondence is implicit. The three topics that cover the air interface of each system are radio transmission, corresponding to layer 1; logical channels, corresponding to layer 2; and messages, corresponding to layer 3. Armed with the details of a system's communication techniques, we proceed in each chapter to give examples of how these techniques are used to perform the key network control operations presented in Section 2.4.

The final section of each system description is a status report that complements the introductory section. The introductory describes system history, and the concluding section is a brief report on the status of the system in early 1997 and its outlook for the future. Because the technology of personal communications is evolving so rapidly, each chapter concludes with a description of work in progress to enhance the system described.

2.5.1 Architecture

Each system defines its own network architecture, which consists of *network elements* and *interfaces* that specify communications between elements. In telephone networks, the architecture with the strongest influence on developments in the 1980s and 1990s has been ISDN (integrated services digital network). ISDN has also influenced the design of personal communications systems, especially GSM (see Chapter 7). Section 9.11 presents the ISDN details that have the strongest influence on personal communications systems architecture.

Network Elements

Some systems use the term *standard devices* to refer to network elements. These devices can be understood as functional units that produce specified outputs in response to specified inputs. Each unit, of course, contains many components. However, the nature of these components and their interactions are irrelevant to the network architecture. The degree of detail in the definition of network elements varies from system to system. Entities that are network elements of some systems are embedded in larger network elements in other systems. Base stations are prominent examples. In some systems, base stations are individual elements. In others, a base station is a collection of network elements, each with interfaces to other elements. It is important to understand that the network architecture specifies distinct *functions* performed by a system, rather than distinct *physical units* within the system. This means that it is not necessary for each network element to exist as a separate piece of equipment. For example, in a residential cordless telephone system, the base station and the switch are parts of the same physical unit.

In Chapter 1, we encountered three network elements that are essential to any personal communications system: a terminal, a base station, and a switch. Each practical system has its own terminology for network elements. We will use the terminology specific to a particular system when we study that system. The most obvious part of a personal communications system is an information *terminal* capable of wireless communications with the rest of its system. Terminals take many forms. The most popular ones are small handheld cellular telephones that carry their own batteries. There are also vehicular telephones connected to antennas mounted outside the vehicle. These telephones obtain their power from the vehicle's storage battery. In standards documents, the most common nomenclature for a terminal is *mobile station*. For non-voice communications, it is possible to embed the functions of a mobile station in a portable computer or other information device.

Each terminal communicates with radio equipment in a fixed location. This collection of equipment performs the functions of a *base station*. Information originating at a wireless terminal arrives at a base station en route to its destination, which is often in an *external information network.* For telephone calls, the external network is referred to as the *Public Switched Telephone Network* (PSTN). For other types of communication, the external network could be a public or private data network. Similarly, signals traveling to a wireless terminal often originate in an external network, enter the personal communications network at a switch, and then go to a base station for transmission to the terminal. A *switch* is responsible for moving information through the fixed part of the personal communications system. Switches and base stations are components of the personal communications *network infrastructure.*

Figure 2.4 displays the essential network elements of a personal communications system along with an external information network. The external information network can take many forms—such as a private branch exchange, a Public Switched Telephone Network, or a local area network. In many situations the external network shown in Figure 2.4 is only one part of the route taken by information moving to and from personal terminals. To reach its destination, user information may traverse many external networks, including another personal communications system.

Although terminals, base stations, and switches are essential to any personal communications system, many systems specify additional network

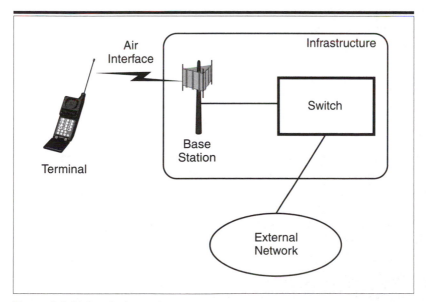

Figure 2.4 Network elements.

elements. Among these additional elements, databases that record the locations of subscribers are the most important.

Interfaces

A network architecture specifies which network elements can communicate with each other. *Interfaces* are the communications protocols for linking network elements. If an interface is *open*, there are published protocols for communication between a pair of network elements. By contrast, a *proprietary* interface is private intellectual property available only to the owner and license holders. A major trend in cellular communications is the move in second-generation systems toward open interfaces. In the original cellular telephone systems introduced in the early 1980s, the only open interface was the air interface that specifies radio communications between terminals and base stations. An open air interface makes it possible for a large number of companies to offer terminals to cellular subscribers. All other first-generation interfaces were proprietary. One consequence is that a single manufacturer controls the supply of base stations, switches, and other infrastructure equipment to each system. Open interfaces, such as those in GSM (see Chapter 7), make it possible for operating companies to assemble a network using elements obtained from a variety of sources.

To facilitate mobility and roaming between service areas covered by different switches, in the early 1990s the North American cellular industry adopted an open interface for inter-switch communications. This interface [TIA, 1991] is the subject of Chapter 4.

Architecture Examples

Personal communications network architectures can be very different from one another. Figures 2.5 and 2.6 show two contrasting examples. Figure 2.5 pertains to first-generation cellular systems. It is hierarchical. Each base station communicates with several telephones—a maximum of 57 in most U.S. systems. Many base stations, on the order of 100 in large systems, are connected to one switch, which in turn has connections to other switches and to the PSTN. The switch maintains tight control over all of the cellular calls in its service area, as will be seen in the following chapters.

Figure 2.6 shows the radically different architecture of first-generation residential cordless telephones. Each telephone has its own infrastructure, consisting of a base station and a simple switch packaged as one unit. The owner connects this unit to the home telephone wires. In contrast with the

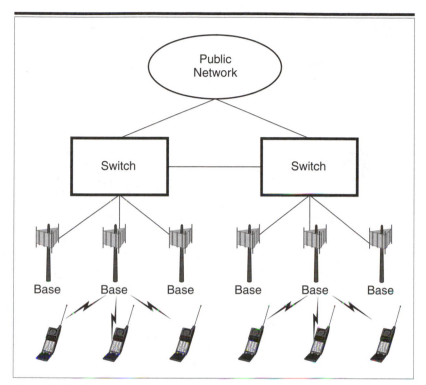

Figure 2.5 Hierarchical cellular telephone architecture. Each switch exercises control over a large number of terminals.

centralized control of the cellular system in Figure 2.5, the cordless architecture is highly distributed. Each cordless phone independently attempts to establish and maintain high-quality communications.

Physical Channels and Logical Channels

The word *channel* has more than one meaning in the context of personal communications technology. Two of the dictionary definitions of the word are "a path along which data passes" and "a band of frequencies." Both these definitions refer to the physical transfer of signals from one network element to another. To refer to this physical transfer we use the term *physical channel*. On the air interface, physical channels can be associated with time slots and codes as well as with frequency bands. In digital systems, physical channels transfer bits from one network element to another.

We also have many occasions to use the term *logical channel*. Regardless of the physical means of signal transfer, logical channels are distinguished

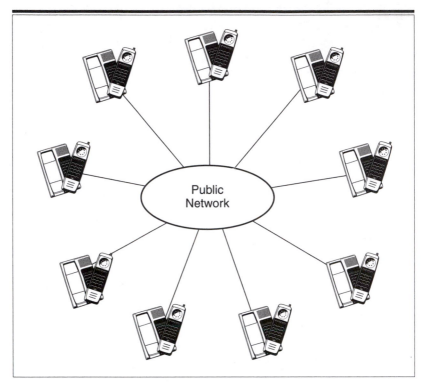

Figure 2.6 Cordless telephone architecture. There is no coordination
of different telephones.

by the nature of the information they carry and by the way in which they
assemble bits into larger data units. Two examples of logical channels in
personal communications are paging channels and synchronization channels. Paging channels inform terminals that the network has information
to transfer to them. The purpose of a synchronization channel is to transfer timing information between base stations and terminals. Each logical
channel has its own way of assembling bit sequences to achieve a specific
purpose.

In examining specific systems, we find that each information transfer
requires the simultaneous use of a logical channel and a physical channel.
To avoid confusion, it is necessary to remember that even though each
one is referred to as a "channel," physical channels and logical channels
play different roles in the task of moving information in a network.

2.5.2 Air Interface Layers

For each system, our air interface presentation consists of three major sections corresponding to the lowest three layers of the seven-layer OSI

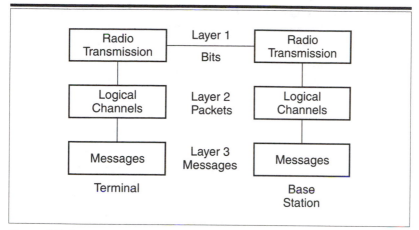

Figure 2.7 Organization of air interface presentations.

(Open Systems Interconnection) model [Halsall, 1992: Sections 1.4, 1.5], described in general terms in Section 9.10. Figure 2.7 displays the relationships of the three sections of each air interface presentation: radio transmission, logical channels, and messages, corresponding approximately to layers 1, 2, and 3, respectively, of the OSI model. The following paragraphs discuss them individually.

2.5.3 Radio Transmission

In the OSI model, the task of the physical layer is to move analog signals or digital symbols from one node in a network to another. In wireless networks, the physical layer plays a critical role in the air interface between a terminal and a base station. It is here that nature places formidable obstacles in the way of information moving over radio channels. These obstacles include

- limited spectrum that must be shared efficiently among many users;
- transmission impairments, such as fading, interference, and multipath propagation, that can change abruptly with time, location, and frequency band;
- interrupted connections associated with handoff procedures; and
- limited power available to portable terminals.

The physical layer, therefore, confronts directly the three technical challenges of personal communications described in Chapter 1: mobility, ether, and energy.

To meet these challenges, personal communications systems deploy an impressive arsenal of techniques including modulation, source coding, channel coding, interleaving, diversity reception, and channel equalization. This battle against nature, with limited-bandwidth weapons, contrasts sharply with many fixed networks, which rely on constant, high-quality transmission facilities.

Each presentation of a specific system begins with the frequency bands assigned to the system by regulatory authorities, and with the range of transmitter power levels. We then address transmission techniques, including modulation, source coding (Section 9.7), forward error correction (Section 9.4), and interleaving (Section 9.5). We also describe the *access* technology (Section 9.1), which is the most conspicuous part of each system. Many people use the access technology designation, such as CDMA or TDMA, as the name for an entire system. If someone knows only one technical fact about a system, this fact is likely to be the access technology.

The access technology determines the nature of the physical channels in a personal communications system. This is a key issue in wireless networks because the signals associated with a large number of communications all meet in the open-air transmission path. The access technology specifies the means of separating the desired signal at a receiver from the multitude of interfering signals that arrive with the desired signal. (In conventional local area networks, access is also a key issue because economics dictate that many terminals share the same physical medium, such as a coaxial cable or a pair of wires, on a dynamic basis.) In personal communications systems, the term *multiple access* refers to transmissions from terminals to base stations, and *multiplex* refers to transmissions from base stations to terminals. OSI protocols refer to the access technology (either multiple access or multiplex) as the *media access* method.

Media access would appear to be sufficiently important to be designated a separate layer of a reference model. Rather than define a new model with an eighth layer, however, each system specifies a *media access sublayer*, which finds its way into layer 1 of some networks and layer 2 of others, as indicated in Figure 2.8. In our system descriptions, we present the access technology in the radio transmission section.

The three access technologies employed by personal communications systems for establishing physical channels are FDMA, TDMA, and CDMA (frequency division multiple access, time division multiple access, and code division multiple access, respectively). These technologies are described in general terms in Section 9.1. While *multiple access* is concerned with the mixture of signals going to and from different terminals,

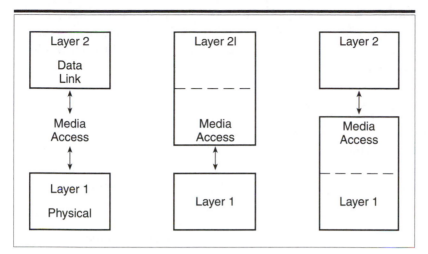

Figure 2.8 Media access could be a separate layer. In protocols conforming to the OSI model, it is either at the bottom of layer 2 or the top of layer 1.

the *duplex* technique separates signals transmitted by base stations from signals transmitted by terminals. Duplex comes in two forms: time division duplex and frequency division duplex. All of the cellular systems described in this book employ frequency division duplex. The digital cordless telephone systems (see Sections 8.1 and 8.2) and the Personal Handyphone System (see Section 8.3) employ time division duplex.

In each chapter, the section covering radio transmission concludes with a discussion of spectrum efficiency. As discussed in Section 2.3.19, there are several different ways to define spectrum efficiency. The definition adopted in this book is the maximum number of simultaneous conversations per base station per MHz of assigned spectrum, which can be written as

$$E = \frac{\text{conversations}}{\text{cells} \times \text{MHz}}.$$

Spectrum efficiency defined this way is an extremely important figure of merit for cellular network operators. Their revenues come primarily from air-time charges (revenue collected while a subscriber occupies a physical channel). This is reflected in the numerator of the fraction. Network costs are heavily influenced by the number of base stations necessary to provide service, which is the first quantity in the denominator. The amount of traffic that a system can handle is directly proportional to the bandwidth it occupies. By including the second quantity in the denominator, we arrive at a fair basis for comparing the efficiencies of systems with unequal spectrum allocations. Section 9.3 describes the factors that influence the efficiency of TDMA and FDMA systems. Chapter 6 addresses CDMA efficiency.

2.5.4 Logical Channels

Layer 2 of the OSI model addresses the unpredictable quality of the physical layer. The purpose of the data link layer is to ensure that information arrives at the receiving end of a link accurately and promptly. The operational definitions of accuracy and promptness depend on the type of information transferred. For example, in conversational speech, the accuracy requirement is not very stringent because humans are accustomed to conversing in situations that involve substantial impairments to the acoustic waves that reach our ears. On the other hand, we are not at all accustomed to delays in conversational speech, and the situation becomes stressful if we have to wait even one-half second to hear each other's voices. By contrast, many types of numerical data transmission depend on complete accuracy, but are fairly tolerant of delays. Consider the transmission of telephone numbers. We would prefer to wait an extra few seconds to be certain of reaching the correct telephone rather than risk being connected, albeit quickly, to another one.

Personal communications systems move several types of information—for example, conversational speech, telephone numbers, and hand-off commands—between terminals and base stations. To address the fact that different information types have different requirements, systems segregate the information into logical channels. Often the name of a logical channel conveys its purpose. For example, a traffic channel carries user information and a paging channel carries paging messages. Each logical channel incorporates its own means of promoting the appropriate levels of accuracy and promptness in information delivery.

To transmit information obtained from layer 3, a layer 2 protocol assembles a collection of bits into a *packet*. The packet is divided into *fields*, with each field serving a special purpose. Some of the fields contain layer 3 information or user information to be transmitted, while other fields contain information added by the layer 2 protocol, such as sequence numbers, packet length indications, indications of the beginning and end of the packet, and parity check bits. The parity bits enable the layer 2 protocol at the receiving end of the link to verify the accuracy of the received information. Depending on the logical channel, the receiver incorporates a means of dealing with packets that arrive with transmission errors. One approach is to send a message back to the transmitter to acknowledge packets received without errors. When a receiver detects errors in an arriving packet, the layer 2 protocol specifies the action to be taken. In many cases the receiver discards the received information and sends no acknowledgment. Failing to receive an acknowledgment, the transmitter sends the packet again. The combination of error detection and retransmission is the most prominent layer 2 mechanism for promoting accurate packet delivery.

In addition to the structure of their packets, logical channels are also characterized by their topology. Traffic channels are typically "one-to-one," carrying information between one base station and one terminal. Other channels carry "one-to-many" information, such as synchronization signals transmitted by a base station to all terminals in a cell. *Broadcasting* is a synonym for one-to-many transmission.

The third type of channel is "many-to-one." This type of logical channel carries information from several terminals to the same base station. An example is the logical channel used by terminals to transmit call setup request messages to a base station. To coordinate the arrival of these messages at the base station, many-to-one channels employ *random access protocols*. Most of them resemble protocols devised for computer networks such as ALOHA, carrier sense multiple access, and busy tone multiple access. An important property of a random access protocol is the sequence of actions taken by a terminal when it determines that the base station has failed to receive one of its packets. Often the cause of the failure is a collision with a packet from another terminal, and the remedy includes retransmission of the packet after a random time delay.

2.5.5 Messages

The third layer of our examination of an air interface focuses on messages. By examining the messages that a system can transmit between base stations and terminals, we learn the system's capability of performing the operations described in Section 2.4. Functionally, the messages arrive at one network element from another network element that performs one of the operations. For example, the call management function generates a request to set up a call. This request causes the terminal to transmit a message labeled "service request" in Figure 1.13 to a base station. Layer 3 sends that message to layer 2, which divides it into packets and adds its own information, depending on the logical channel that carries the message to the terminal.

When we study the messages contained in a specific system, we examine two things: message structure and message content. The structure indicates the nature of the information carried in the message. An example is the message type, which distinguishes a handoff message from a power control message. Other categories of information are the message length and data that the system needs to perform the operation stimulated by the message. For example, a call setup message contains the phone number of the called party. A handoff message indicates the new channel to use and how much power to transmit.

2.5.6 Network Operations

The topics listed in Figure 2.7 and described in Sections 2.5.3 through 2.5.5—radio transmission, logical channels, and messages—comprise the tools that a system uses to move user information to and from terminals and to manage its resources in the battle against the challenges of mobility, ether, and energy. Following the presentation of these tools, each system description presents examples of the way the system uses these tools to accomplish specific tasks, with the emphasis on operations that represent innovations introduced in that particular system.

2.5.7 Status

Each system description concludes with a status report that indicates the deployment of the system in early 1997 and plans for introducing new technology to enhance the system.

Review Exercises

1. Consider the figures of merit described in Section 2.3. Give examples of pairs of goals that are contradictory in the sense that working hard to achieve one of them will make the other one more difficult to achieve (for example, having an exciting social life and getting high grades).

2. Give examples of pairs of figures of merit that are consistent in the sense that working hard to achieve one of them will make the other one easier to achieve (for example, getting good grades and not spending much money on entertainment).

3. If it is your aim to make money by offering a short message service using a personal communications system, which design goals (figures of merit) are most important? Which ones are least important?

4. What is the purpose of power control in a personal communications system? Describe some ways in which power control can be implemented.

5. Why does a terminal register its location with a network? What is the advantage of frequent registration? What is the disadvantage? List some criteria for deciding when a terminal should register its location.

6. Why does a personal communications system sometimes deny service to a terminal that attempts to set up a call? List some criteria for deciding whether a network should allow a terminal to begin a call.

7. Compare fixed channel allocation with dynamic channel allocation. What are some advantages and disadvantages of the two approaches?

8. List some conditions under which a handoff should occur. How can a system decide when to perform a handoff?

9. With the system architecture shown in Figure 2.5, what are some system operations that have to take place when there is a handoff from one base station to another? What is the effect of having the two base stations connected to different switches?

10. This book adopts as a measure of spectrum efficiency "conversations per cell per MHz" (Section 2.5.3). What are some other measures? Compare each one with the measure adopted here.

CHAPTER THREE

Analog Cellular Communications AMPS System

3.1 Background and Goals

Compared with the 120-year history of the telephone, cellular systems are newcomers. Pioneering experiments took place in the 1970s in the United States [Jakes, 1974]. The earliest commercial systems went into service in 1980 and 1981 in Japan and Scandinavia. The 1980s and 1990s have seen rapid expansion of geographical coverage and subscriber populations in most parts of the world. Cellular communications originated in prosperous industrial countries with advanced telephone networks. Most people already had telephones at home and at work. The original purpose of cellular systems was to add motor vehicles to the list of places with telephones. The target customers were a small prosperous minority of the population with special needs.

Not only have these original aims been fulfilled, they have been surpassed in several ways. Cellular telephones are by now familiar parts of expensive cars as anticipated by their developers, and they are even more popular in the form of small, lightweight, portable units. With its own electronic directory of names and numbers, a cellular phone is personal. It belongs to a person rather than a residence, office, or vehicle. The other surprise of cellular service is the mass market appeal. Even though prices are high compared with conventional telephony, cellular service and equipment are popular consumer items. In common with other countries with well-developed cellular services, market penetration in the United States exceeds 15 percent of individuals and 30 percent of households. In addition to their popularity in industrial countries, cellular telephones have attracted markets in countries at all stages of economic development.

The popularity of the original cellular systems has been a principal stimulus for the development of the new technologies described in the chapters following this one. This chapter covers the original, *first-generation* cellular technology, focusing on AMPS (Advanced Mobile Phone System) [Electronic Industries Association, 1989; *Bell System Technical Journal*, 1979]. AMPS and its first-generation relatives are important as precursors of the newer technology. In addition, the existence in 1997 of more than 40 million AMPS subscriber units and supporting infrastructure ensure that AMPS systems will be in service for many years to come, regardless of the relative merits of new systems. In fact, the first digital cellular standards in North America specify "dual-mode" operation, with each terminal capable of both analog and digital voice transmission.

AMPS is one of several first-generation cellular systems. All are mutually incompatible in the sense that terminals conforming to one standard cannot operate with base stations conforming to another standard. Prominent differences between systems include operating frequencies and channel bandwidths [Rappaport, 1996: 548; Mehrotra, 1994]. On the other hand, all analog cellular systems share many characteristics. The most prominent one is voice transmission by means of frequency modulation. Their network architectures are all similar to AMPS and they have similar signaling systems. AMPS systems are in place throughout the United States and Canada as well as several Latin American and Pacific countries.

Referring to the services and design goals of Sections 2.1 and 2.2, AMPS delivers basic telephony and supplementary services of which voice mail and call forwarding are the most popular. Although it is possible to transmit digital data over AMPS channels, service quality is vulnerable to channel impairments and handoffs. The main design goals of AMPS and other first-generation systems were wide area geographical coverage, low probabilities of call blocking and call dropping, high transmission quality, high user mobility, high spectrum efficiency, and early deployment.

3.2 Architecture

AMPS is an American National Standard with the title "Mobile Station, Land Station Compatibility Specification." This title is significant, not only for the words it contains but also for what it omits. The AMPS standard says nothing about communications between base stations and switches. These communications conform to proprietary protocols specific to individual equipment vendors. This makes it impossible for a service provider to use base stations from one supplier with a switch from a competing supplier. It also inhibits coordination of operations between cellular switches produced by different manufacturers. With limited coordination between switches, cellular communications in the United States began as a collection of local services. Each subscriber was able to initiate and receive calls within a home subscription area. Roaming services, which make it possible to use a cellular phone outside of the subscriber's home area, were spotty and inconsistent from company to company. In the mid-1990s, the American cellular operating industry made major advances toward making cellular a national service, allowing everyone within range of a base station to initiate and receive phone calls. A major step toward making this possible was the adoption of a standard interface, described in Chapter 4, for communications between AMPS switches.

3.2.1 Network Elements

The system architecture [*Bell System Technical Journal*, 1979: 11], displayed with the AMPS terminology in Figure 3.1, is the one presented in Figures 1.12 and 2.5. The AMPS specification refers to terminals as *mobile stations* and to base stations as *land stations*. People in the industry often use the terms *mobiles* for terminals and *cell sites* for base stations. Although the AMPS specification does not refer to the cellular switch, this network element plays an essential role in all AMPS communications. The common terminology for an AMPS switch is *mobile telephone switching office* (MTSO). Although the communication links between the base stations and switch are labeled *land lines,* this terminology can be misleading. In many cases, these links are one or more copper wires or optical fibers, carrying digitally multiplexed groups of signals, leased from the local tele-

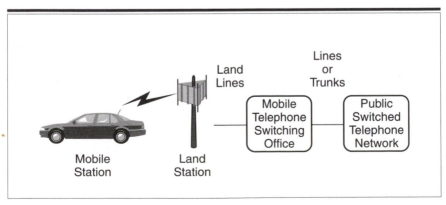

Figure 3.1 AMPS architecture and terminology.

phone company. In many areas, cellular service providers operate private microwave systems to connect cell sites to an MTSO. The connections between the MTSO and the public telephone network can also take a variety of forms. Usually these facilities are the property of a local or long-distance telephone company. Depending on the size of the MTSO, the links to the local telephone company can be in the form of subscriber lines terminating in a central office switch (small MTSO) or trunks terminating in tandem switch (large MTSO).

3.2.2 Identification Codes

AMPS specifies several identification codes for each mobile station. The *mobile identification number* (MIN) is a ten-digit telephone number, stored in a 34-bit binary representation. In the United States, this number has the same format as a conventional telephone number. The first three digits comprise the area code associated with the subscriber's home service area. This is followed by a seven-digit telephone number consisting of an exchange number (three digits) and a subscriber number (four digits). The exchange number is assigned to the cellular operating company. When a subscriber changes operating companies, it is necessary to change cellular phone numbers. In contrast to U.S. practice, many countries assign special prefixes (corresponding to area codes) exclusively to mobile telephone numbers. This practice makes it possible for callers to distinguish calls to mobile telephones from calls to conventional telephones.

Another identification code is a 32-bit *electronic serial number* (ESN) assigned permanently to each terminal. As a permanent characteristic of a physical unit, the ESN is similar to the engine number of a car. The MIN is analogous to the car's registration number, which, in the United States, changes when the car changes owners, or when the owner moves to a different state. A third identification code is the 4-bit *station class mark* (SCM), which describes the capabilities of the terminal. Station class marks indicate whether the terminal has access to all 832 AMPS channels, as described in Section 3.3.1, or whether it is an old model with only 666 channels. Another property conveyed by the SCM is the maximum radiated power of the terminal. This could be either 600 mW or 4 W. As the AMPS system evolves, the industry specifies new station class marks to identify mobile stations with special properties that influence network control operations.

The *system identifier* (SID) is an important 15-bit code stored in all base stations and all mobile stations. In the United States, the Federal Communications Commission issues an SID to an operating company when it issues a license to offer service in a specific area. *System* is AMPS terminology for cellular operations provided by one company in a specific area. Thus, each base station is part of a system. In many places, there is one MTSO per system. However, two or more systems with relatively small numbers of subscribers can share a single MTSO. Conversely, a large system is likely to operate with two or more MTSOs.

Each mobile station stores the identifier of the system that administers its subscription. This is the *home system* of the terminal. When the mobile station performs an initialization procedure (as in Figure 1.5), it compares its own SID with the SID broadcast by the local cell site. Identical SIDs indicate that the mobile station is using its home system. If SIDs are not identical, the mobile station is a roamer in another system. In this event, the terminal indicates, on its display, that it is in a roaming area. This alerts the subscriber to the possibility of incurring special roaming charges.

In addition to the SID assigned by regulatory authorities to each base station, the local operating company assigns two identifiers, the *digital color code* (DCC) and the *supervisory audio tone* (SAT), which help mobile stations distinguish neighboring base stations from one another. The SAT assigned to a base station is one of three analog sine waves. Neighboring base stations operate with different SATs. The 2-bit digital color code serves a similar purpose. Table 3.1 is a summary of AMPS identification codes.

Table 3.1 AMPS Identifiers

Notation	Name	Size (bits)	Description
MIN	Mobile identifier	34	Directory number assigned by operating company to a subscriber
ESN	Electronic serial number	32	Assigned by manufacturer to a mobile station
SID	System identifier	15	Assigned by regulators to a geographical service area
SCM	Station class mark	4	Indicates capabilities of a mobile station
SAT	Supervisory audio tone	*	Assigned by operating company to each base station
DCC	Digital color code	2	Assigned by operating company to each base station

* One of three sine wave signals.

3.3 Radio Transmission

3.3.1 Frequency Bands and Physical Channels

AMPS operates in the frequency bands shown in Figure 3.2. The original allocation in the United States covered a bandwidth of 40 MHz (Figure 3.2a). The bandwidth was later extended to 50 MHz (Figure 3.2b). Frequency division duplex separates signals traveling to a mobile station from signals transmitted by the mobile station. The band for *forward* transmissions, from cell site to mobile station, is 870–890 MHz. The *reverse* band, for transmissions by mobiles, is 45 MHz lower. An AMPS physical channel occupies two 30 kHz frequency bands, one for each direction. There are 666 channels in the original AMPS spectrum allocation, corresponding to the ratio of the entire AMPS bandwidth (per direction), to the width of a physical channel: 20 MHz/30 kHz. AMPS channel numbers begin with 1 at the bottom of the original band and continue to 666. The carrier frequency corresponding to channel C is

$$f(C) = 825,000 + 30\,C\,\text{kHz} \tag{3.1}$$

for transmissions in the reverse direction. In the forward direction the carrier frequency is $f(C) + 45,000$ kHz.

Soon after AMPS entered commercial service in the United States and Canada, regulatory authorities responded to the industry's request for

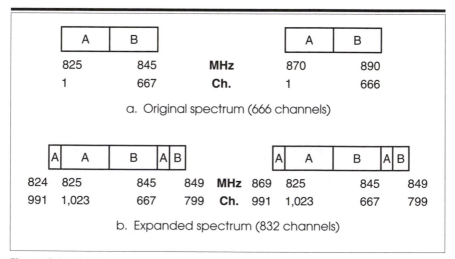

Figure 3.2 AMPS spectrum and channel numbers.

additional radio spectrum by adding 10 MHz to the original 40 MHz allocation. For each direction of transmission, the expanded spectrum contains a 1 MHz band just below each band of the original spectrum and a 4 MHz band just above each original band. There are 832 channels in the expanded spectrum, with channel numbers 1 to 799 related to carrier frequencies according to Equation 3.1. The other 33 channels, in the 1 MHz band below the original band, have the numbers 991 to 1,023. The carrier frequency of one of these reverse channels is

$$f(C) = 825,000 + 30\,(C-1,023);\ \text{kHz};\ 991 \le C \le 1,023. \quad (3.2)$$

Although these channel numbers may look strange, they correspond to negative numbers in two's complement arithmetic, and thereby facilitate the computation of carrier frequencies. Equipment in an AMPS terminal delivers the result of this computation to a frequency synthesizer, which tunes the radio to the assigned frequency.

Figure 3.2 divides the AMPS spectrum into two (equal, but not contiguous) regions labeled A and B. In the United States, regulators issue two cellular operating licenses in each geographical area. One license authorizes a company to operate in the 416 channels of the A-band. The other license applies to the 416 channels in the B-band. There are 1,466 operating licenses in the United States corresponding to two licenses in each of 305 metropolitan statistical areas and 428 rural service areas. The result of this licensing procedure is that each subscriber can choose between two operating companies in any given area. At most locations, a cellular terminal is within the operating range of an A-band cell site and a B-band cell site. The two cell sites have different system identifiers (SIDs). All systems operating in the A-band have odd SIDs (least significant bit = 1). B-band systems have even SIDs. All terminals have access to all 832 AMPS channels (except for the oldest terminals, which can tune only to the original 666 AMPS channels). A particular terminal is programmed with a preference for, or with a restriction to, the band (A or B) in which its home system operates. If it tunes to a control channel at a cell site operating in the other band, the mobile station appears as a roamer in the competing system, even though it is present in the service area of its home system.

Among the 832 AMPS channels, there are 42 channels (21 channels in each band) that carry only system control information. They are channels 313–354, in the center of the original AMPS band. To establish contact with an AMPS system, the receiver at a mobile station tunes to one of these channels. In areas with a high density of cellular subscribers, operating companies may designate additional channels as system control channels.

All other channels (up to 395 channels per operating company) are traffic channels, available to carry user information, which usually takes the form of conversational speech.

3.3.2 Radiated Power

An AMPS terminal is capable of radiating signals at six or eight different power levels depending on the nature of the terminal. A command from the base station establishes the actual power radiated by the terminal. The radiated power levels range from 8 dBm[1] (6 mW) to 36 dBm (4 W) in steps of 4 dB, so that each possible power level is 2.5 times higher than the next lower one. A Class I mobile station, usually powered by a vehicle battery, has access to all eight power levels. A Class III mobile station, typically handheld, ranges over the six lower levels, to a maximum of 600 mW. (The AMPS standard also refers to Class II mobiles with a maximum radiated power of 1.6 W. However, there are no commercial products operating at this limit.) The radiated power at a base station is typically 25 W per channel, for wide area coverage, and lower in cells with small service areas.

Messages from the base station control the transmitted power level of active terminals. Some terminals are designed for *discontinuous transmission* (DTX). During a conversation, it is possible for these terminals to alternate between two power levels, corresponding to ON and OFF states, under the control of a speech activity detector. In the ON state, when a terminal detects a speech input, it transmits at the power level commanded by the base station. In the OFF state, it transmits at a reduced level to conserve battery power. This also reduces the interference to other conversations.

Each AMPS channel can carry signals in an analog format for conveying user information or a digital format for system control information. The following paragraphs describe these signal formats that were designed to promote reliable information transfer in the presence of transmission impairments.

3.3.3 Analog Signal Processing

In Figure 3.3 the four operations prior to the modulator serve to maintain high signal quality and to limit *adjacent channel interference* to transmissions in neighboring physical channels. Compression and pre-emphasis are established techniques for audio signal transmission.

---

[1] The definition of dBm is decibels relative to 1 mW.

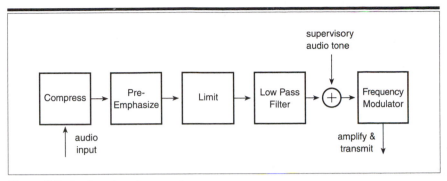

Figure 3.3 Analog signal processing.

The purpose of the compressor and a corresponding expandor at the receiver is to raise transmission quality when the input signal exhibits a large range of amplitudes. Human speech has a high dynamic range. For one speaker, the energy in loud sounds (typically vowels) is 16 times (12 dB) stronger than the energy in weak sounds (unvoiced consonants). Magnifying this range of amplitudes is the difference in average sound level between loud speech and quiet speech. This high dynamic range makes a transmission system vulnerable to degradation of strong sounds by nonlinear distortion and degradation of weak sounds by noise. The compressor reduces this vulnerability by compressing the overall dynamic range (measured in decibels) by a factor of two. At the receiver, the expandor restores the original dynamic range. The result is higher speech quality than there would be without companding. Outside of cellular telephony, we experience the benefits of companding when we listen to tape cassettes with Dolby noise reduction.

The AMPS pre-emphasis filter, with frequency response shown in Figure 3.4 and a complementary de-emphasis filter at the receiver, also improves sound quality. Together they amplify high-frequency sounds (up to 3,000 Hz), which tend to be weaker than low-frequency sounds, prior to transmission, and restore them to their original level after reception.

An amplitude limiter confines the maximum excursions of the frequency modulated signal to 12 kHz on either side of the carrier frequency. Finally, the baseband signal goes through a lowpass filter with the transfer function in Figure 3.5. This filter attenuates signal components at frequencies above 3,000 Hz. It ensures that energy more than 15 kHz away from the carrier frequency is attenuated by at least 28 dB. This energy contributes adjacent channel interference to signals carried in neighboring frequency channels. Note also the notch at 6 kHz relative to the center frequency. This notch removes signal energy at the frequencies associated with the three supervisory audio tones (SAT) of the AMPS system.

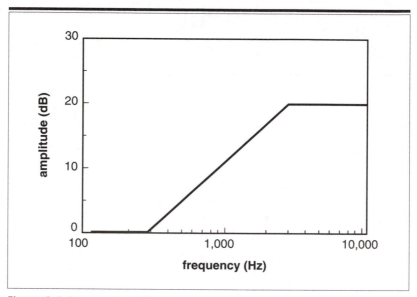

Figure 3.4 Pre-emphasis filter.

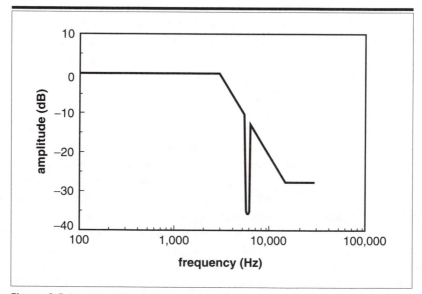

Figure 3.5 Lowpass filter with a notch at 6 kHz.

3.3.4 Supervisory Audio Tone and Supervisory Tone

In AMPS, the SAT transmitted with user information serves to identify the base station assigned to a call. Each base station has its own SAT—at 5,970 Hz, 6,000 Hz, or 6,030 Hz. In the process of setting up each call, the system transmits to the mobile station the identity of the SAT at the base station initially assigned to the call. In transferring the call to another base station, the system informs the mobile station of the identity of the SAT at the new base station. During a call, the mobile station and the base station both inject the same SAT into the signal processed by the frequency modulator. This procedure helps mobile stations and base stations distinguish the desired received signal from interfering signals transmitted in other cells that use the same physical channel.

Nearby cells with the same channel assignments transmit different SATs. Radio propagation conditions can cause the signal from an interfering cell to be stronger than the desired signal. When this happens, the FM receiver demodulates the interfering signal instead of the desired signal. When the mobile station or base station detects an incorrect SAT in the demodulated signal, it mutes the received audio signal. The effect is speech interrupted by silence, which is undesirable but not as objectionable as speech interrupted by a voice from another conversation.

In terms of the channel reuse principles described in Section 9.3.2, the reuse factor of a SAT on a given carrier frequency is three times the reuse factor of traffic channels. Thus, for the typical configuration with a reuse factor of seven cells per cluster, the cells that use the same physical channels have a separation of 4.62 ($\sqrt{21}$) relative to one cell radius. Cells with the same physical channel and the same SAT form a cluster of 21 cells. The distance between two base stations using the same SAT and the same physical channels is 7.94 ($\sqrt{63}$) relative to a cell radius. This means that the distance from a terminal to an interfering base station with the same SAT is at least 7.94 times the distance from the correct base station. Even with this spacing, the signal from an interfering cell with the same SAT from time to time arrives at a higher power than the desired signal. In this situation, the cellular telephone user hears parts of another conversation. In response to this problem, North American digital cellular systems (see Chapters 5 and 6) have the equivalent of at least 256 SATs in their digital channels. This greatly extends the distance between two cells with the same physical channel and the same base station code (the digital equivalent of an SAT).

In addition to the SAT, analog signals from AMPS terminals can also contain a 10 kHz sine wave referred to as a *supervisory tone* (ST). The system uses this tone for signaling in a manner similar to the on-hook/off-hook indications of conventional telephones.

The processed input signal, with SAT and/or ST added as appropriate, frequency modulates the carrier. The terminal then amplifies the FM signal and sends it to the antenna. A terminal uses a single antenna for transmission and reception. A duplexing filter separates the incoming and outgoing signals.

3.3.5 Digital Signals

Although the principal purpose of AMPS is to transport conversational speech to and from mobile stations, it also transmits important network control information in digital form. AMPS digital signals are sine waves either 8 kHz above or 8 kHz below the carrier. The signal format is Manchester coded binary frequency shift keying at a rate of 10 kb/s, or 1 bit per 100 μs. In this format a logical 1 is represented by a transition, in the middle of the 100 μs bit interval, from −8 kHz to +8 kHz relative to the carrier. A logical 0 is represented by the complementary transition. The Manchester code facilitates receiver synchronization by preventing transmissions at a constant frequency for more than 100 μs. Thus, a long sequence of 1s results in the frequency pattern shown in Figure 3.6a, while Figure 3.6b shows the pattern corresponding to a 101010101010 digital signal.

3.3.6 Spectrum Efficiency

Listening tests with juries of potential subscribers have determined that the AMPS transmission technology (frequency modulation in 30 kHz physical channels) requires a received signal-to-interference ratio of at least 18 dB for high-quality sound reproduction at a receiver. In the notation of Figure 9.7, the system has to operate with

$$\left(\frac{S}{I}\right) \geq \left(\frac{S}{I}\right)_{req} = 18 \text{ dB.} \tag{3.3}$$

To meet this requirement with high probability, most AMPS systems operate with reuse factor $N = 7$. This reuse plan is illustrated in Figure 9.9. In this diagram, two cells that use the same physical channels are labeled with the same number. Along with $\left(\frac{S}{I}\right)_{req}$, the nature of the cell site antennas has a strong influence on the reuse factor. To operate with $N = 7$,

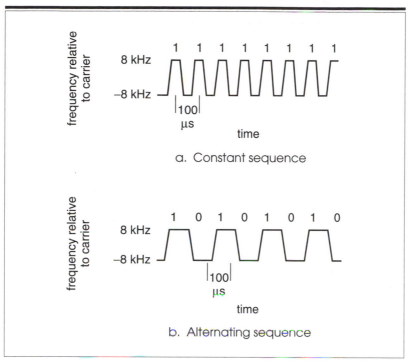

Figure 3.6 Manchester coded frequency shift keying.

AMPS cells require three sets of directional antennas. Each antenna covers 120 degrees.

If a system operates with 395 traffic channels, the average number of traffic channels per cell is 56 4/7 with seven-cell reuse. (Recall that of the 416 assigned channels, at least 21 operate as control channels, leaving a maximum of 395 traffic channels. Assigning these channels as equally as possible to seven cells in a cluster places 56 channels in four cells and 57 channels in the other three cells.) With the total spectrum assignment to an AMPS system 25 MHz, the spectrum efficiency, as defined in Section 2.2.19, is

$$E = \left(\frac{395}{7 \times 25}\right) = 2.26 \text{ conversations/cell/MHz.} \tag{3.4}$$

3.4 Logical Channels

This section describes AMPS information formats designed to foster accurate transfer of network control information in the presence of an imperfect physical connection. These information formats appear in the definitions

Table 3.2 AMPS Logical Channels

Channel Name	AMPS Notation	Purpose	Topology
Reverse traffic channel		User information	Dedicated (one-to-one)
Reverse control channel	RECC	Signaling	Random access (many-to-one)
Reverse voice channel	RVC	Signaling	Dedicated (one-to-one)
Forward traffic channel		User information	Dedicated (one-to-one)
Forward control channel	FOCC	Signaling	Broadcast (one-to-many)
Forward voice channel	FVC	Signaling	Dedicated (one-to-one)

of four logical channels. Table 3.2 lists a total of six one-way logical channels. The term *forward* denotes information transfer from base stations to mobile stations. Less formally, this direction is sometimes referred to as the *downlink*. Conversely, *reverse* (also called *uplink*) channels carry information from mobile stations to base stations. Table 3.2 refers to the pair of channels that carry user information in an analog format (see Section 3.3.3) as *traffic channels*. In addition to the traffic channels, there are four formats for signaling information, as indicated in Table 3.2. The forward and reverse *control channel* formats are used on physical channels reserved exclusively for network control information. These logical channels are sometimes referred to as *common control channels* because they are shared by many mobile stations. As discussed in Section 3.3.1, physical channels 313–354 always carry forward and reverse control channels. In busy systems, operating companies use additional physical channels as forward and reverse control channels. The system uses these control channels to establish calls.

AMPS uses the term *voice channel* to denote the format of system control information carried on a physical channel that also carries user information. A forward voice channel carries system control information from a base station to a terminal when a call is in progress. A reverse voice channel carries system control information from a terminal to a base station when a call is in progress. To transmit information over the forward and reverse voice channels, AMPS uses a technique appropriately referred to as *blank-and-burst*. To send a control message over a voice channel, the system interrupts the flow of user information and inserts a control message, typically of duration around 100 ms. The effect is to time-multiplex a physical channel between user information (traffic) and network control (signaling) information.

When the system inserts a signaling burst, the listener hears a click, not especially obtrusive if it occurs infrequently (once or twice per minute). However, frequent transmissions of control information during a call, by causing a lot of clicks, can seriously undermine a conversation. This impairment limits the amount of AMPS control information that can move between terminals and the system infrastructure during a call.

Owing to the interference and fading on the physical channels, AMPS control signals encounter high binary error rates. Therefore, AMPS protects its control information with robust error-detecting and error-correcting codes.

3.4.1 Logical Channel Categories

Table 3.2 indicates the topologies of the logical channels. A *forward control channel* (FOCC) carries the same information from one base station to all of the terminals in a particular cell that have their power turned on and do not have a call in progress. Similarly, a *reverse control channel* (RECC) carries information from many mobiles that do not have voice channels assigned. To make this possible, AMPS specifies a *random access protocol* that determines how mobiles contend for the attention of the base station receiver. The forward and reverse voice channels are one-to-one links between a base station and a terminal with a call in progress.

3.4.2 Block Codes

All of the logical channels protect the control information with a concatenated pair of block codes, as indicated in Figure 3.7. This figure contains a considerable amount of information about the codes, expressed in a nomenclature consisting of three integers $(n,k;d_{min})$, associated with a block code. The second integer, k, is the number of information bits carried by each code word. The total number of transmitted bits per code word is n. The third quantity, d_{min}, the minimum distance between all pairs of code words, is a measure of the block code's ability to detect and/or correct transmission errors. A high value of d_{min} implies a high immunity to transmission impairments. Section 9.4.1 contains a general description of block codes.

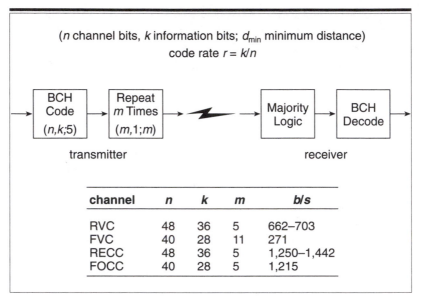

Figure 3.7 Channel coding in AMPS.

Figure 3.7 is a summary of the block codes on the four AMPS logical channels. In each channel, the outer code, which is first applied to a control message, is a shortened (63,51;5) Bose-Chaudhuri-Hocquenghem (BCH) block code [Clark and Cain, 1981: 188–194, 394]. The two mobile-to-base channels (RECC and RVC) carry messages divided into code words of length $k = 36$ bits. The BCH code adds 12 parity check bits to each code word. The result is a transmitted code word with $n = 48$ bits. In the forward direction, the message word length is only $k = 28$ bits and the transmitted code words are $n = 40$ bits long. In the reverse channels, the outer code is thus a (40,28;5) block code. To provide even more protection against binary errors, AMPS employs, as an inner coder, a repetition mechanism that transmits each BCH code word at least five times.

On the FVC, the repetition mechanism transmits each word 11 times! This extremely robust error-control mechanism is warranted because the FVC carries the handoff command that directs an AMPS terminal to establish communication with a new base station after it crosses a cell boundary. This is a critical communication. When it fails, AMPS drops a call and almost invariably inconveniences and irritates the two people who had been speaking to each other. Not only does the FVC carry critical information, but also it does this under difficult circumstances. The event triggering the handoff is a decline in the quality of the physical channel that has to carry the handoff message. As a consequence, the FVC transmits

$$nm = 40 \times 11 = 440 \text{ bits}$$

to convey 28 information bits.

The receiver generates a bit stream that is first processed by a decoder of the inner (repetition) code, and then by a BCH decoder. The operation of the decoders is not part of the AMPS specification. Each base station and terminal manufacturer can decide whether to operate each decoder to:

(a) correct two binary errors,

(b) correct one binary error and detect up to three errors, or

(c) detect up to four binary errors with no error correction.

One approach is to perform majority logic decoding of the inner code [alternative (a) above] and single-bit error correction of the BCH code [alternative (b)]. With this approach, the inner code decoder examines the detected versions of a bit that was transmitted five times (11 times in the case of the FVC). It then employs majority logic by deciding 1 was transmitted if it counts more 1s than 0s. After the terminal or base station performs this operation $n = 40$ times (forward channels) or $n = 48$ times (reverse channels), the inner code decoder delivers an n-bit code word to the outer decoder. If this code word is identical to, or within 1 bit of, a valid transmitted code word, the decoder delivers the corresponding k-bit code word to the message layer of the controller at the terminal or base station.

3.4.3 Logical Channel Formats

Figures 3.8, 3.10, 3.11, and 3.12 show in detail the four different signaling formats of the logical channels. Each format begins with an alternating binary sequence (101010 . . .) that enables the receiver to establish and maintain bit synchronism. On a radio channel, this sequence produces the pattern of frequency shifts shown in Figure 3.6b with the length of the sequence specific to each logical channel. The bit synchronizing sequence is shortest, 10 bits, on the FOCC because this channel transmits continuously, with the synchronizing sequence appearing predictably every 46.3 ms. This makes it an easy matter for a terminal to acquire and hold synchronism on the FOCC. On the other three channels, transmissions take place in bursts, which make it necessary for a receiver to acquire synchronism at the start of each message. On the RECC, the bit synchronizing sequence contains 30 bits, while on the two voice channels, each transmission begins with an alternating sequence of 101 bits. The voice channels also insert a 37-bit alternating sequence before each repetition of a BCH code word. The other synchronization pattern common to all four channels is an 11-bit Barker sequence, labeled "word sync" in Figures 3.8, 3.10, 3.11, and 3.12. When a receiver detects the Barker sequence, it learns that the synchronization transmission has ended and that control information, in the form of protected code words, is about to arrive.

Forward Control Channel

In Figure 3.8, the notation *word A* and *word B* indicates that the channel carries two multiplexed message streams. Word A carries messages for mobiles with even phone numbers (MIN), and word B carries messages for mobiles with odd phone numbers. Transmissions occur continuously on the FOCC in frames containing 463 bits (46.3 ms duration). Each frame carries one 28-bit code word to terminals with even telephone numbers and one code word to terminals with odd phone numbers. Figure 3.8 indicates that the arriving information rate for each terminal is 28 bits per 46.3 ms, or 604.75 b/s.

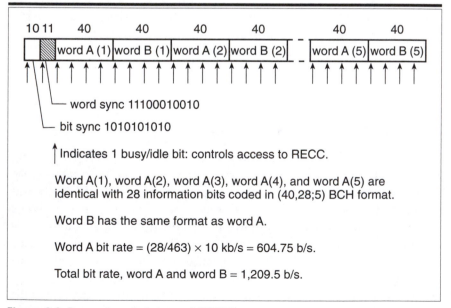

Figure 3.8 Forward control channel (FOCC). (Reproduced under written permission from Telecommunications Industry Association.)

Reverse Control Channel Access Protocol

Expanding Figure 3.8 to show all ten code words would reveal 42 vertical arrows. Each arrow corresponds to 1 busy/idle bit that controls the random access of mobiles to an RECC. The control mechanism is necessary because many terminals use the same RECC. The random access protocol coordinates transmissions from dispersed terminals with the aim of preventing multiple simultaneous transmissions from different terminals. When two or more terminals transmit at the same time on an RECC, their mutual interference usually prevents the base station from detecting any of them. The random access protocol for the RECC uses the FOCC that shares the same two-way physical channel with the RECC. Before transmitting information on the RECC, a terminal examines the state of the busy/idle bits on the corresponding FOCC. These bits are in the idle (1) state when the base station is not in the process of receiving information on the RECC.

Observing the idle state, a cellular terminal with information to transmit initiates a burst in the format shown in Figure 3.10. It continues to observe the FOCC, expecting the busy/idle bits to change to busy (0) within a certain time window. If the transition from idle to busy occurs too soon (in less than 5.6 ms), the mobile station turns off its transmitter to avoid interference with another mobile station that caused the transition. If the mobile station observes no idle-to-busy transition within 10.4 ms, the mobile station turns off its transmitter, assuming that the base station failed to detect the beginning of the burst. If a terminal initially observes busy bits in the FOCC, or if it fails in an attempt to transmit an RECC message, it pauses for a random time interval between 0 and 200 ms and begins the process again. It continues in this way and counts the number of times it observes busy bits (NBUSY) or the number of failed attempts to "seize" the RECC (NSZTR). When one of these numbers exceeds a limit (MAXBUSY or MAXSZTR), the terminal abandons its task. The quantities MAXBUSY and MAXSZTR are system variables broadcast by the FOCC.

After transmitting a message on the RECC, a terminal waits for a response from the system. If the expected response does not arrive within 5 seconds, the terminal returns to the initialization mode. Figure 3.9 is a flowchart of the RECC access protocol.

Reverse Control Channel

On the RECC, terminals transmit network control information to the system in bursts that convey between one and five code words, depending on the control message. Each code word appears on the physical channel

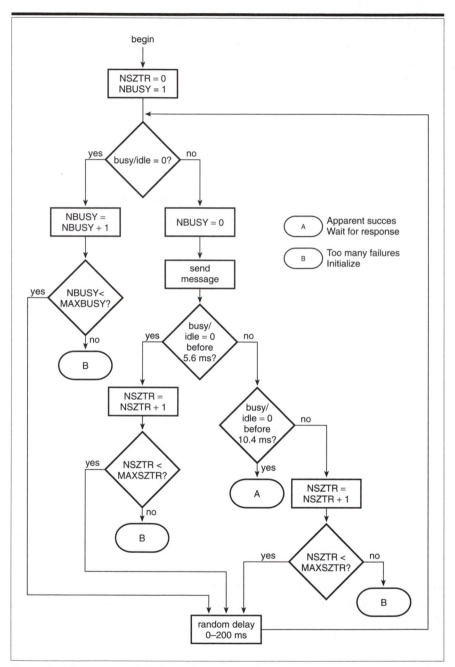

Figure 3.9 Reverse control channel access protocol.

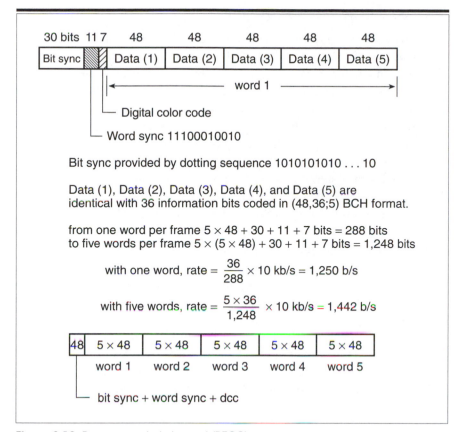

Figure 3.10 Reverse control channel (RECC). (Reproduced under written permission from Telecommunications Industry Association.)

as a sequence of 240 bits (a 48-bit BCH sequence repeated five times). Each message begins with a sequence of 41 synchronization bits, 30 alternating bits for bit synchronism and the 11-bit Barker sequence for frame synchronism. This is followed by a 7-bit *digital color code*. The digital color code plays the same role that the SAT plays in voice channels. Each base station has its own digital color code, broadcast in the FOCC information stream. A terminal echoes this code when it sends a message on the RECC. It is possible for an RECC burst to reach more than one base station tuned to the same physical channel. Base stations ignore RECC signals containing the wrong digital color code. There are four digital color codes in AMPS. The 7-bit RECC transmission corresponds to a (7,2;4) block code with minimum distance $d_{min} = 4$.

Forward and Reverse Voice Channels

To convey system control information between a base station and a terminal when a call is in progress, AMPS relies on in-band signaling over the forward and reverse voice channels. It interrupts user information and sends a control burst in the format of Figure 3.11 (base to terminal) or Figure 3.12 (terminal to base). Network control transmissions on a voice channel begin with an alternating synchronization sequence of duration 101 bits. On detecting this alternating pattern, the base station silences its transmission of the received signal to the MTSO. Similarly, the terminal blanks the audio signal relayed to the loudspeaker in the handset. The base or terminal then waits to detect the 11-bit Barker code, which indicates that the control information is about to arrive.

Before each repetition of a 40-bit or 48-bit BCH code word, there is a 37-bit alternating sequence followed by the 11-bit Barker code. This enables the base station or the terminal to recover from a complete loss of signal, due to a deep fade, during the transmission of a control message over the voice channel. As discussed earlier, each coded message is repeated 11 times on the FVC to provide a high likelihood that the message will be received at the terminal. The FVC carries handoff messages that are essential to preventing call dropping as a subscriber crosses cell boundaries. Handoff messages are transmitted when the signal at the serving base station is weak and therefore requires extra error protection.

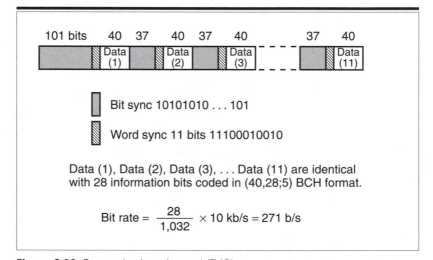

Figure 3.11 Forward voice channel (FVC). (Reproduced under written permission from Telecommunications Industry Association.)

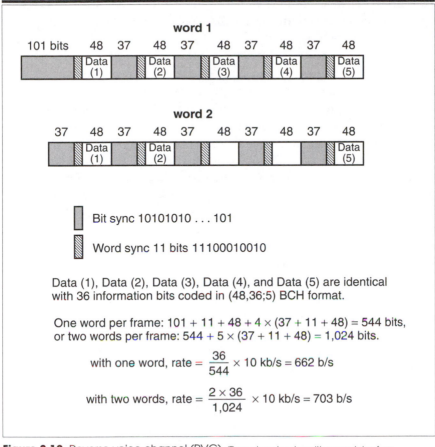

word 1

101 bits   48   37   48   37   48   37   48   37   48

| Data (1) | | Data (2) | | Data (3) | | Data (4) | | Data (5) |

word 2

37   48   37   48   37   48   37   48   37   48

| Data (1) | | Data (2) | | | | | | Data (5) |

Bit sync 10101010 . . . 101

Word sync 11 bits 11100010010

Data (1), Data (2), Data (3), Data (4), and Data (5) are identical with 36 information bits coded in (48,36;5) BCH format.

One word per frame: $101 + 11 + 48 + 4 \times (37 + 11 + 48) = 544$ bits, or two words per frame: $544 + 5 \times (37 + 11 + 48) = 1{,}024$ bits.

$$\text{with one word, rate} = \frac{36}{544} \times 10 \text{ kb/s} = 662 \text{ b/s}$$

$$\text{with two words, rate} = \frac{2 \times 36}{1{,}024} \times 10 \text{ kb/s} = 703 \text{ b/s}$$

Figure 3.12 Reverse voice channel (RVC). (Reproduced under written permission from Telecommunications Industry Association.)

FVC transmissions consist of a single code word containing 28 information bits. Transmissions on the RVC contain either one or two code words, each 36 information bits long. The effective information rate on the FVC is only 271 b/s on a channel with a binary transmission rate of 10 kb/s. On the RVC, the effective rate is either 662 b/s or 703 b/s.

3.5 Messages

In aggregate, the contents of Sections 3.3 and 3.4 describe the techniques used by AMPS to transmit network control code words between terminals and base stations. Each code word is part (or all) of a message that influences the operation of a cellular system. This section begins by examining the structure of AMPS messages. It then describes the system operations associated with specific messages. Table 2.3 lists six categories of network operations. Of these, mobility management, authentication, call management, and radio resources management play a role in every phone call, stimulating an exchange of messages between the terminal and a base station. Each message moving from a base station to a terminal consists of a sequence of 28-bit words transmitted on a forward control channel (FOCC or FVC). Messages transmitted by terminals are sequences of 36-bit words transmitted on a reverse control channel (RECC or RVC).

Table 3.3 is a list of AMPS messages, with an indication of the logical channel that carries each one and the categories of network control operations related to the message. A high proportion of the messages in the AMPS repertory travel from base stations to mobile stations. This reflects the hierarchical nature of AMPS. The MTSO takes major responsibility for the quality and efficiency of AMPS communications. In Table 3.3, the majority of messages are *mobile station control orders*. These are commands from the MTSO to specific terminals. Each command is relayed through a base station and transmitted to a specific terminal by an FOCC (when a terminal is idle or in the process of setting up a call), or an FVC (when a call is in progress). In contrast to the mobile station control orders, the first four messages (each marked with an asterisk) in Table 3.3 are broadcast control messages that provide the same information to all active terminals in a cell.

Table 3.3 AMPS Messages

Message	Network Operations
Forward Control Channel Messages	
SYSTEM PARAMETER*	Call management
	Radio resources management
GLOBAL ACTION*	Radio resources management
REGISTRATION IDENT*	Mobility management
CONTROL-FILLER*	Radio resources management
PAGE	Call management
INITIAL VOICE CHANNEL	Radio resources management
REORDER	Call management
INTERCEPT	Call management

(continued)

Table 3.3 AMPS Messages *(Continued)*

Message	Network Operations
Forward Control Channel Messages	
SEND CALLED-ADDRESS	Call management
DIRECTED RETRY	Radio resources management
RELEASE	Call management
CONFIRM REGISTRATION	Mobility management
Forward Voice Channel Messages	
ALERT	Call management
STOP ALERT	Call management
MAINTENANCE	Operations administration and maintenance
RELEASE	Call management
SEND CALLED-ADDRESS	Call management
HANDOFF	Radio resources management
CHANGE POWER LEVEL	Radio resources management
Reverse Control Channel Messages	
ORIGINATION	Call management, Authentication
PAGE RESPONSE	Call management, Authentication
REGISTRATION	Mobility management
Reverse Voice Channel Messages	
CALLED-STATION ADDRESS ORDER CONFIRMATION	Call management

* Indicates a broadcast message. All other messages are mobile station control orders directed at a specific terminal.

3.5.1 Message Structure

Before addressing the system actions stimulated by specific messages, we give a few examples of the structure of AMPS messages. A striking property of the AMPS system is the lack of uniformity in the formats of different messages. In each of the other systems examined in this book, all messages have a common structure. Within a given system, all code words have the same length. Each message contains a field of fixed length that identifies the message and other fields that transmit variable parameters. In AMPS, code words contain either 28 or 36 bits and each message has its own way of specifying the message type and conveying variable parameters.

As examples, we examine the structures of two messages transmitted on a forward control channel, and the terminal operations stimulated by the messages. Table 3.4 displays a *HANDOFF* message and Table 3.5 displays a *CHANGE POWER LEVEL* message, which has an entirely different structure. The only fields common to the two messages are the preamble, in bit positions 1 and 2, and the present-channel SAT indication, in bit positions 5 and 6. On receiving either a *HANDOFF* message or a *CHANGE POWER LEVEL* message on the FVC, the terminal verifies that the SAT indication in the message corresponds to the SAT (5,970 Hz, 6,000 Hz, or 6,030 Hz) of the cell occupied by the terminal. If the SAT indication does not correspond to the SAT of the present base station, the terminal ignores the message. This occurs when the terminal detects a message from a distant base station communicating with another terminal using the same physical channel.

The terminal learns that the message in Table 3.4 is a handoff command by recognizing a valid SAT identifier in bit positions 3 and 4. If these 2 bits are 11, the terminal has to analyze the message further to determine the control action it conveys. The remainder of the *HANDOFF* message carries three variable parameters: the channel number of the new channel, the power level of the new channel, and the SAT of the present channel. After receiving the *HANDOFF* message, the mobile station turns off its transmitter

Table 3.4 Contents of a 28-Bit *HANDOFF* Message Carried on the FVC

Bit Position	Information
1–2	10 preamble indicates start of message
3–4	SAT of new channel (00, 01, or 10)
5–6	SAT of present channel (00, 01, or 10)
7–14	Not used
15–17	Power level of new physical channel (VMAC)
18–28	New physical channel number

and tunes its transmitter and receiver to the center frequencies corresponding to the new channel number. It then generates the SAT tone corresponding to the new SAT indication in bit positions 3 and 4 of the *HANDOFF* message. Finally, it turns on its transmitter to emit a signal at the power level specified in the *HANDOFF* message.

In Table 3.5, the bits 11 in positions 3 and 4 are not a valid SAT identification. This tells the mobile station that the control message does not command a handoff. To determine the nature of the command, the terminal examines bit positions 24–28. In this case, 01011 indicates that the action is change power level. This causes the terminal to examine bit positions 21–23 to determine the new power level. It then adjusts its transmitter power accordingly.

Tables 3.4 and 3.5 are examples of commands transmitted to terminals on a forward voice channel (FVC). Because this is a (one-to-one) dedicated control channel (Table 3.2), there is no need to identify the terminal that is the target of the command. By contrast, the terminals in a cell that do not have a call in progress receive messages transmitted on a forward control channel (FOCC). These messages must, therefore, identify the terminal that has to take the action specified. To do so, they contain, at the beginning of a message, the 34-bit mobile station identifier (MIN). Therefore, with only 28 bits per code word, mobile station control orders on an FOCC occupy multiple code words.

3.5.2 Message Content

The first four messages in Table 3.3 are broadcast messages that contain information for all of the terminals in a cell. They are referred to as *overhead messages* in AMPS. The FOCC periodically transmits a sequence of

Table 3.5 Contents of a 28-Bit *CHANGE POWER LEVEL* Message Carried on the FVC

Bit Position	Information
1–2	10 preamble indicates start of message
3–4	11 indicates that this is not a handoff message
5–6	SAT of present channel
7–20	Not used
21–23	New power level (VMAC)
24–28	01011 indicates power control message

overhead messages in an *overhead message train*. The information in these messages pertains to a single cell and prepares a terminal for communications with the AMPS infrastructure.

The first message in an overhead message train is a *SYSTEM PARAMETER* message consisting of two 28-bit words. This message contains the first 14 bits of the 15-bit system identifier (SID). (The final bit is determined by the system numbering convention described in Section 3.3.1. If the message arrives on an A-channel, the least significant bit is 1; otherwise it is 0.) The terminal compares the received SID with the SID of its home system (stored in the terminal's memory) in order to determine whether it is tuned to its home system or roaming in another system. If the system identifier does not correspond to the home system of the terminal, the terminal activates an indicator on the terminal's visual display. This indicates to the subscriber that he cannot be certain of access to the local system. If he can use the system, it is possible that service charges will be higher than in the home system. The *SYSTEM PARAMETER* message also indicates the number of forward control channels that carry paging information in the current cell and the number of reverse control channels available to terminals for sending call setup and registration messages. AMPS specifies that each FOCC broadcast a *SYSTEM PARAMETER* message at least every 1.1 seconds and at most two times per second.

In addition to a *SYSTEM PARAMETER* message, an overhead message train can carry one or more *GLOBAL ACTION* messages. These messages contain parameters of the RECC access protocol (Section 3.4.3). Two of these parameters are MAXBUSY and MAXSZTR, which control the maximum number of attempts to transmit an RECC message. Figure 3.9 indicates that if the RECC is busy after MAXBUSY examinations of the busy/idle bits in the FOCC, the terminal abandons its attempt to send the message. It also abandons the attempt if, after MAXSZTR transmissions of a message, the terminal fails to observe the expected response from the base station. Global action messages contain two pairs of values for MAXBUSY and MAXSZTR. MAXBUSY-PGR and MAXSZTR-PGR control the transmission of page response messages and MAXBUSY-OTHER and MAXSZTR-OTHER control the transmission of all other messages.

The third broadcast message that may appear in an overhead message train is a *REGISTRATION IDENT* message. This message contains a 20-bit number (REGID) that controls the frequency with which terminals transmit *REGISTRATION* messages to the system.

AMPS specifies continuous transmission on each FOCC. The transmitter always radiates energy. The system uses *CONTROL FILLER* messages to fill the gaps between necessary control messages. *CONTROL FILLER* messages contain a

3-bit number, CMAC (control mobile attenuation), that specifies the transmit power level for messages transmitted by terminals on an RECC.

Except for the first four messages (those with an asterisk) in Table 3.3, each FOCC message is a mobile station control order directed to a specific terminal. Each control order contains the address of the terminal (34-bit mobile identification number) to which the message is directed. Many of the message names clearly indicate the purpose of the message. A *PAGE* message informs the terminal of an incoming call. An *INITIAL VOICE CHANNEL* message directs the terminal to tune to a traffic channel in order to begin a new call. The message contains the voice channel number (CHAN) and the transmit power level (VMAC) on the voice channel. A *REORDER* message causes the terminal to emit an audible signal to the subscriber. This takes the form of a fast busy signal that indicates that the system is unable, usually because of congestion, to meet the person's call setup request. Similarly, an *INTERCEPT* message causes the terminal to produce a different audible signal to indicate that the subscriber has issued a request (number sequence) that the system cannot interpret. A *SEND CALLED-ADDRESS* message causes the terminal to transmit the telephone number that the subscriber is trying to reach.

Directed retry is a radio resources management procedure. A *DIRECTED RETRY* message commands a terminal to try to gain access to the system through another base station. An AMPS system issues *DIRECTED RETRY* messages when there is an uneven demand for service in a cluster of cells. Before setting up a call, each terminal tunes to the control channel with the strongest received signal. This is likely to come from the nearest base station. It may be that communications are also possible through other base stations. If the original base station is too busy to accommodate a new call request, it can command a terminal to attempt to gain access to the system through one or more other base stations. The *DIRECTED RETRY* message specifies the FOCC channel numbers at adjacent base stations. The terminal, if possible, tunes to one of these channels and attempts again to gain access to the system.

A *RELEASE* message received on the FOCC causes the terminal to abandon its current operation and return to monitoring the FOCC. A *CONFIRM REGISTRATION* message acknowledges receipt of a *REGISTRATION* message on an RECC.

Turning to transmissions on an FVC, the first five messages in Table 3.3 play a role in call management operations. An *ALERT* message causes the mobile station to generate an audible tone (beep) to inform the user of an arriving call. While alerting the user, the terminal transmits an SAT tone and a 10 kHz supervisory tone (see Section 3.3.4) on the forward traffic channel. When the subscriber answers the call by pressing a button on the terminal, the terminal turns off the supervisory tone and the base station reacts by sending a *STOP ALERT* message that commands the terminal to stop beeping. AMPS terminals contain a 65-second timer that controls the duration of the alerting process. If a call is not answered after 65 seconds, AMPS abandons the attempt to reach the mobile subscriber. When this happens, the terminal releases the voice channel and returns to an FOCC. This mechanism limits the amount of time that a voice channel is occupied by an unsuccessful call setup attempt.

A *MAINTENANCE* message allows the system to check the operation of a terminal. The terminal responds to this message in the same way it responds to an *ALERT* message, except that it does not emit an audible beep. When the MTSO learns that the party communicating with a mobile subscriber has ended a call, it commands the base station to send a *RELEASE* message to the terminal. This causes the terminal to leave the voice channel and tune once again to an FOCC. The *SEND CALLED-ADDRESS* message on the FVC stimulates the terminal to transmit a stored telephone number to the system. The other two FVC messages, *HANDOFF* and *CHANGE POWER LEVEL*, play an important role in radio resources management as discussed in Section 3.5.1.

The upstream control messages on the RECC play a vital role in call management and mobility management. In response to the mobile subscriber pressing the SEND button on the terminal, the terminal sends an *ORIGINATION* message to set up a call. This message contains the called-party number and three identifiers of the mobile terminal (see Section 3.2.2): the telephone number (MIN), the electronic serial number (ESN), and the station class mark (SCM). The mobile station transmits a *PAGE RESPONSE* message on the RECC when it detects its MIN in a *PAGE* message on the FOCC. Like the *ORIGINATION* message, the *PAGE RESPONSE* message contains the MIN, ESN, and SCM of the mobile terminal. AMPS uses *REGISTRATION* messages to keep track of the locations of terminals before a call is set up. When a terminal is in the service area of its home system, *REGISTRATION* messages can reduce the number of *PAGE* messages necessary to deliver calls to cellular phones. In the absence of registration, every cell in the home system sends a *PAGE* message in an attempt to set up a call to a mobile terminal. In a large system, with hundreds of base stations and hundreds of thousands of subscribers, the volume of *PAGE* messages can overwhelm the capacity of the system to transmit them. When terminals register their locations, the system can restrict the transmission of *PAGE* messages to cells in the vicinity of the cell that received the most recent *REGISTRATION* message, and greatly reduce the volume of *PAGE* messages.

Registration is essential to deliver calls to terminals that are roaming outside of their home service areas. On receiving a REGISTRATION message from a roaming terminal, a system informs the terminal's home system of the terminal's present location. Chapter 4 describes Interim Standard 41 [TIA, 1991], which specifies procedures that coordinate the operations of home systems and visited systems.

The two messages that the terminal can send on the RVC during a conversation are both responses to messages received on the FVC. This reflects the hierarchical nature of AMPS, with control operations concentrated in the MTSO. A CALLED-STATION ADDRESS message contains a telephone number entered into the terminal's memory by the user. This message is a response to a SEND CALLED-ADDRESS message received from the base station. An ORDER CONFIRMATION message acknowledges the receipt of a message sent to the terminal such as a power control command.

3.6 AMPS Protocol Summary

Figure 3.13 summarizes the AMPS transmission technologies presented in Sections 3.3, 3.4, and 3.5. All information leaves a terminal or base station in the form of a frequency modulated carrier confined to a bandwidth of 30 kHz. The transmitted signal can convey analog user information, network control messages, and signaling tones, including a supervisory audio tone (SAT) at 5,970 Hz, 6,000 Hz, or 6,030 Hz. Traffic channels can also carry, as an on-hook indication, a 10 kHz supervisory tone (ST).

Each network control message is a sequence of from one to five code words. The message is carried on one of four types of logical channels. Each logical channel has its own code-word length, channel-coding techniques, and added synchronization codes. The modulation technique for all four logical channel types is frequency shift keying with a deviation of ±8 kHz from the carrier.

3.7 Tasks Performed by AMPS Terminals

Thus far Chapter 3 has described the capability of AMPS to move network control messages between base stations and mobile stations. We now examine how AMPS uses these messages to establish and maintain telephone calls. To do so, we first look inside a terminal and observe that at any instant we can identify a "task" being performed by the terminal. The AMPS specification defines a large number of tasks. Each can be viewed as one state of a finite-state machine. The terminal moves from

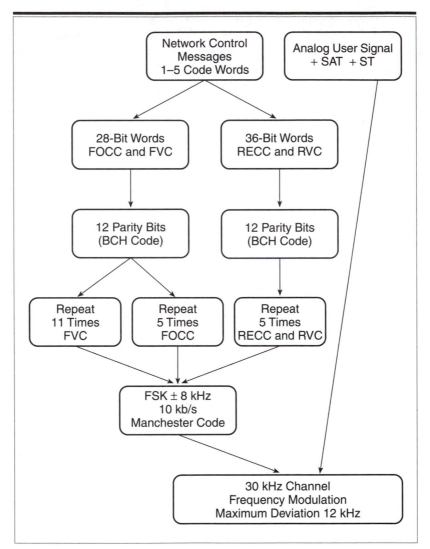

Figure 3.13 Summary of AMPS transmission protocols.

one task to another in response to a specific stimulus such as the completion of a task, a message received from the base station, an action on the part of the subscriber, or a measurement performed by the AMPS terminal itself.

As indicated in Figure 3.14, there are four modes of operation: initialization, idle, access, and conversation. Each mode consists of a sequence of tasks. When a successful communication takes place, the terminal cycles through the four modes, following the heavy lines in Figure 3.14.

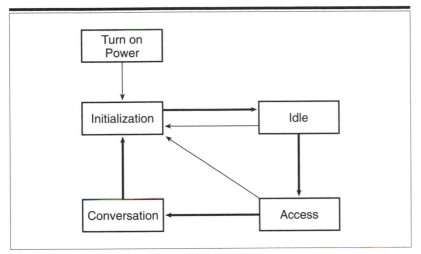

Figure 3.14 Cellular terminal operating modes.

However, at any point during the normal sequence of operations it is possible for the terminal to lose contact with the base station. If this occurs, or if the terminal cannot complete a specific task successfully (such as access to the RECC), the terminal returns to the initialization mode, as indicated by the light lines in Figure 3.14. If this happens while the terminal is in the access mode, a call attempt or a registration attempt fails. If it happens prematurely during a conversation, the system drops a call in progress.

3.7.1 Initialization

Several conditions place the terminal in the initialization mode, including:

- the user turns the power on,
- a conversation ends, or
- the terminal loses contact with the current base station.

To begin the initialization process, the terminal scans either 21 control channels (channel numbers 313–333) in the A-band or 21 channels in the B-band (channels 334–354). Each terminal begins with a preference for either the A-band or the B-band (Figure 3.2). In general, the preferred band is the frequency band used by the subscriber's operating company. However, the subscriber can use the terminal keypad to override this preference and program the terminal to set either A or B as the preferred band. Each base station continuously broadcasts information in the FOCC format on one of the 21 control channels. In most systems, the control channels operate with omnidirectional antennas and a reuse factor of 21 (see Section 9.3.2). This implies that the distance between two cells with the same physical control channels is approximately eight times the cell radius.

The receiver scans the 21 channels in the preferred band and locks on to the strongest one. If no channel in the preferred band is strong enough for accurate reception, the terminal can scan the other band in search of an adequate control channel. The user can program the terminal to perform this search. If the telephone is programmed to remain in the preferred band, the terminal continues to scan the preferred band, in hopes of eventually arriving at a location with an adequately strong control channel. If the terminal cannot find a usable control channel, it turns on a visible "no service" display. With cellular telephony becoming a mature service, companies generally provide coverage throughout their service areas, so that a large majority of initialization procedures result in the terminal tuning to an FOCC broadcast by the nearest base station.

With its receiver tuned to an FOCC, the terminal performs an "update overhead information" task in order to extract important information from the overhead message train broadcast on the FOCC. As described in Section 3.5.2, overhead messages contain the 15-bit identifier of the local cellular system and information about active paging channels in the cell occupied by the terminal. On interpreting these broadcast messages, the terminal decides whether or not to turn on a visible roaming indication. It then tunes to the strongest paging channel operating in the current cell. In all but the busiest cells, there is only one channel transmitting FOCC information, and the paging channel is identical to the original FOCC monitored by the terminal. On completing the initialization tasks, the terminal enters the idle mode. Typically the terminal is in the initialization mode for 5 to 10 seconds.

3.7.2 Idle

In the idle mode, the terminal monitors a paging channel, which is a physical channel selected on the basis of information obtained during initialization. The paging channel transmits, in the FOCC format, system status information. The terminal records this information and uses it to perform mobility management and call setup procedures. Some of the broadcast messages contain registration parameters that determine how often the terminal transmits a message to indicate its location to the system. Other broadcast messages monitored by the system indicate the physical channels used as RECCs in the current cell. In addition to this global information, the paging channel also broadcasts messages directed at specific terminals. These are the eight FOCC messages in Table 3.3 that are not marked with asterisks.

There are several conditions that move the terminal from the idle mode to the access mode. The most important of these are

- a call initiated when the terminal user presses the SEND button,

- an incoming call request detected when the terminal recognizes its MIN in a page message, and

- a registration event stimulated by the value of the parameter REGID received in a *REGISTRATION IDENT* message.

A terminal can remain in the idle mode indefinitely. It moves to the access mode in response to one of the events listed previously. It returns to the initialization mode when it fails to receive accurate information on the current FOCC. Usually, a weak signal on the FOCC is due to the fact that the terminal has entered a new cell. The terminal responds to the weak signal by returning to the initialization mode, which begins with a search for a strong FOCC signal.

3.7.3 Access

In the access mode, the terminal attempts to transmit a message in the RECC format (Figure 3.10) to a base station. To do so it uses a physical channel selected according to information received in the idle mode. This information indicates a set of physical channels available in the present cell for RECC transmissions. The terminal scans the FOCC transmissions on these channels and tunes to the strongest one. It then monitors the global information transmitted by the FOCC on this physical channel in order to extract two access attempt parameters (MAXBUSY and MAXSZTR) that control the access protocol (Figure 3.9). These parameters control the maximum number of times the terminal can attempt to send a message on the RECC.

Upon entering the access mode, a terminal follows the procedure specified in Figure 3.9. When it enters state A in Figure 3.9, indicating that the transmission has apparently succeeded, the terminal waits for a response to the message. In a call setup situation (origination or page response), this system response takes the form of an *INITIAL VOICE CHANNEL* message, which orders the terminal to tune to a physical channel used for voice transmission. On receipt of this message, the terminal tunes to the designated physical channel and enters the conversation mode. If the terminal has entered the access mode to register its location, it waits for a *CONFIRM REGISTRATION* message and eventually returns to the initialization mode.

Figure 3.9 indicates that the access protocol can direct the terminal to return to the initialization mode (state B) if the number of access attempts exceeds one of the limits specified by the system. This occurs when the RECC is congested or the signals transmitted by the terminal encounter too much attenuation or interference to be accurately received at the base station.

When it enters the access mode, the terminal sets a timer with a duration of 12 seconds for an origination and 6 seconds for any other task. If this timer expires, the terminal returns to the initialization mode, regardless of the current status of the access protocol.

3.7.4 Conversation

With a mobile station in the conversation mode, the system serves its ultimate purpose—to connect a mobile station to any other telephone in the worldwide Public Switched Telephone Network. The technical name for the conversation mode is *mobile station control on the voice channel,* and, in fact, various control operations have to take place before a conversation can begin. On entering the conversation mode, the terminal first indicates to the system that it has properly complied with the order to tune to a traffic channel. The SAT transmitted by the terminal (see Section 3.3.3) provides this confirmation.

The terminal enters the conversation mode in response to an *INITIAL VOICE CHANNEL* message. Like the *HANDOFF* message (Table 3.4), this message contains an SAT color code indicator that specifies the SAT (5,970 Hz, 6,000 Hz, or 6,030 Hz) of the assigned base station. The base station transmits the specified SAT in the forward direction and listens for the SAT on the corresponding reverse channel. The mobile station confirms that it has properly selected a voice channel by transmitting the SAT received from the base station. The tuning procedure is complete when the base station detects the correct SAT transmission from the mobile station.

3.7.5 Phone Call Examples

If the call originates at the terminal, the base station confirms to the MTSO that the tuning procedure has succeeded and the MTSO completes the call through the public network. Figure 3.15 displays the sequence of messages and control operations that set up and release a call originating at a mobile station.

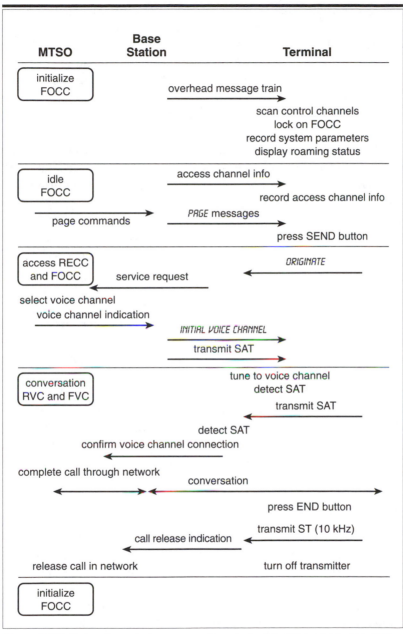

Figure 3.15 Network control sequence for a call originating at a terminal. The call ends when the user presses the END button.

If the call originates in the public network, the terminal enters the conversation mode after it transmits a *PAGE RESPONSE* message and receives an *INITIAL VOICE CHANNEL* message. The base station, on confirming that the terminal has properly tuned to the correct voice channel, sends an *ALERT* message in the FVC format. This stimulates the terminal to emit an audible tone (beep) that prompts the user to respond to an incoming call. Until the user responds by pressing one of the keys on the terminal, the terminal transmits a 10 kHz supervisory tone indicating that the terminal is on hook. When the user responds, the terminal stops transmitting the 10 kHz tone, an event that tells the base station that the call can begin. The base station signals this information to the MTSO, which then completes a voice path to the person who placed the call. The base station also sends a STOP ALERT message to command the mobile station to turn off the audible tone. This sequence of events is shown in Figure 3.16, which begins with the terminal in the idle mode. Initialization procedures conform to those shown in Figure 3.15.

During the call, the base station can send a CHANGE POWER LEVEL message (see Table 3.5) on the forward voice channel to command the mobile station to adjust its transmitter power to one of the eight levels defined by AMPS. The terminal acknowledges receipt of this message by sending an ORDER CONFIRMATION message in the RVC format. The base station can also command a handoff and signal the end of the conversation.

As in a conventional telephone call, either party can conclude the call. If the remote user hangs up first, the base station sends, in the FVC format, a RELEASE message to the terminal. The terminal acknowledges this message by transmitting the 10 kHz supervisory tone for 1.8 seconds. It then turns off its transmitter and returns to the initialization mode. This sequence appears in Figure 3.16. If the mobile user hangs up first, by pressing the END button, the terminal signals this event to the base station by sending the 10 kHz tone for 1.8 seconds. As indicated in Figure 3.15, the terminal then turns off its transmitter and returns to the initialization mode.

3.8 Network Operations

The earlier sections of Chapter 3 present the techniques available for performing the six categories of operations, listed in Table 2.3. Section 3.3 describes in detail AMPS technologies for user information transport and Section 3.7 provides a comprehensive survey of call management in AMPS. This section summarizes the AMPS procedures that contribute to the three sets of network management operations that play a critical role in wireless communications: mobility management, authentication, and radio resources management.

3.8.1 Mobility Management

In the manner described in Section 2.3.2, idle AMPS terminals periodically transmit REGISTRATION messages on a reverse control channel to indicate

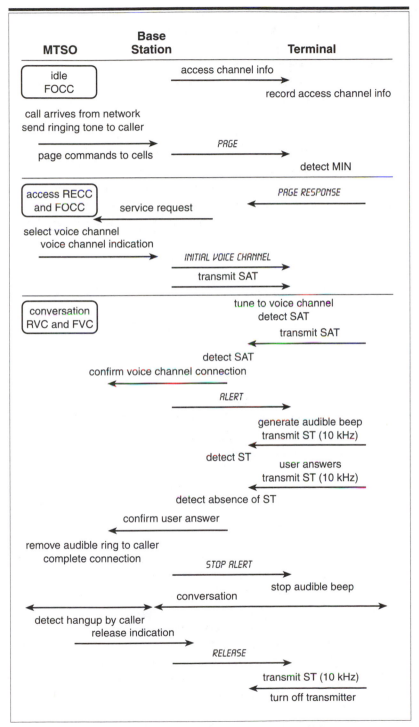

Figure 3.16 Network control operations for a call originating in the fixed network. The call ends when the remote caller hangs up.

their locations. The time intervals between *REGISTRATION* messages are controlled by *REGISTRATION IDENT* messages broadcast on a forward control channel. By transmitting these messages at frequent intervals, a system causes terminals to register their locations frequently and thus provide accurate information about their locations. This information allows the system to restrict the number of cells in which it pages terminals that receive phone calls from the network. Each system adjusts the rate of registration to balance the burdens placed on control channels by *REGISTRATION* messages and *PAGE* messages. The AMPS system gives network operators considerable flexibility in adopting paging and registration strategies. One sophisticated approach is to monitor *REGISTRATION* messages in order to determine subscriber mobility patterns [Madhavapeddy, Basu, and Roberts, 1995]. The system then refers to these patterns to devise a sequential paging strategy, in which it sends *PAGE* messages first to the cells most likely to be occupied by a terminal. If it receives no response to the initial *PAGE* messages, the system pages the terminal in the remaining cells. This has the effect of reducing the number of *PAGE* messages relative to simpler approaches. To gain this information, the system has to acquire, store, and analyze information about mobility patterns.

3.8.2 Authentication

The electronic serial number (ESN, Table 3.1) is at the heart of the network security procedures built into AMPS. To gain access to AMPS services, a terminal transmits both its mobile identifier (MIN in Table 3.1) and its ESN. As the subscriber's telephone number, the MIN is considered public information. The ESN, installed electronically in the terminal, is considered private information, belonging to the cellular operating company. A database in the subscriber's home MTSO records the ESN. Before granting a terminal access to the network, the MTSO verifies that the transmitted ESN is the correct one for the subscriber's MIN. In this sense, the ESN is like a computer password or a personal identification number (PIN) used at automatic teller machines. Just as computer security depends on the secrecy of passwords, network security in AMPS depends on the secrecy of the ESN. This turns out to be a major weakness of AMPS and other first-generation cellular systems. Because mobiles transmit their ESNs through the air, the MIN/ESN pair is subject to interception and fraudulent use. In response to this vulnerability, the systems we study in later chapters all incorporate authentication procedures that are more secure, and more elaborate, than those of AMPS.

3.8.3 Radio Resources Management

As described in Section 2.3.5, personal communications systems deploy several procedures to manage their radio channels.

Call Admission

In AMPS, the call admission policy is part of the MTSO software. The simplest policy is to accept any service request that arrives when there are inactive physical channels in the cell occupied by the terminal requesting service and to deny service when all physical channels are in use. This procedure minimizes call blocking (see Section 2.3.9), but makes the system relatively vulnerable to call dropping (see Section 2.3.14). Since call dropping is far more annoying to users than call blocking, systems adopt *channel reservation* schemes to reduce call dropping at the expense of higher blocking rates. To do so, the system denies service to new calls when there is a small number of inactive channels in a cell, and reserves these channels to satisfy handoff requests.

Channel Assignment and Power Control

With respect to base station and channel assignment, AMPS generally assigns a new call to an available channel at the nearest base station. However, to balance the load over a group of cells, systems can employ a *directed retry* procedure, by which the nearest base station commands a terminal to attempt to gain service through a nearby base station that is less congested than the nearest one. AMPS has the capability for dynamic power control over transmissions from terminals. It does so by commanding each terminal to transmit at one of eight power levels (see Section 3.3.2) listed in the system specification. The system uses CONTROL FILLER messages on the forward control channel to specify the power level (parameter CMAC) for transmissions on a reverse control channel (see Section 3.5.2). The initial power level for transmissions on traffic channels (VMAC) is specified in INITIAL VOICE CHANNEL messages and HANDOFF messages (see Table 3.4). As the terminal changes location within a cell, the system can command it to change its power level by transmitting a CHANGE POWER LEVEL message (see Table 3.5) on a forward voice channel.

Handoff

Perhaps the most impressive property of a cellular telephone system is its ability to maintain calls as mobile stations move from cell to cell or into different sectors of cells operating with directional antennas. In AMPS, handoff from one base station to another is controlled by the MTSO, which assembles measurements of received signal strength from the current base station and surrounding base stations. Each system has its own proprietary handoff algorithm, which typically consists of a set of signal strength thresholds, referred to as *RSSI (received signal strength indication) levels.* One threshold, typically around −100 dBm, is the level at the current cell that causes the system to initiate a handoff. Below this level, AMPS may be unable to maintain adequate voice quality. Another, higher, threshold, perhaps −90 dBm, is the signal strength required at the new cell. The difference between these two thresholds introduces hysteresis, which inhibits repeated handoffs as a mobile station moves along the boundary between two cells. The handoff algorithm may also be designed to limit call dropping due to overload by considering the number of active channels in candidate cells before directing a call in progress to a new cell.

In addition to handoff from one cell to another, the base station can also initiate an "intracell" handoff to another channel in the same cell. Typically, this type of handoff takes place in response to the mobile station moving to a new sector, served by a different directional antenna.

When the system control software decides to initiate a handoff, the sequence of messages in Figure 3.17 begins. First, the base station transmits, over the current physical channel, a *HANDOFF* message in the FVC signal format. This message includes the new channel number, an indication of the SAT in the new cell, and the initial transmitter power level (VMAC). The terminal acknowledges this message by transmitting the 10 kHz supervisory tone for 50 ms. It then turns off its transmitter, tunes to the new channel, generates the SAT of the new cell, turns on its power, and resumes voice transmission.

3.9 AMPS Status

With respect to both technology and commerce, AMPS is a major success story. However, since the late 1980s, the cellular industry has recognized the need for improvements in several areas including capacity, roaming, security, and support for non-voice services. All of these issues are addressed by new technology described briefly in the following paragraphs and in detail in Chapters 4, 5, and 6.

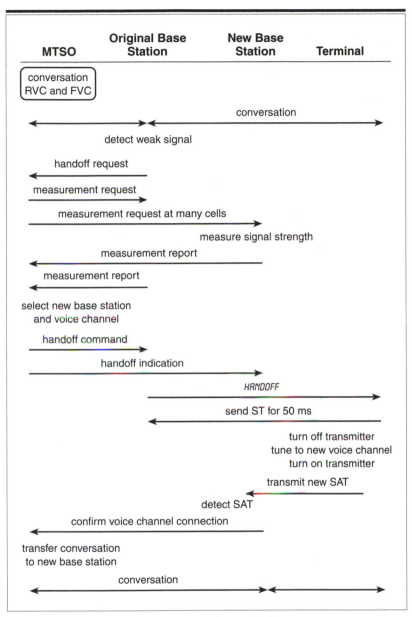

Figure 3.17 Network control sequence for handoff.

3.9.1 Capacity

There are three ways to increase the capacity of cellular systems:

- operate with smaller cells,
- obtain additional spectrum allocations, and
- introduce new technology to improve spectrum efficiency.

While cell-splitting to increase capacity is fundamental to the cellular idea [MacDonald, 1979], this approach has its limitations. Using the original, high-power, high-elevation base stations of cellular systems, the practical lower limit on cell radius is around 1.5 km [Mehrotra, 1994]. To achieve smaller dimensions, companies install low-power, low-elevation *microcells* [Steele, 1992: 24–41] in densely populated areas. In many instances, the microcells are within the service areas of larger, conventional cells. Thus, many terminals in microcells have access to at least two base stations, one serving a low-power microcell and the other operating in a conventional high-power mode. This situation raises a host of challenging radio resources management issues.

Beyond the practical problems of operating with small cells, cell splitting is the most expensive of the three approaches to increasing capacity. With respect to the preferred approach, obtaining new spectrum, the U.S. Federal Communications Commission, in the 1980s, added 10 MHz to the original 40 MHz allocation of radio spectrum to cellular services (see Figure 3.2). The FCC then announced that no additional cellular bandwidth would be available. However, the FCC also issued new rules [FCC, 1990a] to make the third approach to capacity enhancement, deploying new technology, available to license holders. The FCC encouraged operating companies to adopt new transmission technologies by permitting each operating company to transmit signals in any format, providing the signals do not interfere with the signals of other license holders. The industry response to these rules is embodied in three transmission technologies that have higher spectrum efficiency than AMPS. Two of them transmit speech in digital format, one using time division multiple access [TIA, 1996d] and the other using code division multiple access [TIA, 1993b]. They are presented in detail in Chapters 5 and 6, respectively. The third new technology, NAMPS, based on analog speech transmission, closely resembles AMPS and is described briefly in the following paragraphs.

Motorola developed Narrowband AMPS [TIA, 1993a] in response to uncertainties about the relative merits of the two digital standards and uncertainties about when they would be available in commercial products. Operating companies use NAMPS technology to provide a short-term solution to capacity problems and then introduce the preferred digital standard after uncertainties are resolved. NAMPS gains its capacity advantage by dividing an original AMPS channel into three narrowband channels: one with a carrier frequency equal to an AMPS carrier (Equations 3.1 and 3.2), and the other two offset by ±10 kHz relative to an AMPS carrier. The modulation technique on a narrowband channel is FM with a maximum deviation of 5 kHz from the carrier.

NAMPS is a dual-mode system, so that all NAMPS terminals are capable of operating with 30 kHz channels as well as 10 kHz channels. In a system not equipped for NAMPS operation, the dual-mode terminal functions as a conventional AMPS terminal. NAMPS employs the AMPS control channels for call setup and the call management sequence conforms to Figure 3.14. In the access mode, an *INITIAL VOICE CHANNEL* message on a forward control channel directs a NAMPS terminal to either a wide traffic channel (30 kHz bandwidth) or a narrow traffic channel (10 kHz bandwidth). NAMPS systems are capable of four types of handoff: wide channel to wide channel, wide to narrow, narrow to narrow, and narrow to wide.

In the conversation mode, network control in NAMPS differs significantly from AMPS. Instead of the blank-and-burst operation on forward and reverse voice channels, NAMPS transmits out-of-band control information continuously in both directions over narrow traffic channels. It conveys network control information in logical *associated control channels*, which transmit 200 b/s (non-return-to-zero) signals and 100 b/s (Manchester coded) signals in the "sub-audible" (low-frequency) portion of the traffic channel input signal spectrum. Figure 3.18 shows the contents of an AMPS channel divided into three narrow physical channels. There are four types of network control information:

- messages similar in format or identical to AMPS messages,

- synchronization sequences that replace the dotting sequences and Barker codes of AMPS control channels (see Section 3.4.3),

- digital versions of the AMPS supervisory audio tone (seven possible codes), and

- a digital replacement for the AMPS supervisory tone.

In addition to transmitting AMPS call management messages and radio resources management messages, NAMPS control channels are capable of

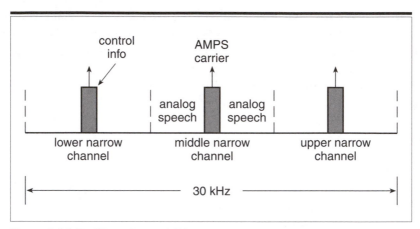

Figure 3.18 Partition of one AMPS channel into three narrow channels, each carrying analog user signals and digital control signals.

operating with an extended protocol that brings special features and network services to subscribers, including calling-number identification, voice mail control, and short message services (see Section 2.2.2).

The principal purpose of NAMPS is to achieve higher spectrum efficiency than AMPS. In cellular systems, spectrum efficiency depends on the bandwidth of each signal and also the sensitivity of the signal to interference (see Section 9.3). With its smaller bandwidth, a narrow traffic channel is more vulnerable to interference than a wide traffic channel and would normally require a higher reuse factor (N in Section 9.3.2) than AMPS signals have. To control the interference to signals in narrow channels, NAMPS introduces a radio resources management procedure referred to as *mobile reported interference*. To perform this procedure, a terminal measures the received signal strength on a forward narrow traffic channel and the binary error rate of the control signals on the associated control channel. When the measurements go outside of a range specified in a message sent by the base station, the terminal reports these measurements to the base station by means of a message on the reverse associated control channel. Based on this report, the system can initiate a handoff in order to improve signal quality.

3.9.2 Roaming

In the United States, the cellular industry operates with separate licenses in 305 metropolitan statistical areas and 428 rural service areas. In this situation, the ability of a subscriber to obtain cellular service outside of her home service area depends on two things. It requires business arrangements between her cellular telephone company and a network operator in the visited area. It also requires technology to transfer information between the home system and the visited system. For many years this technology was primitive, requiring considerable intervention (dialing special codes, for example) by the subscriber and/or the person calling her. These inconvenient, inconsistent arrangements were far removed from the ideal of using the same procedures to place a call to or from a cellular phone, regardless of the phone's location. To address this problem, the industry developed Interim Standard 41 [TIA, 1991], which specifies standard protocols for communications between MTSOs. The principal uses of IS-41 are to deliver calls to roaming subscribers and to hand off calls in progress from one system to another as a subscriber travels. IS-41 is the subject of Chapter 4.

3.9.3 Network Security

The AMPS authentication procedures are weak. They rely on the secrecy of each terminal's electronic serial number (ESN). The terminal transmits this number on a reverse control channel each time it initiates a call, responds to a *PAGE* message, or registers its location. As a consequence, the ESN can be intercepted by radio receivers tuned to AMPS control channels. People who operate these receivers illegally use the serial numbers to gain unauthorized access to cellular systems. This activity, referred to as *cloning*, is highly prevalent and a matter of great concern to the cellular industry. To address this problem, network operators have introduced a variety of measures. A common one is to require each subscriber to key in a personal identification number each time she makes a phone call. The terminal transmits this number on a reverse voice channel, making it somewhat harder to intercept than the ESN transmitted on a common control channel.

In addition, the industry has devised robust network security technology based on encryption and secure key distribution. These measures, which are integral parts of the digital systems presented in Chapters 5, 6, and 7, achieve a significantly higher level of network security. In 1996, these cryptographic authentication techniques were introduced to analog systems. However, they have to be implemented at terminals, as well as at base stations and switching offices. Therefore, they are available only to subscribers with new terminals that incorporate the secure authentication technology. Tens of millions of existing terminals remain vulnerable to cloning.

3.9.4 Non-Voice Services

A growing proportion of the population uses telephone lines to gain access to a wide variety of digital information services such as facsimile, electronic mail, the World Wide Web, and a large collection of specialized services. The data protocols that link fax machines and personal computers to these information services in many situations suffer severe performance degradation in the presence of the interference levels on cellular channels as well as the signal interruptions caused by handoffs and blank-and-burst transmission of signaling information. To cope with these problems, advanced cellular systems apply several approaches. One approach is to convey short text messages through special logical channels as in NAMPS (see Section 3.9.1) and North American TDMA (see Chapter 5). Another approach is to introduce a separate packet data network, cellular digital packet data [CDPD Forum, 1995] that transmits its own signals through AMPS logical channels. A third approach is to incorporate special signal processing methods for signals moving to and from telephone data modems and fax machines. This approach is especially relevant to the dual-mode digital systems described in Chapters 5 and 6.

Review Exercises

1. What is the purpose of the bit synchronization patterns in the AMPS logical control channels (dotting sequence = 10101010 . . .)? Account for the different lengths of the bit sync patterns in AMPS logical channels: FOCC (10 bits), RECC (30 bits), RVC (101 bits), and FVC (101 bits).

2. What is the purpose of the word sync (Barker sequence)? What are desirable characteristics of the Barker sequence?

3. What is the purpose of the busy/idle bits in the FOCC? Why are they not used in the other control channel formats?

4. Explain how the AMPS system uses supervisory audio tones (SAT) and a digital color code (DCC). Why are both required?

5. How does AMPS use the 10 kHz supervisory tone? Indicate other ways in which the same effects can be achieved.

6. The RECC transmits each 48-bit code word five times, for a total of 240 bits. The same information could be sent by transmitting the code word once with each bit sent five times in a row. Why is the AMPS approach more effective? (See diagram on page 123.)

7. Explain the use in AMPS of the blank-and-burst technique for transmitting network control messages when there is a call in progress. What is another way of sending control messages? What are the advantages and disadvantages of blank-and-burst?

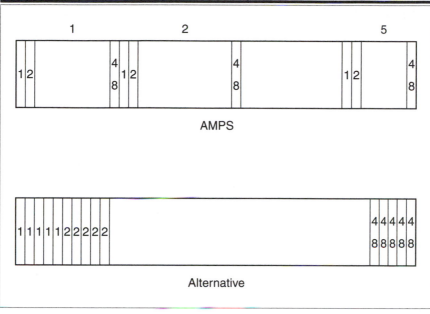

AMPS

Alternative

Question 6.

8. Why do cellular telephones have displays to indicate roaming status? How does an AMPS terminal determine whether to display a roaming indication?

9. Explain why it is sometimes desirable for the AMPS system to set up a call through a base station that is not the nearest base station to the terminal. How does the AMPS system achieve this effect?

10. AMPS specifies only two messages to be transmitted on a reverse voice channel. Suggest some other messages to transmit and explain how they would improve the system.

CHAPTER FOUR

North American Intersystem Operations

4.1 Background and Goals

With AMPS ubiquitous in the United States and Canada, every cellular phone in the two countries is compatible with every base station. However, because the AMPS standard is confined to the air interface between terminals and base stations, AMPS by itself does not contain the technology necessary to deliver to the public the full benefit of this widespread compatibility. As described in Section 3.9.2, service to roaming subscribers requires business arrangements between operating companies and technology to coordinate network management operations between AMPS systems.

With respect to business arrangements, cellular licensing regulations in the United States initially produced a fragmented cellular industry that inhibited roaming between systems. A symptom of this problem was the practice of people who had regular travel habits to obtain separate cellular telephone subscriptions in different locations. The practice became so widespread that the telephone manufacturing industry produced terminals capable of working with two or more different cellular phone numbers. To overcome the many inconveniences that initially confronted roaming subscribers, the operating industry, through mergers, acquisitions, and alliances in the early 1990s, created wide area cellular networks that offer subscribers convenient access to full service in the largest possible geographical area. The technology that makes this possible is embodied in Telecommunications Industry Association Interim Standard 41 (TIA IS-41), the subject of this chapter.

IS-41 makes it possible for roamers to originate phone calls, receive phone calls, and maintain calls in progress as they cross system boundaries. Each of these operations presents its own technical problems. When a roaming subscriber initiates a call, the local system has to verify that the caller is authorized to use the system. For calls originated by a roaming subscriber and calls directed to a roaming subscriber, the local system requires information about the services that the subscriber has selected. When a call arrives for a roaming subscriber, the home system requires information about the subscriber's location and the means to direct the call to that location. When a subscriber moves from the service area of one system to the service area of another system, the two systems have to exchange information about radio transmission conditions. They also require facilities for transmitting user information between the two systems.

4.1.1 The Situation Prior to IS-41

Prior to widespread implementation of IS-41, the cellular service industry improvised a variety of schemes for serving roaming subscribers. In the absence of technology for immediately verifying that a roaming subscriber is authorized to initiate a phone call, companies subscribed to a database, referred to as a *clearing house,* that recorded information about delinquent phone numbers. A common practice for dealing with call requests from roamers was for a visited system to accept the first call request of a new roamer and simultaneously consult the clearing house database. When it found that a roamer's number was in that database, the system would deny subsequent call requests.

The situation was considerably more complicated with respect to phone calls directed to roamers. One approach was to have the traveling cellular subscriber provide potential callers with a special telephone number connected to a cellular system in a specific geographical area. After dialing this number, callers would then dial the subscriber's cellular phone number. Upon receiving this number, the local system would initiate AMPS call setup procedures. This was clearly a burdensome process of limited utility. As an alternative, many operating companies subscribed to a GTE proprietary service, referred to as *Follow Me,* in which roamers key in a special number when they turn on their phone or enter a new area. This indicates their current location and sets up a call forwarding procedure to be invoked when someone calls the roaming subscriber. This is a burden on the subscriber, who must be aware of entering each new service area and then make the effort to register his presence.

The situation was better when the home system and visited system employed infrastructure equipment from the same manufacturer. Then it was possible for the operating companies in the two locations to serve roamers by using AMPS autonomous registration in connection with proprietary signaling links between two MTSOs. The visited MTSO used the signaling link to inform the home MTSO of the subscriber's location. When a call arrived, the home MTSO directed the visited MTSO to page the subscriber.

In addition to setting up a call to a roaming subscriber, intersystem procedures are also necessary for maintaining a call in progress when a subscriber crosses system boundaries. In the absence of standard signaling technology, handoff of a call from one system to another could take place only when both systems used infrastructure equipment from the same source. This equipment employed proprietary protocols to coordinate handoff. Another wrinkle stemmed from telecommunications industry regulation in the United States that prevented some cellular companies from providing long-distance service. As a person travels, a local call can turn into a long-distance call, which would have to be discontinued under the original regulations. A U.S. court finally waived this rule in order to allow cellular phone companies to maintain calls in progress as subscribers travel across service area boundaries. Although this removed the legal obstacle to intersystem handoff, technical problems remained.

In summary, the North American cellular industry, for many years, employed a variety of techniques to serve roamers. Often the availability of service to a roamer and the actions required to initiate or receive phone calls were unknown in advance by the traveler.

4.1.2 The Role of IS-41

To replace this collection of techniques with a national standard that requires no subscriber intervention, the Telecommunications Industry Association (TIA) produced Interim Standard 41. IS-41 is an open communications interface between a pair of AMPS systems. The TIA published a preliminary form of the standard (designated "Revision 0") in 1988. Revision A appeared in January 1991. Revision B, designated IS-41B and dated December 1991, has been implemented widely [TIA, 1991]. IS-41B is the subject of this chapter. IS-41C, which adds authentication and handoff procedures performed by digital cellular telephones, will be published in 1997. In producing IS-41, the North American industry benefited from the results of the GSM (see Chapter 7) development effort that took place a few years before IS-41 was standardized [Mouly and Pautet, 1992]. As a consequence, IS-41 borrows terminology, network, architecture, and protocol details adopted by GSM.

4.1.3 Services and Design Goals

The purpose of IS-41 is to deliver AMPS services to roaming subscribers. A subscriber has roaming status when the system identifier (SID, Table 3.1) of his home system differs from the system identifier of the base station currently communicating with the subscriber terminal. IS-41 includes mobility management functions that allow a roaming terminal to register its location in a system that is not its home system. It specifies call management functions of call setup, call release, and delivery of special features to roamers. It also specifies handoff from one AMPS system to another. The mobility management and call management functions involve communication between the subscriber's home system and the system visited by the subscriber. Intersystem handoff requires communication between two or more switches that serve adjacent areas. These switches may or may not include a switch in the subscriber's home system. In terms of the design goals of Section 2.3, IS-41 promotes roaming and mobility.

4.2 Architecture

Formally, IS-41B consists of five related, but separate, standards. All have titles that begin with the words "Cellular Radio-Telecommunications: Intersystem Operations." The remainder of each title describes the scope of its document as follows:

- IS-41.1-B: Functional Overview

- IS-41.2-B: Intersystem Handoff

- IS-41.3-B: Automatic Roaming

- IS-41.4-B: Operations, Administration, and Maintenance

- IS-41.5-B: Data Communications

With respect to the approach taken in this book, our presentation of IS-41 differs from the way we look at other systems because communications between terminals and base stations, a central issue in every other system described here, does not play a role in IS-41. Thus, in contrast to the other chapters on specific systems, there are no sections in this chapter on radio transmission or logical channels. The other topics that we examine—architecture, messages, and system operations—appear in IS-41 as follows. Documents IS-41.1 and IS-41.5 describe network architecture, the

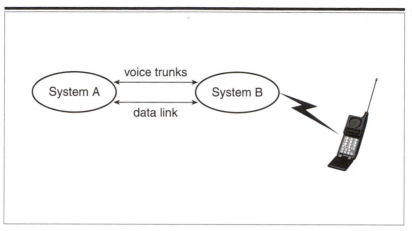

Figure 4.1 Two adjacent systems connected with voice trunks and a data link.

subject of this section. Messages (see Section 4.3) appear in IS-41.5. Network operations, including call management, mobility management, and handoff, are covered in IS-41.2 and IS-41.3.

IS-41 specifies two types of communication, illustrated in Figures 4.1 and 4.2, between a pair of cellular systems. In the case of handoff, the two systems, labeled "A" and "B," are adjacent geographically and have dedicated connections for signaling (*data link*) and for transport of user information (*voice trunks*), as shown in Figure 4.1. For automatic roaming and for operations, administration, and maintenance, the two systems can be separated geographically, and they use an external data network to exchange messages, as shown in Figure 4.2. In the case of automatic

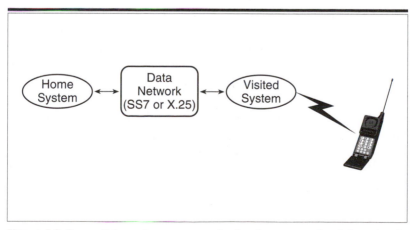

Figure 4.2 Two cellular systems communicating by means of a data network.

roaming, one of the two communicating systems is the subscriber's home system and the other one is a visited system. When IS-41 performs hand-off procedures, A and B can both be visited systems, or one of them could be the home system.

Formally, IS-41 is a seven-layer protocol conforming to the OSI model (Section 9.10). At layer 7 (applications), IS-41 employs the transactions capabilities applications part (TCAP) of Signaling System Number 7 (SS7) [Ramteke, 1994: Chapter 14]. SS7 is a layered network protocol designed to perform telecommunications network management. It plays a major role in fixed telephone networks throughout the world. Before it was incorporated in IS-41, it was adopted for communications between GSM network elements. Section 9.12 provides a brief introduction to SS7.

Below the applications layer, IS-41 specifies no operations at layers 6 and 5 (presentation and session), and a "null transport protocol" at layer 4, which consists of a small number of addressing conventions that conform to either X.25 or SS7 practices depending on the data carriage network employed. At the lowest three layers of the OSI protocol, IS-41 functions with two types of data networks: SS7 networks and X.25 networks. X.25 is a general-purpose packet data protocol that is older and simpler to implement than SS7. However, with the widespread adoption of SS7 in fixed telephone networks, SS7 is the preferred protocol for cellular operations. Our presentation of IS-41 focuses mainly on messages defined at the applications layer. When we refer to network layer operations, we confine our attention to SS7.

IS-41 adopts the functional network architecture of the Pan-European digital cellular system, GSM (Chapter 7), which has the flavor of the ISDN architecture (Section 9.11). The IS-41 architecture defines standard devices and interfaces, designated by letters, between pairs of standard devices. In addition to mobile terminals, base stations, and switches, IS-41 identifies, as standard devices, several databases. The two types of database that play a role in IS-41B are home location registers (HLR) and visitor location registers (VLR), which together coordinate roaming. Figure 4.3 shows the complete IS-41 network model, including the standard interfaces between network elements. The nomenclature for the most part is identical to that of GSM. The main departures from AMPS are the designation *mobile switching center* (MSC), instead of *mobile telephone switching office,* for a switch, and *cellular subscriber station* (CSS), instead of *mobile station,* for a terminal.

U_m is the air interface, which plays the role of the U interface in ISDN. It is the subject of the AMPS standard described in detail in Chapter 3

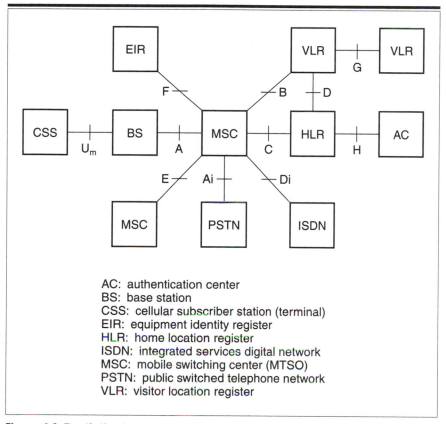

Figure 4.3 The IS-41 reference model. (Reproduced under written permission from Telecommunications Industry Association.)

[Electronic Industries Association, 1989]. The majority of IS-41 specifies the flow of information among switches and databases through interfaces B, C, D, and E. As discussed earlier, these interfaces all conform to the SS7 protocol at the applications layer and either SS7 or X.25 at layers 1–3. Although IS-41 defines the EIR (equipment identity register) and AC (authentication center) as standard devices, their status is "reserved" in IS-41B. IS-41B defines no network control operations that include either the EIR or the AC.

Although the network elements specified in Figure 4.3 are functionally distinct, it is not necessary that they be physically separate. Thus, it is a common, but not exclusive, practice to incorporate a home location register or a visitor location register, or both, in a mobile switching center.

4.3 Messages

The IS-41 message structure conforms to the TCAP (transactions capabilities applications part) of SS7. There are four types of messages: *INVOKE, RESULT, ERROR,* and *REJECT.* Each network operation begins with an *INVOKE* message transmitted from one network element to another. The network element that receives the *INVOKE* message responds with one of the other types. If it is possible for the operation to proceed, the receiving element sends a *RESULT* message to the entity that began the operation. If a network element receives an *INVOKE* message and cannot perform the task requested, it returns either an *ERROR* message or a *REJECT* message. An *ERROR* message includes a specific reason that the network element cannot perform the operations requested in the *INVOKE* message. For example, the receiving signal element may not support the operation requested, or the resources needed to perform the operation may not be available. *REJECT* messages indicate protocol errors, which prevent the system from performing the task requested in the *INVOKE* message. Each message contains fields that indicate the message length, the message type, and variable parameters that depend on the message.

Table 4.1 displays the three sets of messages conforming to the three categories of network control operations defined in IS-41: intersystem handoff; automatic roaming; and operations, administration, and maintenance. The following sections give examples of roaming protocols and intersystem handoff protocols, each consisting of a sequence of messages. These examples include only successful operations with *INVOKE* messages followed by *RESULT* messages such as *HANDOFF MEASUREMENT REQUEST INVOKE* and *HANDOFF MEASUREMENT REQUEST RESULT.*

4.4 Automatic Roaming

Automatic roaming makes it possible for a subscriber to initiate and receive calls in the service area of a system that is not her home system. It is "automatic" in the sense that the subscriber and the caller do not have to make any special effort to set up calls. The subscriber and the other party to the call act as if the subscriber is using the home system. Although automatic roaming is transparent to users, cellular systems require considerable coordination to make it possible. The visited system has to determine whether the subscriber is currently authorized to obtain service. It has to learn the subscriber's service profile, including the special features she has subscribed to. The home system has to know which system the subscriber is visiting and how to direct a call to her in that system.

Table 4.1 IS-41B Messages

Intersystem Handoff	Automatic Roaming	Operations, Administration, and Maintenance
HANDOFF MEASUREMENT REQUEST	QUALIFICATION REQUEST	BLOCKING
FACILITIES DIRECTIVE	QUALIFICATION DIRECTIVE	UNBLOCKING
MOBILE ON CHANNEL	REGISTRATION NOTIFICATION	RESET CIRCUIT
HANDOFF BACK	REGISTRATION CANCELLATION	TRUNK TEST
FACILITIES RELEASE	LOCATION REQUEST	TRUNK TEST DISCONNECT
HANDOFF TO THIRD	ROUTING REQUEST	UNRELIABLE ROAMER DATA DIRECTIVE
FLASH REQUEST	REMOTE FEATURE CONTROL REQUEST	
	SERVICE PROFILE REQUEST	
	SERVICE PROFILE DIRECTIVE	
	TRANSFER TO NUMBER REQUEST	
	CSS INACTIVE	
	REDIRECTION REQUEST	
	CALL DATA REQUEST	

The automatic roaming capabilities of IS-41 include three types of system operations described in Section 2.4: mobility management, authentication, and call management. Mobility management operations take place before a call setup request arrives. Authentication procedures protect the system against unauthorized access, while call management procedures coordinate call setup, release, and delivery of special features. All of the procedures involve communications between MSCs and two types of databases defined in Figure 4.3: home location registers (HLRs) and visitor location registers (VLRs).

An HLR contains the subscription information of a set of terminals. In addition, IS-41 procedures make it possible for the HLR to record information about the terminal's current location and status. Except for the information about subscribers' current locations, the information in an HLR changes infrequently. The information about each terminal is accessed by means of the terminal's directory number (MIN). A call request for a terminal arrives at the terminal's home MSC (MTSO, in AMPS terminology). The home MSC interrogates the terminal's HLR in order to determine the location of the terminal. The home MSC then coordinates the actions necessary to page the terminal in the visited system.

A VLR stores information about terminals that are currently in its service area. It uses this information to set up calls initiated by cellular subscribers and to deliver calls directed to cellular subscribers. The service area of a VLR can span the coverage areas of one or more MSCs. When a terminal registers its location or initiates a phone call, the MSC that controls the terminal communicates with the VLR. Therefore, the information in the VLR changes frequently as subscribers change locations. The VLR plays a central role in all of the system operations associated with automatic roaming.

4.4.1 Mobility Management

Mobility management procedures begin when a system detects the presence of a visiting terminal. This can occur when the terminal sends an AMPS registration message or an origination message to a base station. The base station informs its MSC of the terminal's action. In an internal database, the MSC records the fact that the terminal is in its operating area. The MSC also sends this information to its VLR. If the VLR is physically part of the MSC, this notification does not have to follow IS-41 specifications. However, there are situations in which it is desirable to associate one VLR with many MSCs and to use IS-41 messages to coordinate their operations.

On learning of the presence of a terminal in its service area, the VLR notifies the terminal's HLR. If the HLR has previously recorded the presence of the terminal in the service area of another VLR, the HLR notifies the old VLR that the terminal is no longer under its jurisdiction. Figure 4.4 shows this sequence of operations and the associated IS-41 messages. In this example, the mobility management procedures are triggered by the arrival of a registration message at a base station.

The sequence of operations begins when the local base station informs the MSC that it has received a registration message from the terminal. With the interface between base station and MSC proprietary in AMPS, the nature of this information exchange depends on the equipment in place. It is not standardized. When the MSC receives this information, it examines an internal database to determine whether the terminal is already registered in its service area. If it is not registered, the MSC stores

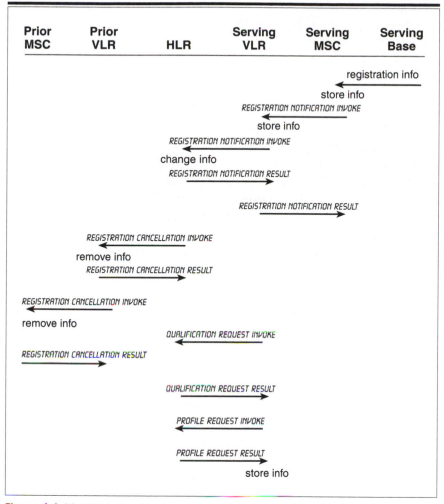

Figure 4.4 Message sequence and system operations for registration of a terminal in a visited service area.

the information that the terminal is present and sends an IS-41 *REGISTRATION NOTIFICATION INVOKE* message to its VLR. This message contains the directory number and electronic serial number (MIN and ESN in Table 3.1) of the terminal and the system identifier (SID) of the MSC. The VLR then determines whether it has already registered the presence of the terminal in its database. This can happen when the VLR covers more than one MSC. If the terminal is not already registered in the VLR, the VLR stores the registration information in its database and also sends this information to the

HLR of the terminal in another *REGISTRATION NOTIFICATION INVOKE* message. In addition to information about the terminal and the serving MSC, this message contains the Signaling System Number 7 address of the VLR.

In this example, the terminal was registered in another VLR prior to initiating the sequence of operations shown in Figure 4.4. In this situation, the HLR records the new location information of the terminal and sends a confirmation to the new VLR in a *REGISTRATION NOTIFICATION RESULT* message. This VLR sends a corresponding message to the serving MSC. In addition, the HLR informs the previous system that the terminal is no longer present in its service area. It does so by sending a *REGISTRATION CANCELLATION INVOKE* message to the prior VLR. This message causes the prior VLR to delete from its database information about the terminal. The VLR acknowledges this action by transmitting a *REGISTRATION CANCELLATION RESULT* message to the HLR. Similarly, the prior VLR and the prior MSC exchange *REGISTRATION CANCELLATION INVOKE* and *RESULT* messages that cause the MSC to delete from its database information about the terminal.

4.4.2 Authentication

Using AMPS procedures (see Section 3.8.2), the terminal transmits its electronic serial number (ESN) and the subscriber's telephone number (MIN) in a registration message. The serving MSC sends this information to its VLR and the VLR asks the subscriber's HLR if the pair of identifiers is valid. IS-41 provides a few alternatives for performing this procedure. In Figure 4.4, authentication begins with a *QUALIFICATION REQUEST INVOKE* message initiated by the VLR. The response from the HLR is contained in the *QUALIFICATION REQUEST RESULT* message in Figure 4.4. This message indicates whether the HLR accepts or denies the authentication request. If it accepts the request, the HLR can authorize the VLR to provide service to the terminal for a limited time period, or indefinitely. If it denies the request, the HLR can use a data field in the *QUALIFICATION REQUEST RESULT* message to indicate the reason for the rejection, such as invalid serial number or stolen unit.

Upon receiving authorization to serve the mobile unit, the VLR, in Figure 4.4, sends a *PROFILE REQUEST INVOKE* message to the HLR. The response is a *PROFILE REQUEST RESULT* message containing details of the subscriber's service features, which may include unconditional call forwarding, call forwarding when busy, call waiting, or three-way calling capability. In Figure 4.4, registration, authentication, and profile delivery to the VLR all take place in separate exchanges of messages. It is also possible to group these operations in a single message exchange. In this event, the *REGISTRATION NOTIFICATION RESULT* message contains information about qualification acceptance or denial, and service features available to the subscriber.

4.4.3 Call Management

Following the completion of the operations in Figure 4.4, the VLR is prepared to set up calls to and from the roaming subscriber. Subscriber call origination takes place by means of AMPS protocols. IS-41 specifies procedures for processing requests to set up calls to roaming subscribers. Which of several different procedures is used depends on the subscriber's service profile and the terminal's status (busy or idle, powered up or turned off, etc.). Figure 4.5 shows the simplest situation in which the terminal is turned on in the service area of the serving VLR and is not busy with an existing call. Here the call begins when someone dials the cellular phone number of the roaming subscriber. The telephone network transmits a call setup request to the subscriber's home system, typically by means of an SS7 procedure. The home MSC interrogates the subscriber's HLR to determine the present location of the subscriber. In order to respond to the location request, the HLR asks the serving system for a *temporary local directory number* to be used to set up a call to the subscriber. This is a phone number that will route the call to the serving MSC. The HLR sends this phone number to the home MSC, which uses it to set up a call to the serving MSC. On receiving the call setup request, the serving MSC pages the terminal using the terminal's directory number (MIN). The call setup then follows the procedures described in Section 3.7.5 (see Figure 3.16).

The sequence of operations shown in Figure 4.5 begins when the home MSC receives a request to set up a call to the roaming subscriber. Typically, this request arrives from the Public Switched Telephone Network in an SS7 message. The MSC establishes a billing identifier, which is a numeric code unique to this phone call. It then sends a *LOCATION REQUEST INVOKE* message to the subscriber's HLR. This message identifies the subscriber and supplies the billing identifier. The HLR interrogates its internal database and determines the address of the serving VLR. Recall that the HLR received this address in a *REGISTRATION NOTIFICATION INVOKE* message from the VLR (Figure 4.4). Using this address, the HLR sends a *ROUTE REQUEST INVOKE* message to the serving VLR. The VLR examines its database to determine the serving MSC and forwards the *ROUTE REQUEST INVOKE* message to this MSC. The serving MSC assigns a temporary telephone number to this call and forwards this number to the home MSC in the

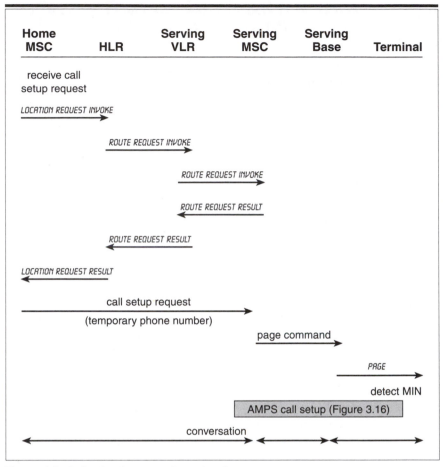

Figure 4.5 Call setup to a roaming subscriber.

data fields of confirmation messages (*ROUTE REQUEST RESULT* and *LOCATION REQUEST RESULT*) transmitted to the serving VLR, the subscriber's HLR, and the home MSC.

When the home MSC receives the temporary phone number, it places a call to this number. The serving MSC receives the call setup request and retrieves from its database the terminal identifier (MIN) associated with the temporary directory number. It then performs the sequence of operations in Figure 3.16 that begins with the transmission of paging commands to a set of base stations. All of these base stations broadcast AMPS *PAGE* messages on their forward control channels. The terminal detects its MIN in a *PAGE* message and sends an AMPS *PAGE RESPONSE* message to the serving base station, which informs the serving MSC that the call setup can proceed. The remaining operations in Figure 3.16 establish a voice path between the serving MSC and the terminal. The serving MSC also uses standard SS7 procedures to set up a voice path to the home MSC, which forwards the call to the serving system.

Figure 4.5 represents an example of a successful call setup procedure: call delivery to an idle roaming terminal. IS-41 also specifies protocols for several other call management conditions. Figure 4.6 indicates seven conditions, including the one shown in Figure 4.5. Each condition depends on the service profile and status of the called subscriber. For example, the called terminal may be busy with a call in progress when the call request arrives. In this event, the serving MSC informs its VLR of this condition in the *ROUTE REQUEST RESULT* message. This information is relayed through the HLR to the home MSC in a *ROUTE REQUEST RESULT* message and a *LOCATION REQUEST RESULT* message, respectively. If the called subscriber does not use call forwarding, the home MSC then sends a busy indication to the telephone that originated the call. If, on the other hand, call forwarding is in effect, for example to a voice mailbox, the *LOCATION REQUEST RESULT* message from the HLR to the home MSC contains a forwarding telephone number.

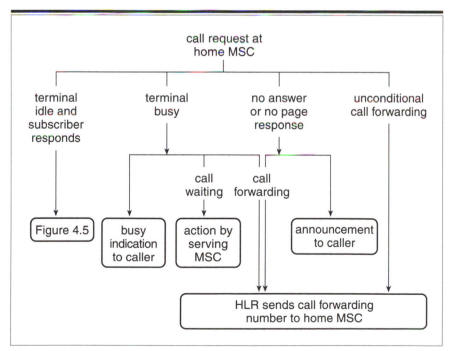

Figure 4.6 Examples of call management scenarios.

The home MSC then sets up a call from the originating telephone to the forwarding number.

The other conditions covered in Figure 4.6 are:

- delivery of call waiting service to a busy subscriber;

- forwarding a call (perhaps to a voice mailbox) when the subscriber does not respond to an incoming call;

- transmitting an announcement to the calling party when the subscriber does not respond; and

- unconditional call forwarding to another telephone number, in which case the system makes no attempt to set up the call to the mobile terminal.

The system procedures that take place during these conditions involve the *REDIRECTION REQUEST* and *TRANSFER TO NUMBER REQUEST* messages in Table 4.1. The other automatic roaming messages in Table 4.1 that do not appear in Figures 4.4 and 4.5 are used in other call management conditions covered by IS-41, including

- special feature requests by mobile subscribers (*REMOTE FEATURE CONTROL REQUEST*, *CALL DATA REQUEST*, and *SERVICE PROFILE DIRECTIVE* messages),

- changes in the subscription conditions of a roaming subscriber (*QUALIFICATION DIRECTIVE* message sent by the HLR), and

- information that the subscriber is unable to receive a call in the visited system (*CSS INACTIVE* message from the VLR).

4.5 Intersystem Handoff

Section 3.8.3 and Figure 3.17 present a sequence of operations that take place in an AMPS handoff. The sequence begins when the received signal level at the base station falls below a threshold. The base station sends a message to its switch and the switch asks neighboring base stations to measure the signal arriving from the mobile phone. Intersystem handoff procedures, specified in IS-41, begin when the terminal is in a cell at the boundary of the serving area of the switch. If another system serves the adjacent area, and there is a direct communication path to the other system, the switch handling the call can request information from the adjacent system and eventually request that the adjacent system handle the call.

IS-41 uses four terms to describe the status of MSCs involved in an intersystem handoff procedure: *anchor, serving, candidate,* and *target.* The *anchor* MSC is the one in which a call begins. This MSC connects the call to the Public Switched Telephone Network. It retains the status of anchor MSC for the duration of the call. At any time, the *serving* MSC is the one that currently handles the call. An intersystem handoff procedure moves the status of serving MSC from one MSC to another. During the procedure, the serving MSC asks a neighboring MSC to measure the strength of the signal received from the terminal. The MSC that obtains the measurements is a *candidate* MSC. If the serving MSC decides that the call should move to the candidate MSC, the candidate MSC becomes the *target* MSC of the handoff. Finally, after the handoff is complete, the target MSC becomes the serving MSC. Table 4.2 shows how the status changes as a terminal moves from its original service area, controlled by MSC-A, to a new service area, controlled by MSC-B.

The table shows that the call remains anchored in MSC-A. During the handoff procedure, MSC-B becomes first the candidate MSC, then the target MSC, and finally the serving MSC.

4.5.1 Handoff Categories

IS-41 specifies three handoff protocols: *handoff forward, handoff back,* and *handoff to third,* as described in the following paragraphs.

As shown in Figure 4.1, intersystem handoff requires dedicated communication links between a pair of MSCs. These links consist of voice trunks for carrying user information in calls handed from one MSC to another, and data links for carrying control messages between the two switches. Intersystem handoff usually results in one MSC forwarding a call to another. Figures 4.7 and 4.8 show the simplest situation, referred to

Table 4.2 MSC Status Before, During, and After a Handoff Procedure

	Anchor	Serving	Candidate	Target
Call begins	MSC-A	MSC-A		
Terminal approaches service area of MSC-B	MSC-A	MSC-A	MSC-B	
MSC-A decides to transfer call to MSC-B	MSC-A	MSC-A		MSC-B
Handoff complete	MSC-A	MSC-B		

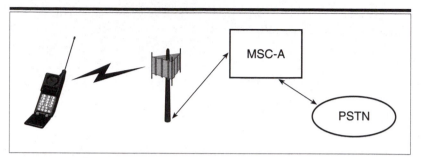

Figure 4.7 A call originates in System A (anchor system).

as *handoff forward*. In Figure 4.7, the call begins in System A. The terminal moves into the service area of System B causing MSC-A and MSC-B to perform a handoff to a base station connected to MSC-B. Figure 4.8 shows the result. The conversation passes through voice trunks connecting MSC-A to MSC-B and through land lines within System B that connect MSC-B to the cell occupied by the terminal. In this situation, MSC-A is the *anchor MSC*. It is responsible for routing the call to the remote party. MSC-B is the *serving MSC* because it currently has control of the call.

After a handoff from System A to System B, there is no guarantee that the terminal will remain in the service area of System B. IS-41 identifies two possibilities. It can return to the service area of System A, or it can move into the service area of a third system, System C. When the terminal returns to the service area of System A, MSC-B recognizes that the call

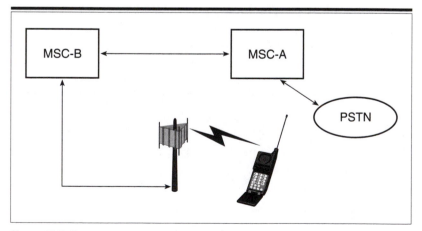

Figure 4.8 The situation after a handoff forward from System A (anchor system) to System B (serving system).

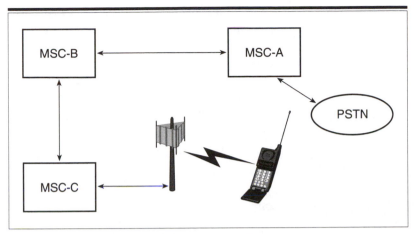

Figure 4.9 Call path after handoff forward to System C.

arrived from System A and it initiates a handoff back protocol, which releases the voice circuit between MSC-A and MSC-B and restores the signal path of Figure 4.8. Without this protocol, the systems would tie up two voice trunks (one taking the call from System A to System B, and the other taking it from System B back to A) when none is necessary.

It is also possible that the terminal will move from System B to a third system, System C. This produces two possibilities illustrated in Figures 4.9 and 4.10. In Figure 4.9, MSC-B and MSC-C perform a handoff forward procedure like the one that moved the call from System A to System B.

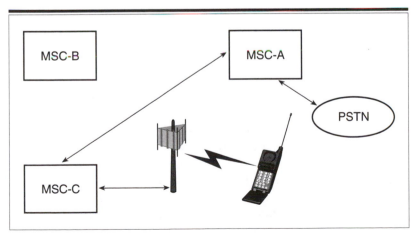

Figure 4.10 If there are circuits connecting MSC-A and MSC-C, the system can perform handoff to third with this result.

Now System B simply provides a path from MSC-A to MSC-C. The situation can continue, adding more and more MSCs to the chain, up to a limit established by the anchor system. An alternative occurs when there is a direct connection between System A and System C. To economize on transmission and switching facilities in this situation, IS-41 includes a protocol referred to as *handoff to third*, which establishes a direct link between MSC-A and MSC-C and releases the link between A and B. Figure 4.10 shows the result. The call now uses one inter-MSC voice trunk, from MSC-A to MSC-C. Without handoff to third, it would use two circuits (A to B and B to C), as in Figure 4.9.

4.5.2 Handoff Protocols

There are two phases to every handoff procedure. In the *location phase*, the serving MSC collects measurement reports from cells in the neighborhood of the cell presently occupied by a terminal. The operations in this phase are independent of the nature of the handoff protocol. When measurements are required from one or more cells in a system adjacent to the serving system, the adjacent system becomes a *candidate system*. During the location phase, the serving MSC and a candidate MSC exchange handoff measurement request messages. A HANDOFF MEASUREMENT REQUEST INVOKE message, transmitted by the serving MSC, includes:

- information about the terminal (station class mark, in Table 3.1),

- information about the serving base station (SAT and a base station identifier), and

- information about the radio channel carrying the call (channel number).

Based on the identity of the serving base station, the candidate MSC selects one or more candidate cells and transmits a HANDOFF MEASUREMENT REQUEST RESULT message to the serving MSC. This message contains identities of candidate cells and associated signal strength measurements. After collecting measurement reports from several cells, the serving MSC selects a *target cell* for the handoff. If the target cell is served by a candidate MSC, this MSC becomes the *target MSC* for the handoff.

The handoff procedure then moves from the location phase to the handoff phase. In the handoff phase, the serving MSC determines the type of handoff to initiate (forward, back, or handoff to third).

Handoff Forward

For a handoff forward, the serving MSC sends a *FACILITIES DIRECTIVE INVOKE* message to the target MSC. This message contains several parameters that will allow the target MSC to process the call. These parameters include:

- information about the terminal (SCM, MIN, ESN, Table 3.1);

- information about the call;

  - billing ID (established by the anchor MSC at the beginning of the call, as described in Section 4.4.3);

  - inter-MSC circuit (identifies the voice trunk that will carry the call from the serving MSC to the target MSC);

  - inter-switch count (the total number of MSCs through which the call will pass after the handoff);

- information about the call status (serving cell, serving channel); and

- target cell identifier (based on measurement reports from the target MSC).

If the target MSC accepts the handoff, it selects a channel to handle the call in the new cell and then sends a *FACILITIES DIRECTIVE RESULT* message to the serving MSC. This message contains information about the new channel, including the channel number, SAT, and VMAC (transmit power level). On receiving this message, the serving MSC sends an AMPS *HAND-OFF* message (see Table 3.4) to the terminal through the serving cell. The terminal performs the operations described in Section 3.8.3. When the target base station detects the SAT, it sends a message to the target MSC, which completes the handoff forward operation by sending a *MOBILE ON CHANNEL INVOKE* message to the prior serving MSC. This message requires no response. Figure 4.11 is a message flow diagram for the handoff forward protocol.

Handoff Back

If the location phase results in a determination by the serving MSC that the call would best be handled in the system previously occupied by the terminal, the serving MSC initiates a handoff back procedure. It sends a

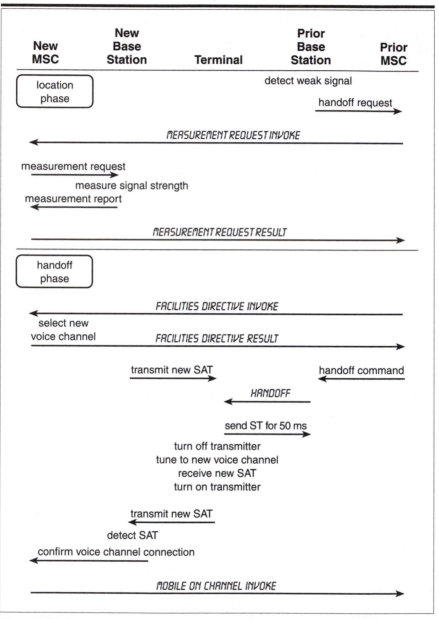

Figure 4.11 Message sequence and system operations for handoff forward.

HANDOFF BACK INVOKE message to the previous MSC, which is now the target MSC of the handoff protocol. This message plays the same role as the *FACILITIES DIRECTIVE INVOKE* message in Figure 4.11, but contains only a subset of the parameters required for handoff forward. The remaining information about the call is available at the target MSC because the call passes through this MSC prior to the handoff. The *HANDOFF BACK RESULT* message from the target MSC to the serving MSC contains the same information as the *FACILITIES DIRECTIVE RESULT* message in the handoff forward protocol. This message provides the information necessary for the serving MSC to send (through the serving base station) a handoff command to the terminal. When the target MSC learns that the terminal has arrived on the assigned channel at the target base station, it sends a *FACILITIES RELEASE INVOKE* message to the serving MSC. This message identifies the voice trunk that carries the call between the two MSCs. On receiving this message, the serving MSC releases the voice trunk and sends a *FACILITIES RELEASE RESULT* message to the target MSC. This message completes the handoff back protocol and causes the target MSC to release the inter-MSC voice trunk. Figure 4.12 shows the sequence of operations in the handoff phase of a handoff back protocol.

Any two MSCs in a chain can perform the handoff back protocol. For example, one can be the anchor MSC, in which case the call path changes from the one shown in Figure 4.8 to the call path shown in Figure 4.7, or they can be two other MSCs, such as MSC-B and MSC-C in Figure 4.9. In this case handoff back changes the call path from the one in Figure 4.9 to the one in Figure 4.8.

Handoff to Third

This is the most complicated of the three protocols because it involves at least three MSCs, rather than two as in handoff forward and handoff back. Handoff to third is an example of a path minimization procedure, in which the system reduces the number of voice trunks carrying a call through three or more systems. The simplest example of handoff to third is the one that changes the call path in Figure 4.8 to the one in Figure 4.10. In this example, the serving MSC (MSC-B) initiates the procedure by sending a *HANDOFF TO THIRD INVOKE* message to the anchor MSC (MSC-A). This message contains all of the information conveyed in a *FACILITIES DIREC-TIVE INVOKE* message and one additional parameter: the identity of the target MSC (MSC-C). On receiving the *HANDOFF TO THIRD* message, MSC-A takes control of the handoff procedure. It sends a *FACILITIES DIRECTIVE INVOKE* message to the target MSC. The target MSC responds with a *FACILITIES DIRECTIVE RESULT* message containing the information necessary to move the terminal to a new cell. MSC-A relays this information to the serving MSC in a *HANDOFF TO THIRD RESULT* message. The serving MSC sends this information to the serving base station, which transmits it to the terminal in an AMPS *HANDOFF* message. On receiving this information, the terminal tunes

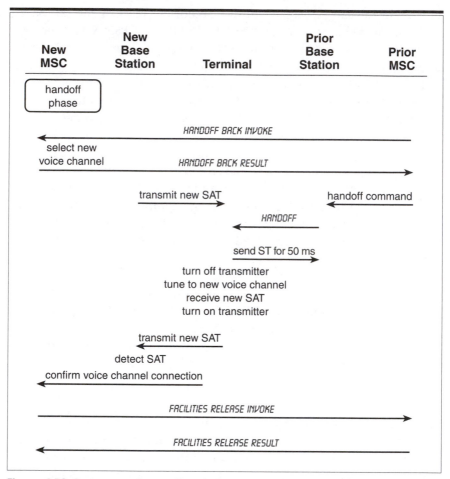

Figure 4.12 Sequence of operations for handoff back.

to the assigned channel and transmits the SAT specified in the handoff message. On detecting the SAT transmission, the target base station informs the target MSC, which sends a *MOBILE ON CHANNEL* message to MSC-A. MSC-A then routes the call through a voice trunk connecting MSC-A to the new serving MSC (MSC-C). It also sends a *FACILITIES RELEASE INVOKE* message to the prior serving MSC (MSC-B), which releases the trunk that carried the call from MSC-A to the prior serving MSC. The prior serving MSC then sends a *FACILITIES RELEASE RESULT* message to MSC-A, confirming that the handoff procedure is complete. Figure 4.13 shows the operations in a handoff to third protocol.

In Figure 4.13, MSC-A, the controlling MSC, is the anchor MSC. In general, the role of MSC-A, as described in the previous paragraph, can be

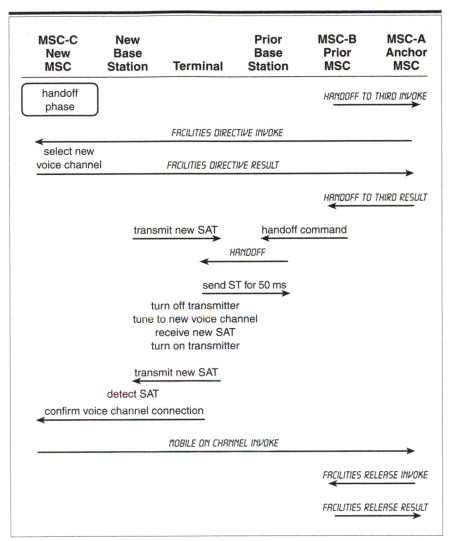

Figure 4.13 Sequence of operations for path minimization (handoff to third).

performed by any MSC in the call path, for example, MSC-X, provided that MSC-X precedes (is closer to the anchor MSC than) the serving MSC.

4.5.3 Other Intersystem Handoff Procedures

In addition to transferring a call from one system to another, IS-41 specifies two other operations that take place after an intersystem handoff has been performed. One type of operation is the release of voice trunks when a call ends. Systems perform this operation by exchanging *FACILITIES RELEASE INVOKE* messages and *FACILITIES RELEASE RESULT* messages. The other operation is the transmission of service requests by a terminal with a call in progress on a TDMA digital voice channel (Chapter 5) that passes through more than one AMPS system. The terminal sends a *FLASH WITH INFO* message to the serving base station. The information in this message affects the processing of the call by the anchor MSC. The purpose of the IS-41 *FLASH REQUEST INVOKE* message is to relay this call management information from the serving MSC to the anchor MSC, perhaps through a sequence of intermediate MSCs. There is no acknowledgment of the *FLASH REQUEST INVOKE* message.

4.6 Status of IS-41

The material presented in this chapter describes IS-41B, which is in widespread use in the United States and Canada. At the beginning of 1997, the Telecommunications Industry Association was in the process of approving IS-41C. IS-41C will make it possible for roaming calls to use the network operations defined in the TDMA and CDMA North American digital cellular standards (IS-136 and IS-95, respectively). The main operations are cryptographic authentication and handoff to and from digital voice channels.

Review Exercises

1. Why is a cellular phone capable of operating with more than one telephone number?

2. How does AMPS use dedicated communications links between two mobile switching centers?

3. How does AMPS use an external signaling network in intersystem operations?

4. Someone uses a conventional wired telephone in Los Angeles to call a person who subscribes to AMPS service, including voice mail, in Chicago. The subscriber's telephone has recently registered its location in Los Angeles. However, the subscriber is away from the phone when the call is placed. Describe the communications that take place among the fixed telephone network, the home system in Chicago, and the visited system in Los Angeles.

5. During a phone call, explain the conditions that cause a mobile switching center to perform each of the following functions:
   • anchor MSC,
   • serving MSC,

- candidate MSC, and

- target MSC.

6. With respect to the four functions in Question 5, can an MSC perform more than one function at a particular instant? Give all the examples you can of functions that can be performed simultaneously by one MSC.

7. Why does IS-41 specify three different handoff procedures?

8. Prior to performing a handoff to a new MSC, the serving MSC transmits three types of information to the candidate MSC. Which type is essential? Explain why the other two are useful. How does the candidate MSC use the three types of information?

9. The following diagram shows the route taken by a car with a call in progress. The rectangles are the service areas of three different MSCs. What type of handoff operation is necessary as the terminal crosses each boundary between two MSC service areas? When the car is at the position indicated by **+** , indicate which MSCs perform the four functions listed in Question 5.

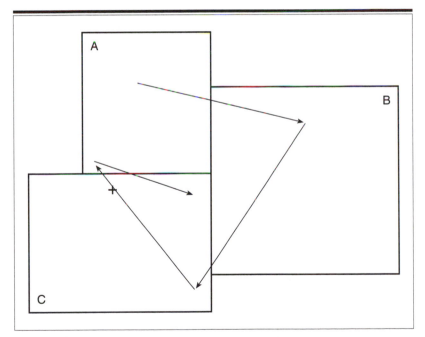

Question 9.

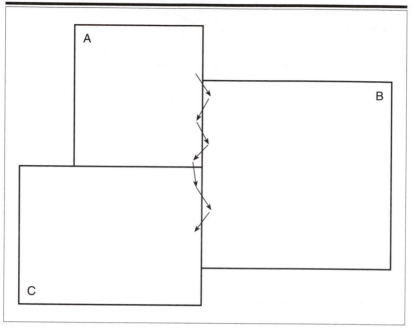

Question 10.

10. Answer Question 9 for the next diagram. This could be a picture of a car driving on one road. The path of the road and local propagation conditions (shadow fading) move it from one service area to another.

CHAPTER FIVE

North American Cellular System Based on Time Division Multiple Access

5.1 Background and Goals

Cellular telephones entered commercial service in North America late in 1983. Their popularity soon became apparent. Within 5 years, this success aroused industry concern that system capacity would be saturated in the 1990s, first in the largest cities, with other locations predictably following. When consumer demand saturates the capacity of a cellular system, there are three ways to expand: move into new spectrum bands, split existing cells into smaller cells by installing new base stations, or introduce new technology to make more efficient use of existing bandwidth and base stations. After expanding the cellular bandwidth allocation from 40 MHz to 50 MHz as shown in Figure 3.2, regulators announced that no new radio spectrum would be available for cellular services. As the geographical density of a cellular system increases, adding base stations becomes an increasingly expensive means of expansion, and beyond a certain point, it becomes impractical. As cell dimensions become smaller and smaller in congested areas, the precise locations of new base stations become critical and the opportunities for acquiring new sites in the necessary locations diminish.

With no new bandwidth available for cellular services and a limit in sight to cell splitting, new technology is the best route to system expansion. To stimulate technology creation, the Federal Communications

Commission declared in 1987 that "cellular licensees may employ alternative cellular technologies in the cellular bands 824–849 MHz and 869–894 MHz . . . provided that interference to other cellular systems is not created" [FCC, 1990a]. This encouraged the cellular industry to search for new transmission techniques that would increase the efficiency, as measured by channels per cell, of existing AMPS systems.

In 1988 and 1989, industry panels considered technology proposals submitted by several different companies. All of the proposals incorporated digital speech transmission. The principal distinguishing feature of the different approaches was the multiple access technique (Section 2.6.2). Many companies advocated TDMA (time division multiple access), while others favored FDMA (frequency division multiple access). Early in 1990, the industry voted to adopt a hybrid frequency-division/time-division approach (Figure 9.3). The industry also decided to retain the AMPS 30 kHz spacing of carrier frequencies by dividing an analog AMPS physical channel into time multiplexed digital channels. The digital TDMA channels are part of a dual-mode system, in which each digital subscriber unit is also capable of analog speech transmission. This gives the operating industry flexibility to install digital channels incrementally in each area as the number of dual-mode telephones increases. In cells with no digital channels, subscribers with dual-mode AMPS telephones use the original AMPS technology described in Chapter 3.

In addition to meeting a growing demand for cellular service with efficient transmission technology (see Section 2.3.19), another pressing industry need was for enhanced network security (see Section 2.3.20). The original AMPS authentication procedure, based on verifying electronic serial numbers, proved vulnerable to fraud. Unauthorized users monitor cellular transmissions in order to detect the mobile identity numbers and electronic serial numbers of system subscribers. They then use these codes to gain access to the system. To meet this challenge, the industry adopted enhanced security measures involving encrypted radio transmission of identity codes. Another design goal for second-generation cellular equipment in North America was to accommodate small, portable terminals with self-contained power supplies (see Section 2.3.3). A large majority of subscribers favors these terminals in preference to vehicle-mounted mobile telephones that use the vehicle's battery. Other goals that influenced system design were early deployment and adaptability to technical change (see Section 2.3.21).

The result of the industry design effort was Interim Standard 54, *Cellular System Dual Mode Subscriber Equipment* published by the Telecommunications Industry Association [TIA, 1992]. Under this standard, new terminals have all of the attributes of first-generation AMPS terminals, with the added ability to transmit user information in digital format over "digital traffic channels." These channels can be installed in existing base stations as replacements for some of the conventional frequency modulation voice channels. With IS-54 in place, analog AMPS cellular telephones continue to have access to all AMPS systems, while network operators encourage subscribers to acquire dual-mode terminals with access to digital channels operating in congested areas. The technology advances that make IS-54 at least three times as efficient as AMPS are mainly in the area of digital signal processing. These advances make it possible for mobile terminals to perform sophisticated speech coding and equalization algorithms.

IS-54 was created after GSM, the Pan-European digital mobile radio system (see Chapter 7), and it incorporates several important GSM network control innovations. Examples are associated control channels, authentication procedures based on encrypted transmissions, and mobile-assisted handoff procedures. Conversely, the Japanese digital cellular system, referred to as Pacific Digital Cellular (PDC) [Kinoshito, Kuramoto, and Nakajima, 1991], was designed soon after IS-54 was published. Many details of PDC radio transmission resemble those of IS-54.

While calls are in progress, IS-54 telephones have access to *associated control channels* that perform some of the functions of AMPS transmissions on forward voice channels (FVC) and reverse voice channels (RVC). However, before an IS-54 system assigns a traffic channel, it uses AMPS forward control channels (FOCC) and reverse control channels (RECC) to perform mobility management, authentication, and call management operations. The RECC and FOCC reflect 1970s technology, which by the mid-1990s limited system performance in various ways. To overcome these limitations, the industry published, in December 1994, the specification of a digital control channel (DCCH). The DCCH is more efficient than the AMPS RECC and FOCC. It also enhances the performance of terminals by introducing a *sleep mode,* in which terminals can turn off their receivers for a significant fraction of the time when they do not have a call in progress. This extends battery life (see Section 2.3.15). The digital control channel also delivers short message services (see Section 2.2.2) to terminals.

The digital control channel specification is TIA/EIA Interim Standard 136.1. Interim Standard 136.2 is a revised version of IS-54. IS-136.2 takes into account the existence of digital control channels. It specifies analog and digital traffic channels and the AMPS out-of-band control channels (FOCC and RECC). The IS-136 nomenclature for a combined FOCC/RECC is *FSK (frequency shift keying) control channel.* The publication of IS-136 opens the way to the production of all-digital TDMA telephones, rather than the dual-mode units specified in IS-54. All-digital terminals will be attractive when there is a sufficient infrastructure of AMPS base stations that incorporate TDMA channels. Meanwhile, TDMA telephones will continue to have dual-mode capability, ensuring their owners of service in all locations covered by AMPS systems. A revised version of IS-136, published in October 1996 [TIA, 1996d], specifies TDMA operation in the North American PCS bands at 1,900 MHz (see Section 1.6.5). In these bands, TDMA networks operate exclusively with digital control channels. In the absence of analog transmissions at 1.9 GHz, there are no AMPS forward and reverse control channels.

As in AMPS, the principal application is telephony (see Section 2.2.1). In addition to basic telephony, IS-136 has provisions that facilitate the delivery of supplementary services including voice mail, call waiting, caller identification, and group calling. IS-136 also specifies a short message service (see Section 2.2.2). While the transmission technologies for voiceband data and facsimile transmissions are specified in other documents [TIA, 1995c; TIA, 1995d], IS-136 contains protocols that make it possible for a network to set up calls that transfer facsimile and data signals. The system can also inform users that the network holds facsimile, voice mail, or text messages addressed to them (see Section 5.5.1).

The remainder of this chapter describes the system specified in IS-136. We refer to the system as *NA-TDMA (North American TDMA)*.

5.2 Architecture

NA-TDMA is an extension of AMPS. IS-136 systems are capable of operating with AMPS terminals, dual-mode terminals, and all-digital terminals. The network architecture, Figure 5.1, is a more general version of the AMPS architecture in Figure 3.1. Corresponding to the AMPS network infrastructure of land stations and mobile telephone switching offices (base stations and switches), NA-TDMA defines a BMI: "Base Station, Mobile Switching Center, and Interworking Function." Because IS-136 is confined to the air interface, it is appropriate to specify, in this general way, the functions performed in the network infrastructure. Each equipment vendor then makes its own decisions on how to allocate functions performed by the BMI to specific pieces of equipment.

In accordance with the goal of a personal communications system to accommodate multiple modes of operation (see Section 2.3.16), NA-TDMA specifies three types of external network: public systems, residential systems, and private systems. Thus, a terminal can function as a

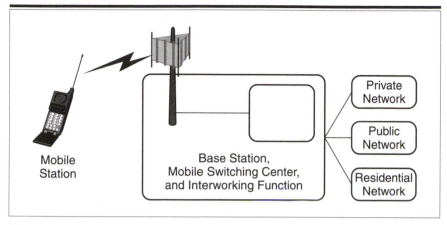

Figure 5.1 NA-TDMA architecture.

cellular telephone with access to the base stations of cellular operating companies (*public network*). It can also be programmed to function as a cordless telephone operating with a specific residential base station (residential network), and as a business phone operating with a specific wireless private branch exchange (private network).

To deliver mobile telephony, cryptographic authentication, and a wide range of service enhancements relative to AMPS, NA-TDMA defines a large number of identification codes, including all of the AMPS identifiers in Table 3.1. Table 5.1 is a partial list. A major addition to the set of identification codes is the 64-bit A-key, assigned to each subscriber by her cellular operating company. This encryption key plays a critical role in promoting network security and communication privacy in a dual-mode TDMA system. Section 5.6.1 describes the use of the A-key in authentication of subscribers and encryption of user information. Another identification code in NA-TDMA is a 12-bit location area identifier, LOCAID. The system can divide its service area into clusters of cells, referred to as location areas. Each base station broadcasts its LOCAID. When a terminal that does not have a call in progress enters a new location area, it sends a registration message to the system. When a call arrives for the terminal, the system pages the terminal in the location area that received the most recent registration message (see Section 5.6.4).

The IMSI in Table 5.1 is a telephone number with up to 15 decimal digits that conforms to an international numbering plan (E.212) published by the International Telecommunication Union. The value of PV reflects the standards document (for example, IS-54 or IS-136) that governs the operation of a base station or terminal. The system operator code (SOC) transmitted

Table 5.1 NA-TDMA Identifiers

Name	Meaning	Size (bits)	Description
MIN	Mobile identifier	34	Directory number assigned by operating company to a subscriber
IMSI	International mobile subscriber identification	50	Directory number conforming to international convention
ESN	Electronic serial number	32	Assigned by manufacturer to a mobile station
A	A-key	64	Secret key for cryptographic authentication
SID	System identifier	15	Assigned by regulators to a geographical service area
SCM	Station class mark	5	Indicates capabilities of a mobile station
PV	Protocol version	4	Indicates capabilities of a mobile station or a base station
SAT	Supervisory audio tone	*	Assigned by operating company to each base station
SOC	System operator code	12	Identifies the company operating a particular system
BSMC	Base station manufacturer code	8	Identifies the manufacturer of the local base station
LOCAID	Location area identifier	12	Identifies the location area of a base station
DCC	Digital color code	2	Assigned by operating company to each base station
DVCC	Digital verification color code	12	Assigned by operating company to each base station with digital channels

* One of three sine wave signals.

by a base station identifies to terminals the company that operates the base station, while BSMC indicates the manufacturer of the base station. The digital verification color code (DVCC) plays the same role in digital traffic channels as the SAT transmitted in analog traffic channels.

5.3 Radio Transmission

IS-136 specifies dual-mode NA-TDMA/AMPS operation in the AMPS frequency bands of Figure 3.2. It also anticipates all-digital operation in six frequency bands near 1.9 GHz allocated in 1995 for personal communications in the United States (see Figure 1.18). In all of these bands, NA-TDMA specifies carriers spaced at 30 kHz. (In the AMPS band, each pair of NA-TDMA carriers corresponds to an AMPS physical channel.) The access technology conforms to the hybrid FDMA/TDMA model of Figure 9.3. As illustrated in Figure 5.2, the frame duration is 40 ms. Each frame contains six time slots. Therefore, the length of each time slot is $40/6 = 6.67$ ms. The time offset between reverse (mobile-to-base) and forward (base-to-mobile) transmissions is important. It implies that a terminal can perform full-duplex, two-way communications without transmitting and receiving simultaneously. This can simplify the equipment in all-digital telephones by eliminating the need for a duplexing filter to separate the strong transmitted signal from the weak received signal. In Figure 5.2, a receive time slot (base-to-mobile) begins approximately 1.9 ms after the end of the corresponding transmit time slot.

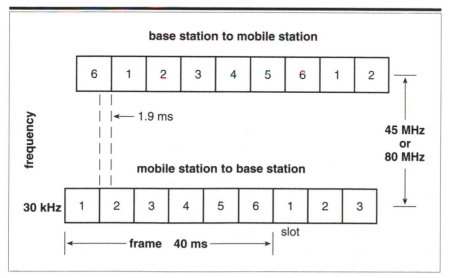

Figure 5.2 Frames and time slots.

5.3.1 Physical Channels

Each time slot carries 324 bits, so that the data rate per carrier is

$$\frac{324 \, \frac{\text{bits}}{\text{slot}} \times 6 \, \frac{\text{slots}}{\text{frame}}}{40 \, \frac{\text{ms}}{\text{frame}}} = 48.6 \text{ kb/s.} \tag{5.1}$$

NA-TDMA defines four types of physical channels, distinguished by the number of slots per frame that they occupy. Initial implementations carry information in *full-rate* physical channels that occupy two slots per frame. A full-rate channel can occupy slots 1 and 4, slots 2 and 5, or slots 3 and 6. Therefore, the bit rate of a full-rate physical channel is

$$\frac{324 \, \frac{\text{bits}}{\text{slot}} \times 2 \, \frac{\text{slots}}{\text{frame}}}{40 \, \frac{\text{ms}}{\text{frame}}} = 16.2 \text{ kb/s,} \tag{5.2}$$

one-third of the transmission rate.

Reflecting the flexibility of TDMA, IS-136 also defines

- half-rate channels (8.1 kb/s) consisting of one slot per frame,

- double full-rate digital channels (32.4 kb/s) with four slots per frame, and

- triple full-rate digital channels that occupy an entire carrier (48.6 kb/s).

Figure 5.3 shows a full-rate physical channel and a half-rate physical channel as regions in the time-frequency plane of the AMPS spectrum band. Figure 5.3 shows only the reverse direction of transmission. Each physical channel simultaneously occupies corresponding regions in the forward band: 869 MHz to 894 MHz.

In contrast to AMPS, which always operates physical channels 313–354 as forward and reverse control channels (see Section 3.3.1), NA-TDMA has no fixed assignment of physical channels to digital control channel operation. As a consequence, on arriving in a new cell, a TDMA terminal without a call in progress performs a scanning procedure to search for a digital control channel.

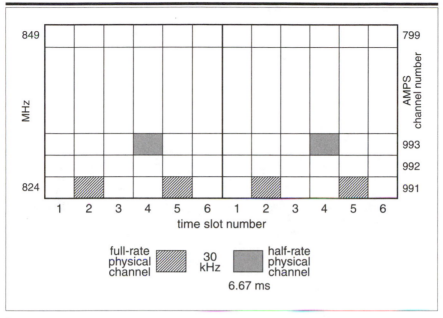

Figure 5.3 Half-rate physical channel (one slot per frame) and full-rate physical channel (two slots per frame).

5.3.2 Modulation

The modulation format for the 48.6 kb/s transmitted signal is $\pi/4$ shifted DQPSK (differential quaternary phase shift keying) [Feher, 1991]. At any instant, the modulator is capable of producing one of eight possible signals. Each signal is a sine wave at the carrier frequency with one of eight possible phase angles that are multiples of $\pi/4$ radians. DQPSK is a four-level modulation scheme, which means that each transmitted signal, referred to as a *channel symbol*, carries 2 bits to the receiver. Therefore, a modulator generates 48.6 kb/s ÷ 2 bits/symbol = 24,300 symbols per second. Symbols are distinguished by the carrier phase shift during a symbol interval. For each symbol, the possible phase changes are odd multiples of $\pi/4$ radians (45°): $-3\pi/4$, $-\pi/4$, $\pi/4$, and $3\pi/4$ radians. Differential PSK is an appropriate modulation technique for cellular radio systems because it reduces the demodulator's job to detecting phase differences, rather than absolute phase. Owing to the nature of the signal fading in a mobile radio system, it is a difficult task to determine the absolute phase of a received signal.

Figure 5.4 displays the possible phase transitions in the NA-TDMA modulation technique. At the beginning of a symbol interval, the signal

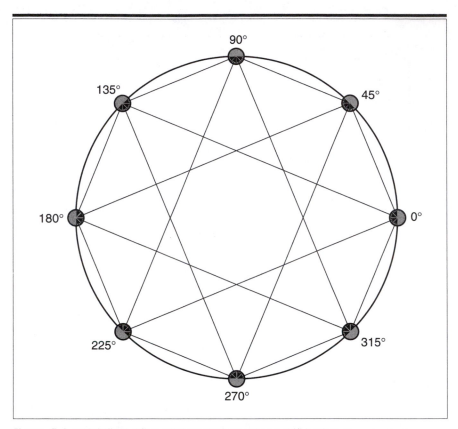

Figure 5.4 π/4 shifted differential quaternary phase shift keying. (Reproduced under written permission from Telecommunications Industry Association.)

phase is at the location of one of the eight small circles on the circumference of the large circle. The four lines leading away from each small circle represent the possible phase changes that can take place during the transmission of a symbol. It is significant that the transition lines cut through the interior of the large circle. This indicates that the signal envelope decreases and then increases again during phase transitions. A constant-envelope form of π/4 DQPSK would preserve the signal envelope at all times by prescribing phase transitions that follow the circumference of the large circle. Relative to constant-envelope modulation, the NA-TDMA approach concentrates signal energy in a narrower bandwidth. The measure of this concentration is the modulation efficiency defined as the number of transmitted bits per second for each Hz of bandwidth occupied by a physical channel. In NA-TDMA, the modulation efficiency is 48.6 kb/s ÷ 30 kHz = 1.62 b/s/Hz. This is the highest modulation efficiency among the systems described in this book.

Although it promotes modulation efficiency (see Section 2.3.19), the departure from constant-envelope modulation is costly in terms of the power drain on telephone batteries (see Section 2.3.15). The nature of the NA-TDMA tradeoff between bandwidth efficiency and energy efficiency reflects the importance of achieving high spectrum efficiency. Other systems, including analog AMPS (see Chapter 3) and GSM (Pan-European digital cellular, Chapter 7), transmit constant-envelope signals. The terminals in these systems contain energy-efficient (Class C) power amplifiers.

5.3.3 Radiated Power

NA-TDMA specifies 11 radiated power levels for terminals, including the eight power levels of AMPS terminals (see Section 3.3.2). The highest power level is 4 W (6 dBW) and the levels differ by increments of 4 dB, ranging to a low of −34 dBW (0.25 mW). In a dual-mode system, the three lowest power levels can be assigned only to digital traffic channels and digital control channels specified in IS-136.1. Analog traffic channels and FSK control channels (FOCC and RECC) that transmit in the AMPS formats are confined to the eight power levels in the AMPS specification. Note that the transmitter of a terminal using a TDMA full-rate physical channel is active only one-third of the time. Thus, the average transmitted power is 10 log (3) = 4.8 dB lower than the specified radiated power level. This extends battery life of a terminal operating on a digital traffic channel relative to a terminal operating at the same radiated power on an analog traffic channel.

5.3.4 Spectrum Efficiency

NA-TDMA operates with the same frequency plan as an analog AMPS system. The most common reuse factor is $N = 7$ cells per cluster with three antenna sectors (120° antennas) in each cell. An all-digital network occupying half of the AMPS band (25 MHz in all) has 416 carriers and $3 \times 416 = 1,248$ full-rate physical channels. Although IS-136 specifies no minimum number of digital control channels, we can assume that a practical network with $N = 7$ frequency reuse will operate at least 21 digital control channels (corresponding to the 21 antenna sectors in a cell cluster). This leaves a maximum of $1,248 - 21 = 1,227$ full-rate digital traffic channels, which corresponds to an efficiency of

$$E = \frac{1,227}{7 \times 25} = 7.01 \text{ conversations/cell/MHz}, \qquad (5.3)$$

approximately three times the efficiency of AMPS.

It was originally anticipated that the coding and modulation techniques of NA-TDMA would be more robust to co-channel interference than the analog FM modulation of AMPS. Designers expected this robustness to permit systems with 120° antennas to operate with a reuse factor of $N = 4$ cells per cluster, which would provide an additional substantial increase in the number of conversations per cell per MHz. To date, practical systems have not been able to achieve the robustness necessary to reduce the reuse factor from seven to four. The unexpected vulnerability to interference appears in the adaptive equalizers (see Section 9.6) in NA-TDMA receivers. With terminals moving at vehicular speeds, channel conditions change substantially over the 6.67 ms duration of each time slot. In adapting to changing channel conditions, equalizers are vulnerable to interference from nearby transmitters using the same physical channel.

5.4 Logical Channels

As a dual-mode system, NA-TDMA is capable of supporting all of the AMPS logical channels in addition to the digital control channels and digital traffic channels specified in IS-136. Figure 5.5 is a generalized view of the logical channels that operate on NA-TDMA digital physical channels. It contains all of the logical channels defined in IS-136, and, in addition, several data fields (for example, DVCC) that carry information between terminals and base stations but are not defined as logical channels in the standard. Like logical channels, these data fields carry specific types of information and incorporate their own error-control mechanisms. A digital traffic channel transmits information in six formats in the forward direction and five formats in the reverse direction. Forward digital control channels multiplex information in nine distinct formats, including three broadcast control channels and three point-to-point channels. They also broadcast two types of synchronization information (SFP and SYNC) and carry feedback on the results of random access transmissions from terminals (SCF). A random access channel is a many-to-one channel carrying messages from terminals to a base station. The channels listed in Figure 5.5 incorporate a variety of techniques for transmitting multiple information types on a single physical channel:

- The SYNC information, superframe phase information, digital verification color code, slow associated control channel, and digital control channel locator all occupy distinct fields of each time slot of a digital traffic channel or a digital control channel. The remainder of the time slot carries either user information or a system control message.

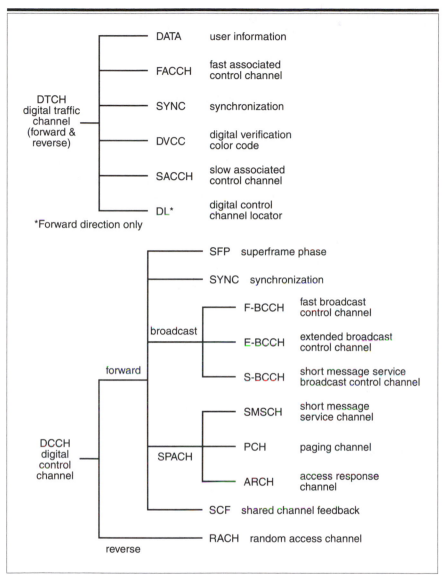

Figure 5.5 NA-TDMA logical channels.

- The FACCH uses a blank-and-burst technique (similar to that of the AMPS forward and reverse voice channels) to transmit information on a digital traffic channel.

- The broadcast control channels and the SPACH all occupy their own time slots on a forward digital control channel.

- Terminals contend for access to the RACH. Shared channel feedback plays an important role in the contention process.

The following sections describe the logical channel information formats.

5.4.1 Digital Traffic Channel (DTCH)

Figure 5.6 displays the contents of each time slot in a DTCH. The format for forward (base-to-mobile) transmissions differs from the format for reverse transmissions. The principal reason for the difference is the fact that NA-TDMA base stations transmit continuously while terminals turn their transmitters on at the beginning of each transmitting time slot and

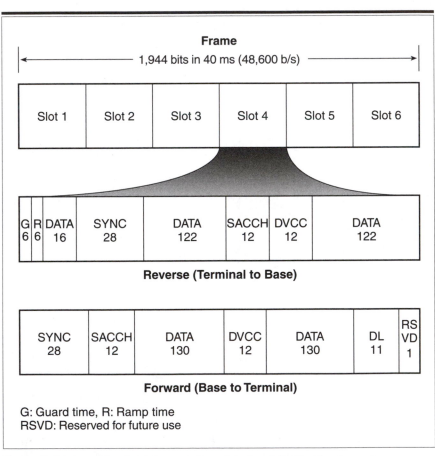

Figure 5.6 Information fields in each time slot of a digital traffic channel.
(Reproduced under written permission from Telecommunications Industry Association.)

turn them off at the end of the slot. Typically, three terminals share the same carrier and it is important to prevent their signals from arriving at the base station simultaneously. The terminals sharing a carrier are at different distances from the base station and they are in motion. As a consequence, it is impossible to predict the exact arrival time at the base station of any transmission. To accommodate this uncertainty, NA-TDMA specifies, at the beginning of each mobile-to-base time slot, a 6-bit guard time (G). This time interval (0.123 ms) prevents the signal transmitted at the beginning of one time slot from interfering with the signal transmitted (by another terminal) at the end of the previous time slot. After the guard interval, the transmitting terminal turns on its transmitter. The 6-bit ramp time (R) allows the transmitter to come up to its full radiated power level. The base, transmitting continuously, requires no guard or ramp interval. In forward time slots, the 11 digital control channel locator (DL) bits and 1 "reserved" bit (RSVD, set to 0) take the place of the guard and ramp intervals in reverse time slots. (In forward DTCH time slots of IS-54, there are 12 reserved bits. IS-136 uses 11 of them for the DL described in Section 5.4.1.)

In addition to using the reverse time slot format in Figure 5.6, it is also possible for terminals to transmit a *shortened burst* when they acquire a new physical channel. A shortened burst has a guard time with a duration of 50 bits (1.03 ms), rather than 6 bits (0.123 ms) as in Figure 5.6. The long guard time prevents the signal from one terminal from interfering with signals from other terminals using adjacent time slots. A shortened burst consists of repeated transmissions of the DVCC and SYNC fields, separated by blank fields (all 0) of various lengths. While the terminal transmits a shortened burst, the base station determines the correct timing for the terminal. This depends on the distance between the terminal and the base station. The base station then sends a PHYSICAL LAYER CONTROL message to the terminal. Based on time alignment information in this message, the terminal adjusts its transmitter timing relative to the nominal timing in Figure 5.2. It then begins transmitting information in the format of Figure 5.6.

Table 5.2 is a summary of the contents and purpose of each information field in a DTCH time slot. The following paragraphs describe them in detail.

Synchronization

The 28-bit SYNC field in each time slot serves two purposes. It contains frame synchronization information and it enables a receiver to train an adaptive equalizer.

Table 5.2 Information Fields in DTCH Time Slots

Field	Length (bits)	Purpose
DATA*	260	transport user information
FACCH*	260	network control message
DVCC	12	lock terminal to correct base station
SYNC	28	provide frame synchronization
		lock terminal to correct time slot
SACCH	12	network control message
DL	11 (forward)	information about location of DTCH
RSVD	1 (forward)	always set to 0

* A DTCH time slot normally carries DATA. The DATA can be replaced by a blank-and-burst FACCH message.

NA-TDMA traffic channels achieve frame synchronization by means of six different 28-bit SYNC sequences, one assigned to each time slot. To tune to a physical channel, a terminal tunes to the assigned carrier frequency and searches the received waveform for the SYNC pattern corresponding to an assigned time slot. On recognizing the SYNC pattern for that time slot, the receiver has an indication of when to receive the information in the time slot and when to transmit reverse DTCH information to the base station. Figure 5.2 indicates that forward and reverse time slots are offset by 1.9 ms. This is the nominal value of the timing offset. During a call, NA-TDMA base stations transmit time-alignment parameters to specific terminals. A terminal uses this information to adjust the timing offset to correspond to the location of the terminal in a cell. Terminals remote from the base station have a large timing advance. They transmit their signals earlier than terminals closer to the base station. This timing adjustment prevents interference between signals in adjacent time slots.

In each time slot, the transmitted SYNC pattern is known at the receiver. An adaptive equalizer in the receiver examines the received waveform in the SYNC field and compares it with the known transmitted waveform. Based on this comparison, the equalizer estimates the characteristics of the transmission channel. It uses this estimate to compensate for distortions introduced by the fading radio channel. Section 9.6 provides a brief description of an adaptive equalizer.

Digital Verification Color Code (DVCC)

The fields labeled DVCC in Figure 5.6 carry a digital verification color code that serves the same purpose as the supervisory audio tone (SAT) in AMPS. Like the SAT, the DVCC informs mobiles and base stations that they are receiving the desired signal and not a signal from another cell using the same physical channel. Interfering co-channel cells have different DVCCs and in NA-TDMA the protection against undetected reception of the wrong signal is very strong because the system has, in contrast to the three SATs of AMPS, a total of 255 different DVCCs. Each DVCC is represented by an 8-bit word, expanded to 12 bits by a (12,8;3) error-correcting block code (see Section 9.4.1). When a digital channel is assigned to a terminal, the terminal learns the DVCC of the base station and transmits that code in each time slot. If the base does not detect the correct DVCC, it does not send the accompanying user information to the switch. Correspondingly, if the terminal detects an incorrect DVCC, it suppresses the user information in the affected time slots.

User Information and Speech Coding

The DATA fields contain 260 bits of user information per time slot. In the forward direction, the DATA bits occupy two fields, each with 130 bits. In the reverse direction, there are three DATA fields, one with 16 bits, and the other two with 122 bits each. The principal user information in digital cellular systems is conversational speech. In NA-TDMA, a full-rate user channel occupies two time slots per frame, which together carry

$$2 \frac{\text{time slots}}{\text{frame}} \times 260 \frac{\text{bits}}{\text{time slot}} \div 0.040 \frac{\text{s}}{\text{frame}} = 13,000 \text{ b/s.} \quad (5.4)$$

The speech-coding technique is vector sum excitation linear prediction (VSELP) at a bit rate of 7,950 b/s. Section 9.7 provides a brief introduction to linear prediction speech coding.

Figure 5.7 is a summary of speech signal processing at a transmitter. At a terminal, an analog-to-digital converter produces digital speech samples at a rate of 8,000 samples per second in a linear format quantized to at least 13 bits/sample. At a base station, digital speech arrives from a switch with 8,000 samples per second coded in an 8-bit μ-law format. These samples are converted to a linear representation. A VSELP encoder, either at a terminal or a base station, processes speech in frames of 20 ms duration (160 samples). The VSELP algorithm produces a 159-bit representation of the 160 samples.

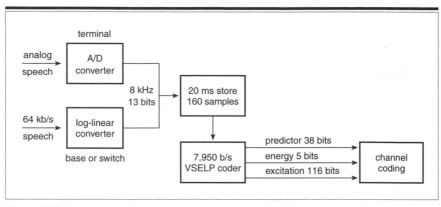

Figure 5.7 Speech-coding summary.

Of the 159 bits per block, 38 bits represent vocal tract information in the form of ten linear prediction coefficients (short-term), and 5 bits represent the energy in the 20 ms block. The remaining 116 bits are divided into four sets, each containing 29 bits. Each set of 29 bits represents a 5 ms segment of the 20 ms speech block. The 29 bits per segment convey pitch information (long-term predictor) in the form of a delay estimate and excitation estimates in the form of code book indices.

Among the 159 VSELP bits in a vocoder frame, there are 77 Class 1 bits that are especially vulnerable to transmission errors. Transmission errors in any of these bits result in highly distorted speech at a decoder. Errors in the other 82 bits (designated Class 2) impair the reconstructed speech, but less severely. Among the 77 Class 1 bits, 12 bits are designated the "Most Perceptually Significant" bits. The NA-TDMA channel coder generates a 7-bit cyclic redundancy check (CRC) error-detecting code (see Section 9.4.1) to protect these 12 bits. Added to the 77 Class 1 bits, this results in a block of 84 bits. This block, with 5 tail bits appended, is processed by a rate 1/2 convolutional coder (see Section 9.4.2), with a constraint length of 6 bits, to produce $2 \times (84 + 5) = 178$ bits of protected speech information. These bits are multiplexed with the 82 unprotected vocoder bits to produce a block of 260 bits to represent 20 ms of speech. Hence, the total transmitted speech bit rate is

$$260 \, \frac{\text{bits}}{\text{speech block}} \div 20 \, \frac{\text{ms}}{\text{speech block}} = 13,000 \text{ b/s.} \qquad (5.5)$$

Figure 5.8 is a summary of NA-TDMA channel coding for VSELP speech.

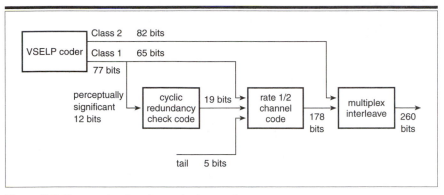

Figure 5.8 Error protection for coded speech.

At a receiver, the decoder performs the inverse operations, first decoding the 178 bits generated by the convolutional channel coder. It then performs a cyclic redundancy check to detect errors in the 12 perceptually significant bits. These are errors that could not be corrected by the convolutional decoder. If the bit stream fails the cyclic redundancy check, the decoder generates a "bad frame" indication, which causes the VSELP decoder to ignore the received information and to repeat the most recent block of speech bits received without a CRC failure. This repetition reduces the distortion in the reconstructed speech signal relative to the two main alternatives: decoding the bit stream containing errors in important bits, or inserting silence in place of the affected speech block.

If the decoder requires three or more consecutive repetitions of a speech block, it attenuates the received signal, and if there are six or more successive bad frame indications, the receiver completely mutes the received signal. The process, referred to as *bad frame masking*, conforms to the state diagram in Figure 5.9.

Before presenting the coded speech to a transmitter, NA-TDMA performs an interleaving process to combat the effects of fading. Section 9.5 describes interleaving in a general way. To obtain the 260 bits transmitted in each time slot, NA-TDMA collects the final 130 bits from an "old" VSELP block and multiplexes them with the first 130 bits from a "new" block. Figure 5.10 illustrates this process. It shows speech coder frames S_6, S_7, S_8, and S_9 generating the contents of time slots T_7, T_8, T_9, and T_{10}, where the subscript is a time index. (S_7 is the seventh speech block in a conversation.) Each time slot T_i contains bits 130–259 from coder block S_{i-1} and bits 0–129 from block S_i. The 26×10 interleaving matrix in Figure 5.11 specifies the transmission sequence for one time slot.

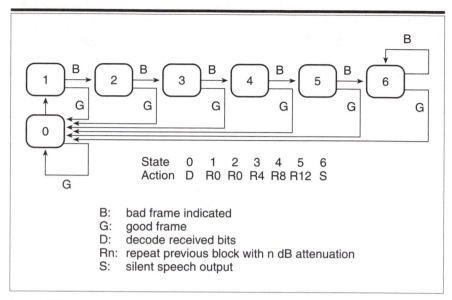

Figure 5.9 Bad frame masking.

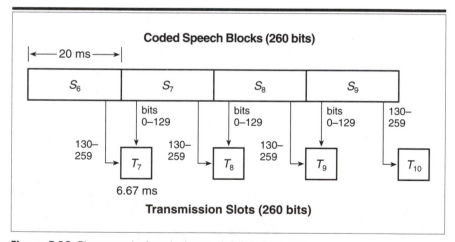

Figure 5.10 Placement of coded speech bits in time slots.

To perform the multiplexing and interleaving operations indicated in Figure 5.11, a transmitter multiplexes on a bit-by-bit basis the final 130 bits (bit numbers [130]–[259]) of the previous (old) block of coded speech bits and the first 130 bits (bit numbers [000]–[129]) of the current (new) block of coded speech bits. It places these 260 bits in a 26 × 10 matrix by

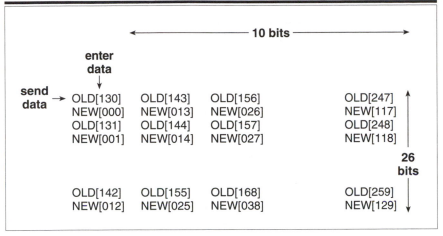

Figure 5.11 Interleaving matrix for one time slot in Figure 5.10.

successively filling the ten columns. It transmits the bits by reading them in sequence from the 26 rows of the matrix. The receiver reconstructs the matrix (with some of the bits possibly inverted by transmission errors). It restores the original order by reading the columns, and then sends the resulting sequence to the decoder. In Figure 5.11, two successive errors on the channel (for example in bits [131] and [144] of the old block) are separated by 13 bits in the sequence presented to the decoder. Thus, if the fade that causes the errors is sufficiently short, incorrect bits appear as isolated errors that can be corrected by the convolutional coding decoder.

Slow Associated Control Channel (SACCH)

Like an ISDN D-channel (see Section 9.11), the SACCH is an out-of-band signaling channel carrying information to and from a terminal while a call is in progress. With 12 bits per time slot, the bit rate of the SACCH is

$$2 \, \frac{\text{time slots}}{\text{frame}} \times 12 \, \frac{\text{bits}}{\text{time slot}} \div 0.040 \, \frac{\text{s}}{\text{frame}} = 600 \text{ b/s.} \qquad (5.6)$$

in a full-rate physical channel and 300 b/s (one time slot per frame) in a half-rate physical channel. The SACCH makes it possible for terminals and base stations to exchange network control information on a continuous basis without affecting the quality of user information. Recall that the corresponding logical channels in AMPS are the FVC and RVC. They are in-band signaling channels that operate in a blank-and-burst mode, interrupting user information each time they carry a message.

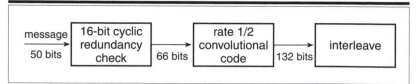

Figure 5.12 Error protection on the slow associated control channel.

In the SACCH, 132 bits, corresponding to the contents of 11 time slots, comprise a code word. The code word contains a 50-bit network control message protected by an error-detecting cyclic redundancy check (CRC) code and an error-correcting convolutional code, as shown in Figure 5.12.

In contrast to the rectangular interleaving arrangement of Figure 5.11 for user information signals, the SACCH uses a diagonal interleaver that spreads the 132 bits in Figure 5.12 over 12 transmission time slots, or 240 ms on a full-rate channel. On a half-rate channel, the transmission time of each word is 480 ms. The interleaver takes a block of 12 bits from the convolutional coder and distributes the 12 bits over 12 consecutive time slots in the manner shown in Figure 5.13. Here, a new block of 12 bits (bits 60–71 of the current word) enters the diagonal of a 12×12 matrix; and then 12 bits leave the first column of the matrix. After transmitting each group of 12 bits, the interleaver shifts the matrix to the left by one column. As a consequence, each transmission time slot contains 1 bit from each of 12 blocks of 12 bits. This interleaving causes two consecutive transmission errors (for example, in bits 5 and 16 in Figure 5.13) to be 11 bits apart in the received signal.

Digital Control Channel Locator (DL)

The DL field in the forward digital traffic channel helps terminals locate a digital control channel. It indicates the location of a carrier that presently contains a DCCH. The 11-bit DL field contains a 7-bit digital locator value protected by an (11,7;3) error-correcting code.

Fast Associated Control Channel (FACCH)

The transmission time of a message on an SACCH associated with a full-rate traffic channel is 240 ms. For an SACCH associated with a half-rate traffic channel, the transmission time is 480 ms. For some network control functions, these delays are acceptable. However, there are some functions (such as handoff) that require faster communications between terminals and base stations. For this reason, NA-TDMA also incorporates an in-band

(1) arriving bits

60	61	62	63	64	65	66	67	68	69	70	71

(2) place new bits on the diagonal of a 12 × 12 matrix
    (71' indicates bit 71 in previous code word)

60											
49	61										
38	50	62									
27	39	51	63								
16	28	40	52	64							
5	17	29	41	53	65						
126'	6	18	30	42	54	66					
115'	127'	7	19	31	43	55	67				
104'	116'	128'	8	20	32	44	46	68			
93'	105'	117'	129'	9	21	33	45	57	69		
82'	94'	106'	118'	130'	10	22	34	46	58	70	
71'	83'	95'	107'	119'	131'	11	23	35	47	59	71

(3) transmit bits in the first column of the matrix

71'	82'	93'	104'	115'	126'	5	16	27	38	49	60

(4) shift the matrix one column to the left

(5) arriving bits

72	73	74	75	76	77	78	79	80	81	82	83

Figure 5.13 Diagonal interleaving on the slow associated control channel.

signaling channel referred to as the fast associated control channel
(FACCH). When the system requires an FACCH, it transmits a 260-bit
code word. The code word replaces the 260 bits from a speech coder block
(Figure 5.8). The information content of an FACCH code word is a message
of length 49 bits that is protected by a 16-bit error-detecting cyclic redundancy check. The 65 resultant bits are processed by a rate 1/4 convolutional coder to produce $65 \times 4 = 260$ channel bits, as shown in Figure 5.14.
These bits go through the same two-slot interleaver as the coded speech
bits (Figures 5.10 and 5.11) and thus their transmission time is 40 ms on a
full-rate channel, and 80 ms on a half-rate channel. The result is that the
FACCH information is transmitted in 1/6 the time of SACCH informa-

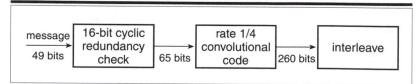

Figure 5.14 Error protection on the fast associated control channel.

tion. Moreover, the reliability of the information is considerably higher because the FACCH incorporates a rate 1/4 code, instead of the rate 1/2 code of the SACCH.

Because it interrupts the digital speech signal, the operation of the FACCH resembles the blank-and-burst operation of the AMPS forward and reverse voice channels. The FACCH is, however, less disruptive than analog blank-and-burst channels. This is because NA-TDMA employs bad frame masking (Figure 5.9) when the FACCH takes over the digital traffic channel. It repeats previously received speech blocks in order to conceal brief interruptions in received speech.

5.4.2 Digital Control Channel (DCCH)

Figure 5.15 displays the DCCH timing structure. In addition to the bits, slots, and frames of the traffic channels, digital control channels multiplex information in *blocks, superframes,* and *hyperframes.* In Figure 5.15, a block, consisting of three slots, is half of a frame. Like a full-rate traffic channel, a full-rate control channel occupies 1 slot per block (2 slots per frame). A half-rate DCCH occupies 1 slot per frame. There are 32 blocks per superframe and two superframes per hyperframe. Thus, a full-rate DCCH occupies 32 time slots per superframe. A half-rate DCCH occupies 16 slots per superframe.

The structure of each DCCH time slot is similar to the corresponding (forward or reverse) DTCH slot (Figure 5.6). In the reverse time slot, there are 40 bits of additional synchronization information relative to the DTCH. This includes a 16-bit preamble (consisting of eight consecutive phase shifts of −45°) that replaces the first 16 DATA bits of the DTCH. Another sequence of 24 bits (SYNC+) replaces the DVCC and SACCH fields of a reverse DTCH. The reverse DCCH requires this extra synchronization information because terminals send isolated bursts of information on a reverse DCCH. This is in contrast to a DTCH, where information arrives periodically (at 20 ms intervals for a full-rate channel) from terminals with calls in progress. Similar to shortened bursts (see Section 5.4.1) on the

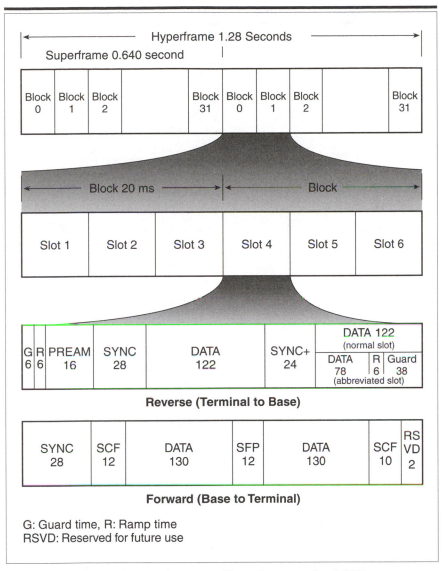

Figure 5.15 Slots, blocks, superframes, and hyperframes on the DCCH.
(Reproduced under written permission from Telecommunications Industry Association.)

reverse DTCH, NA-TDMA specifies an "abbreviated slot format" for the reverse DCCH. In this format, the second DATA field contains 78 bits. The remaining 44 bits of an abbreviated slot consist of a 6-bit ramp interval and 38 bits additional guard time.

Table 5.3 Information Fields in Digital Control Channel Time Slots

Field	Length (bits)	Purpose
DATA	260 (forward) 244 (reverse, normal) 200 (reverse, abbreviated)	network control information or short message service user information
SYNC	28	provide frame synchronization lock terminal to correct time slot
SYNC+	24 (reverse)	additional synchronization information
PREAM	16 (reverse)	additional synchronization information
SFP	12 (forward)	position of current block in the control superframe
SCF	22 (forward)	feedback on RACH access attempts consisting of BRI, R/N, CPE
BRI	6 (forward)	busy/reserved/idle indication
R/N	5 (forward)	received/not received indication
CPE	11 (forward)	part of the address in received data

Table 5.3 displays the information fields in the DCCH. The SYNC field is identical to the SYNC field of a digital traffic channel (Section 5.4.1). SYNC+ and PREAM are fixed bit patterns that provide additional synchronization information on the random access channel as described in the preceding paragraph.

The superframe phase (SFP) field appears in the same part of a time slot as the DVCC in a DTCH. SFP informs terminals of the location of the current block in the 32-block DCCH superframe. The block number is represented as an 8-bit word. (With block numbers ranging from 0 to 31, the first 3 bits are always 0.) The 12 bits of the SFP field comprise the code word of a (12,8;3) error-correcting code. Because this code differs from the code protecting the DVCC, terminals can inspect this 12-bit field to determine whether the current time slot carries a DCCH or a DTCH.

Shared channel feedback (SCF) appears in 22 bits of each forward DCCH time slot. It contains the following three types of information used in the access protocol of the random access channel:

- A busy/reserved/idle (BRI) indication (6 bits) informs terminals of whether the current slot is being used by a random access channel. The three possible code words of the BRI indication are separated from each other by a distance of 4 bits in the 6-bit BRI code.

- A received/not-received (R/N) indication (5 bits) informs terminals of whether the base station has successfully decoded the information transmitted in a time slot on the reverse DCCH. This information appears in a 5-bit repetition code (5,1;5). Five 1s indicate "received"; 0s indicate "not received."

- A coded partial echo (CPE, 11 bits) acknowledges receipt of information on the reverse DCCH. It carries the least significant 7 bits of the directory number (MIN or IMSI, see Table 5.1) of the terminal that transmitted the acknowledged information. An (11,7;3) block code protects this information.

Multiplexed Logical Channels on the Forward DCCH

Each time slot on a forward DCCH carries information from one of the six logical channels shown in Figure 5.5. The logical channels share each superframe in the following prescribed order:

1. fast broadcast control channel (F-BCCH),

2. extended broadcast control channel (E-BCCH),

3. short message service broadcast control channel (S-BCCH), and

4. short message service, paging, and access response channel (SPACH).

Each superframe begins with information from the fast broadcast control channel and ends with information from one of the SPACH channels. The F-BCCH occupies the first three blocks of every superframe. It can occupy as many as ten blocks. At least one, and as many as eight, blocks following the F-BCCH contain E-BCCH messages. If there are any short message service broadcast messages to send, they follow the E-BCCH. The maximum number of blocks to be allocated to the S-BCCH is 15 in a full-rate DCCH, and 11 in a half-rate DCCH. The remainder of the superframe is allocated to SPACH, consisting of messages from the short message service, the paging channel, and the access response channel. Figure 5.16 summarizes the assignment of logical channels to a forward DCCH superframe.

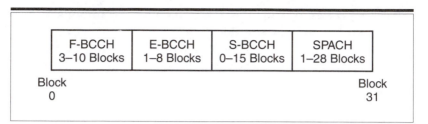

Figure 5.16 Logical channels in a digital traffic channel superframe.

Paging Channel Operation, Sleep Mode

When there is no call in progress, a terminal monitors paging messages in order to detect the arrival of a new call. The power consumed by the radio receiver while it waits for a paging message has a strong influence on the standby time of the terminal's battery. Because an AMPS telephone monitors the control channel continuously, there is a constant drain on the battery. NA-TDMA improves this situation by organizing the paging channels in a way that makes it possible for terminals to operate in *sleep mode* when there is no call in progress. In this mode, a terminal turns off its receiver for a significant fraction of the time. It then wakes up for a short time interval. If there is a paging message for the terminal, the base station schedules the message to arrive during this brief wake-up interval.

Paging messages arrive in the SPACH blocks (Figure 5.16) of each superframe. Paging messages are always transmitted twice, once in each superframe of a hyperframe. NA-TDMA defines paging subchannels that occupy a fraction of the hyperframes on a DCCH. The number of paging subchannels is a parameter, PFN, that ranges from 1 to 96. With PFN = 1, there is only one paging subchannel that occupies all hyperframes. With PFN = 96, a subchannel appears once in 96 hyperframes. There is a hyperframe counter in the BCCH that informs terminals of the current paging subchannel. Each terminal listens to a specific paging subchannel. It sleeps during superframes occupied by the other PFN-1 paging subchannels. The subchannel number (between 1 and PFN) assigned to a specific terminal is determined by a hashing function. This function is a formula that computes the subchannel number from the least significant 16 bits of the mobile station identifier (MIN or IMSI, in Table 5.1). The function is designed to distribute terminals uniformly among the PFN paging subchannels.

RACH Access Protocol

Like the AMPS reverse control channel (see Section 3.4.3), the NA-TDMA random access channel has a many-to-one topology. Dispersed terminals contend for access to the RACH under the control of the shared channel feedback (SCF) information transmitted in forward DCCH time slots. This information serves the same function as the busy/idle bits in the AMPS forward control channel. However, the three types of information in the SCF provide considerably closer coordination between base stations and terminals than the binary information in the AMPS busy/idle bits. This coordination is enhanced by synchronization of transmissions on the reverse DCCH with responses on the forward DCCH. After transmitting information in a reverse DCCH slot, a terminal examines a specific slot in the forward DCCH in order to learn the results of the transmission. To accommodate processing delays, NA-TDMA divides a full-rate reverse DCCH into six RACH subchannels. Since the DCCH occupies two slots per frame, each subchannel occupies one slot per three frames. Figure 5.17 indicates that the information content of an RACH time slot is either 100 bits (normal slot) or 79 bits (abbreviated slot). On gaining access to the RACH, a terminal can transmit information at a rate of

$$\frac{101 \text{ bits}}{\text{slot}} \times \left(\frac{1 \text{ slot}}{120 \text{ ms}}\right) = 842 \text{ b/s (normal slot), or}$$

$$\frac{79 \text{ bits}}{\text{slot}} \times \left(\frac{1 \text{ slot}}{120 \text{ ms}}\right) = 658 \text{ b/s (abbreviated slot).}$$

There are two modes of transmission on the RACH, *random access* and *reserved access*. In the random access mode, a terminal with information to transmit waits for an IDLE indication in the BRI bits of a forward DCCH time slot. The terminal then transmits its information in a specified slot of the reverse DCCH. The base station reports the result of this transmission in a time slot that occurs 120 ms (three frames) after the slot with the IDLE indication that stimulated the transmission by terminal. The base station indicates a successful result by means of a BUSY indication in the BRI bits of the response slot, a RECEIVED indication in the R/N bits, and the final 7 bits of the mobile station identifier in the coded partial echo (CPE) portion of the shared channel feedback. Failing to receive a success indication, the terminal waits a random time and attempts again to transmit its information to the base. The transmission attempts continue until the base station receives the RACH information and successfully transmits an acknowledgment to the terminal, or until the number of attempts exceeds a limit specified in an *ACCESS PARAMETERS* message broadcast on the fast BCCH (see Section 5.5.3).

In the reserved mode, the base station prompts the terminal for a transmission by means of a RESERVED indication in the BRI bits and the last 7 bits of the mobile station identifier in the CPE portion of the shared channel feedback. This prompt from the base station grants the terminal exclusive access to a time slot in the reverse DCCH.

DATA Fields of the DCCH

The purpose of a digital control channel is to carry network control messages and short message service messages between terminals and base stations. These messages are contained in the DATA fields of each slot. The DATA fields are protected by error-detecting codes and error-correcting codes similar to those of the DTCH. In addition to the message content, the DATA fields carry headers that describe the message carried in the slot. Figure 5.17 shows the format of the DATA in a forward DCCH time slot. The CRC is the same cyclic redundancy check that appears in the slow associated control channel (Figure 5.12). The convolutional coder is identical to the one used for speech and SACCH signals.

In Figure 5.17, the numerical table displays bit counts corresponding to the three time slot formats in Figure 5.15. To find the number of message bits carried in a time slot, it is helpful to begin with the total DATA field allocation, for example, 260 bits in a forward DCCH time slot. This implies that there are $260 \div 2 = 130$ bits at the input to the convolutional coder. With a 16-bit CRC and 5 tail bits, this leaves $130 - 16 - 5 = 109$ bits available for the header and message information. Similar reasoning

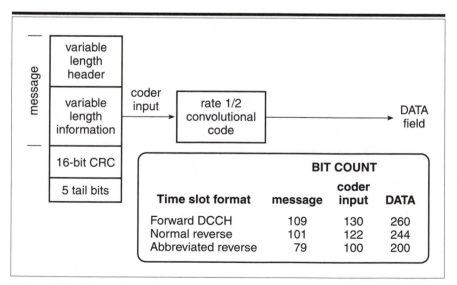

Time slot format	BIT COUNT		
	message	coder input	DATA
Forward DCCH	109	130	260
Normal reverse	101	122	244
Abbreviated reverse	79	100	200

Figure 5.17 Digital control channel DATA field.

leads to the allocation of 101 message bits per slot for a normal reverse time slot and 79 message bits in an abbreviated reverse time slot. The header information indicates whether the message content begins a new message or is a continuation of a message in progress. It also indicates the length of the message portion of the DATA field. Because SPACH messages are point-to-point communications, the header field of SPACH slots contains address information (MIN or IMSI) that specifies the terminal receiving the message. It also indicates the SPACH message type: an access response message, a paging message, or a short message service message.

5.5 Messages

NA-TDMA specifies the following three sets of messages classified by the logical channels that carry them:

- Messages transmitted on AMPS logical channels (FOCC, RECC, FVC, and RVC). This message set is an expanded version of the AMPS messages described in Section 3.5.

- Messages transmitted on the in-band (FACCH) and out-of-band (SACCH) signaling channels associated with TDMA traffic channels. Most of these messages were originally specified in IS-54.

- Messages transmitted on the digital control channels.

5.5.1 Messages on AMPS Logical Channels

As a dual-mode system, NA-TDMA retains the AMPS messages described in Section 3.5. It also introduces new messages and defines new message fields in existing messages on the AMPS control channels and voice channels (FOCC, RECC, FVC, and RVC). The main purposes of the added messages and information fields are to:

- control NA-TDMA authentication procedures, which introduce enhanced network security relative to AMPS;

- direct dual-mode terminals to digital traffic channels, either at call setup or in handoff procedures that move calls from analog traffic channels to digital channels; and

- inform the base station and switch of the capabilities of a terminal.

Table 5.4 is a list of new messages carried on the AMPS control channels. These messages augment the message set in Table 3.3.

Table 5.4 Additional Messages on AMPS Control Channels

Base-to-Mobile Messages (FOCC and FVC)	Mobile-to-Base Messages (RECC and RVC)
Authentication Messages	
SERIAL NUMBER REQUEST CONFIRM BASE STATION CHALLENGE UNIQUE CHALLENGE ORDER SHARED SECRET DATA UPDATE	SERIAL NUMBER RESPONSE BASE STATION CHALLENGE CONFIRM UNIQUE CHALLENGE CONFIRM SHARED SECRET DATA UPDATE
Call Management Messages	
PAGE WITH SERVICE MESSAGE WAITING	ORIGINATION WITH SERVICE PAGE RESPONSE WITH SERVICE
Radio Resources Management Message	
INITIAL DIGITAL TRAFFIC CHANNEL	

Corresponding to the INVOKE and RESPONSE messages in IS-41 (see Chapter 4), the NA-TDMA authentication protocol specifies messages in pairs. Each of the four rows in the authentication message section of Table 5.4 contains an initial message (either from a base station or a terminal) and a corresponding response. Section 5.6.1 describes the use of these message pairs in the authentication protocol.

The new call management messages relate to service enhancements of NA-TDMA relative to AMPS. ORIGINATION WITH SERVICE, PAGE WITH SERVICE, and PAGE RESPONSE WITH SERVICE distinguish voice calls from calls carrying facsimile signals or asynchronous data signals. MESSAGE WAITING informs a terminal that the system has received voice mail, facsimile, or short message service messages addressed to the terminal. An INITIAL DIGITAL TRAFFIC CHANNEL message directs a terminal to a digital traffic channel at the beginning of a call.

In addition to transmitting new messages on the AMPS control channels, dual-mode terminals and base stations are capable of transmitting augmented versions of the original AMPS messages listed in Table 3.3. One example of additional information carried in AMPS messages is a *preferred call mode* indication in call setup messages (*ORIGINATION, PAGE RESPONSE*) transmitted by terminals. Preferred call mode information informs the system of the capabilities of a terminal. Another example of additional information is a location area identifier (LOCAID, Table 5.1) carried in *GLOBAL ACTION* messages. LOCAID plays an important role in area-based registration (see Section 5.6.4), a mobility management protocol available in NA-TDMA.

5.5.2 Messages Carried on Associated Control Channels

In addition to augmenting the message sets on the AMPS control channels, NA-TDMA specifies a completely new message set for network management functions coordinated by means of the in-band and out-of-band control channels associated with digital traffic channels (FACCH and SACCH). Table 5.5 is a list of these messages. With one exception (*CHANNEL QUALITY* message), all of the messages stimulate acknowledgment responses (confirmation) from the receiving network element (terminal or base station). Some acknowledgment messages simply inform the sending base station or terminal that the message was received. These messages indicate only the type of message received. Other acknowledgments contain information specific to the purpose of the original message. For example, they echo important parameters carried in the message or they contain a response to a query. Rather than list all the acknowledgment messages separately, Table 5.5 contains a mark next to each message to indicate whether the acknowledgment contains information specific to the original message (*) or is a general-purpose acknowledgment (†).

Message Structure

Like the original AMPS network control messages, the new messages on the AMPS control channels have formats tailored to their specific purposes. By contrast, messages on the associated control channels share a common format, which is similar to the formats of messages exchanged in other communications systems, such as ISDN (see Section 9.11) and Signaling System Number 7 (see Section 9.12). All of the messages are carried in 49-bit code words.[1] The first bit in each word indicates whether this code word is the final code word in a message (0) or if additional code words follow (1). The final 48 bits in each code word contain network control information.

---

[1] Code words on the SACCH have 50 bits. However, the second bit is a filler bit set to 0. It is inserted in order to create a transmitted word, generated by the channel coder in Figure 5.12, that fills exactly 11 SACCH fields, each with 12 bits.

Table 5.5 Messages Carried on Associated Control Channels

Forward SACCH and FACCH	Reverse SACCH and FACCH
Call Management Messages	
†ALERT WITH INFO †RELEASE *SEND BURST DTMF *SEND CONTINUOUS DTMF *FLASH WITH INFO	†CONNECT †RELEASE *SEND BURST DTMF *SEND CONTINUOUS DTMF *FLASH WITH INFO
Authentication Messages	
*SHARED SECRET DATA UPDATE *UNIQUE CHALLENGE †PARAMETER UPDATE	*BASE STATION CHALLENGE
Radio Resources Management Messages	
*MEASUREMENT ORDER †STOP MEASUREMENT ORDER †HANDOFF *PHYSICAL LAYER CONTROL	CHANNEL QUALITY
User Information Transport Message	
*R-DATA	
Operations, Administration, and Maintenance Messages	
†MAINTENANCE †AUDIT *STATUS REQUEST	

* Confirmation contains message-specific data.
† Confirmation is a generic base acknowledgment message or mobile acknowledgment message.

Each message begins with a 2-bit preamble. NA-TDMA refers to this preamble as a *protocol discriminator*. In early implementations of NA-TDMA, these bits are always 00. The next 8 bits comprise a *message type* field that specifies the nature of the message. The remainder of the message contains variable data specific to the purpose of the message. Table 5.6 shows the contents of a *HANDOFF* message transmitted on the forward FACCH. It corresponds to Table 3.4, which shows the contents of a *HANDOFF* message on the FVC of an analog AMPS system. Data elements in Table 5.6 direct the terminal to the new carrier frequency and indicate the amount of power to be radiated by the terminal. If the handoff is to a digital voice channel, the *HANDOFF* message indicates whether the terminal will operate at half rate (6.5 kb/s) or full rate (13 kb/s). It also specifies the time slot occupied by the new physical channel and the digital verification color code of the new base station. Other information in the *HANDOFF* message directs the terminal to begin transmission with abbreviated time slots, or to begin with standard time slots. If transmissions begin in the standard format, the terminal applies the time alignment correction indicated by the *HANDOFF* message. The final 2 bits specify whether user information or network control information or both are to be transmitted in encrypted form on the new digital traffic channel.

Table 5.6 Contents of a 48-Bit *HANDOFF* Message Carried on the FACCH

Bit Position	Information
1–2	00 protocol discriminator
3–10	11011100 *HANDOFF* message
11–21	AMPS channel number (specifies carrier)
22	full rate or half rate
23–25	time slot
26–35	SAT if handoff to analog channel, DVCC if handoff to digital channel
36–39	transmit power level
40–44	time alignment
45–46	shortened burst indicator
47	voice privacy mode
48	message encryption mode

Acknowledgment and Retransmission

After transmitting a network control message on the SACCH or FACCH, a base station or terminal waits for a confirmation, in the form of an acknowledgment message, that the network control message has been received. The waiting time is 200 ms for messages on an FACCH and 1.2 seconds for SACCH messages. If it receives no acknowledgment within this time, the transmitter sends the network control message again. For transmissions from a mobile terminal, the maximum number of attempts is three, after which the terminal abandons its attempt to send the message. For base station transmissions, there is no standard maximum. The number of attempts depends on the system implementation.

Message Content

Some of the messages, such as RELEASE and AUDIT, correspond directly to messages carried on the AMPS FVC. The ALERT WITH INFO message is similar to the AMPS ALERT message. The "info" it contains can include the telephone number of the calling party, which enables it to deliver caller identification service to the cellular subscriber. The DTMF messages refer to the dual-tone multiple-frequency sounds produced by push-button telephones. These sounds control an increasing number of services delivered to telephone users. Because the speech coders in digital traffic channels distort these tones, NA-TDMA sends messages between base stations and terminals corresponding to the signaling tones. For example, a cellular phone user checking his voice mail will push buttons on the telephone to obtain his messages. These actions will cause the telephone to transmit a sequence of SEND BURST DTMF or SEND CONTINUOUS DTMF messages to the local base station. The base station or the switching center will then generate the tones and send them to their destination. A burst DTMF has a limited duration. Continuous DTMF continues until the subscriber releases the button on the telephone. At that time, the phone sends a special SEND CONTINUOUS DTMF message that causes the system to stop transmitting the tones.

Other messages in Table 5.5 are related to protocols that control the digital communications in NA-TDMA. For example, the PHYSICAL LAYER CONTROL message is a generalization of the AMPS CHANGE POWER LEVEL message. In addition to the radiated power, it can also control the timing of TDMA transmissions. The MEASUREMENT ORDER, CHANNEL QUALITY, and STOP MEASUREMENT ORDER messages are all part of the mobile-assisted handoff protocol described in Section 5.6.2. R-DATA messages are part of the short message service. They convey text messages to and from terminals with calls in progress. Terminals that do not have calls in progress send and receive text messages on a digital control channel (see Section 5.4.2).

5.5.3 Messages Carried on Digital Control Channels

The functions of the DCCH correspond to those of the AMPS forward control channel and reverse control channel. Reflecting the more advanced network control capabilities of NA-TDMA, there are 58 messages defined for the DCCH in contrast to the 16 FOCC/RECC messages in AMPS (see Table 3.3). Table 5.7 contains the complete message list. The three columns indicate which type of control channel (Figure 5.5) carries each message: a broadcast channel (BCCH); a random access channel (RACH); or a short message service, paging, and access response channel (SPACH).

Table 5.7 Digital Control Channel Messages

BCCH Messages	SPACH Messages	RACH Messages
Initialization Messages	Call Management Messages	
DCCH STRUCTURE	REORDER/INTERCEPT	ORIGINATION
CONTROL CHANNEL SELECTION PARAMETERS	PAGE	PAGE RESPONSE
SYSTEM IDENTITY	USER ALERT	
REGULATORY CONFIG	RELEASE	
ALTERNATE RCI INFO	MESSAGE WAITING	
SERVICE MENU	CAPABILITY REQUEST	CAPABILITY REPORT
NEIGHBOR SERVICE INFO		
TIME AND DATE		
	Authentication Messages	
		AUTHENTICATION
	BASE STATION CHALLENGE CONFIRMATION	BASE STATION CHALLENGE ORDER
	UPDATE SSD ORDER	UPDATE SSD CONFIRMATION
	UNIQUE CHALLENGE ORDER	UNIQUE CHALLENGE CONFIRMATION
	PARAMETER UPDATE	SERIAL NUMBER

(continued)

Table 5.7 Digital Control Channel Messages *(Continued)*

BCCH Messages	SPACH Messages	RACH Messages
User Information Transport Messages		
EMERGENCY INFORMATION BROADCAST	SPACH NOTIFICATION	SPACH CONFIRMATION
	R-DATA	R-DATA ACCEPT
	R-DATA ACCEPT	R-DATA
	R-DATA REJECT	R-DATA REJECT
Mobility Management Messages		
REGISTRATION PARAMETERS	REGISTRATION ACCEPT	REGISTRATION
	REGISTRATION REJECT	
	TEST REGISTRATION RESPONSE	TEST REGISTRATION
Radio Resources Management Messages		
MOBILE ASSISTED CHANNEL ALLOCATION	DIGITAL VOICE CHANNEL DESIGNATION	MACA REPORT
ACCESS PARAMETERS	ANALOG VOICE CHANNEL DESIGNATION	
NEIGHBOR CELL	DIRECTED RETRY	
OVERLOAD CLASS	—	
Special Services Messages		
SOC/BSMC IDENTIFICATION	—	—
BSMC MESSAGE DELIVERY	BSMC MESSAGE DELIVERY	BSMC MESSAGE DELIVERY
SOC MESSAGE DELIVERY	SOC MESSAGE DELIVERY	SOC MESSAGE DELIVERY
	Operations, Administration, and Maintenance Messages	
	AUDIT	AUDIT CONFIRMATION

Message Structure

All DCCH messages share a common format. The length of each message is an integer multiple of 8 bits, up to a maximum length of $255 \times 8 = 2{,}040$ bits (255 octets). Like the DTCH messages (see Section 5.5.2), each message begins with a protocol discriminator of length 2 bits, which for IS-136 messages are always 00. The following 6 bits comprise a *message type* that specifies the nature of the message. The remainder of the message contains data specific to the purpose of the message. The data fields begin with mandatory data carried in every message of a specific type. This data is followed by optional parameters, which are carried in some, but not all, messages. Each optional parameter is preceded by a 4-bit field that identifies the parameter to follow. As an example, Table 5.8 shows the contents of a *SYSTEM IDENTITY* message carried on a broadcast control channel.

Like an AMPS *SYSTEM PARAMETER* message, the main purpose of the *SYSTEM IDENTITY* message is to convey the 15-bit system identifier to all of the terminals in a cell. The 3-bit network type field indicates whether the base station is part of a public, private, or residential network (Figure 5.1). The protocol version information indicates whether the base station operates

Table 5.8 Contents of a *SYSTEM IDENTITY* Message on a BCCH

Bit Position	Information
1–2	00 protocol discriminator
3–8	001011 *SYSTEM IDENTITY* message
Mandatory Data	
9–23	system identifier (SID)
24–26	network type
27–30	protocol version
Optional Data	
variable	PSID/RSID set
next 14 bits	mobile country code
variable	alphanumeric system name

according to the IS-136 specification or the older IS-54 standard, or some other standard that may be developed in the future. When a base station is part of a private network or a residential network, the data in the PSID/RSID set provides additional information about the network. Anticipating multinational deployment of NA-TDMA, the *SYSTEM IDENTITY* message is capable of broadcasting a mobile country code to indicate the country in which a system operates. Information carried in the optional alphanumeric system name data field allows a terminal to display the name of the system in which a base station operates. Each of the optional data fields begins with a 4-bit code indicating the data type (for example, 1000 indicates alphanumeric system name).

Message Content

The broadcast control channel carries the same information to all of the terminals in a cell. In Table 5.7, the first seven messages in the BCCH column are mandatory. They are carried on every digital control channel and provide essential information about communications in a cell. The first five messages travel on the fast BCCH (Figure 5.5) and appear in every superframe. The *DCCH STRUCTURE* message describes multiplex arrangements on the control channel. For example, it broadcasts the partition of superframes among F-BCCH, E-BCCH, and SPACH frames (Figure 5.16). It also contains the digital verification color code of the cell as well as information about paging subchannels (see Section 5.4.2). The *ACCESS PARAMETERS* message contains the parameters of the RACH access protocol, including the maximum number of transmission attempts. This message also carries information about authentication procedures to be used when a terminal transmits an RACH message. *CONTROL CHANNEL SELECTION PARAMETERS* messages inform terminals of procedures for searching for a new DCCH, which may be of higher quality than the present DCCH. The search is necessary when a terminal without a call in progress moves into a new cell. Additional information about this search procedure appears in mandatory *NEIGHBOR CELL* messages on the extended BCCH. With the DCCH occupying at most two slots per frame, this search can take place during the other four time slots. The *REGISTRATION PARAMETERS* message informs terminals of mobility management (see Section 5.6.4) procedures in effect.

Among the optional BCCH messages, the *MOBILE ASSISTED CHANNEL ALLOCA-TION* message and the associated *MACA REPORT* message on the RACH contribute to radio resources management. In *MACA REPORT* messages, terminals inform the system of received signal quality (error rate measurements) on a set of channels specified by the base station. The system uses this information to assign a traffic channel to the terminal when a call begins. Several optional messages allow network operators and equipment vendors to introduce their own network control protocols. These protocols are outside of the NA-TDMA standard and require terminals designed to operate with them. *SOC MESSAGE DELIVERY* and *BSMC MESSAGE DELIVERY* messages coordinate these protocols. SOC and BSMC (Table 5.1) identify the system operator and base station manufacturer, respectively. The *SOC/ BSMC IDENTIFICATION* message carries these identifiers to all of the terminals in a cell. Terminals use *CAPABILITY REPORT* messages to describe their ability to perform the nonstandard procedures of system operators and base station manufacturers. Depending on the specific SOC and BSMC in each cell, a fraction of the terminals will be able to operate the nonstandard protocols.

The *EMERGENCY INFORMATION BROADCAST* message and the *R-DATA* messages in Table 5.7 perform user information transport. An *EMERGENCY INFORMATION BROADCAST* message transmits the same text message to all of the terminals in a cell. The *R-DATA* messages are part of the NA-TDMA short message service. They move text messages to and from terminals that do not have calls in progress. Terminals with calls in progress send and receive text messages on the associated control channels (see Section 5.5.2). In order to maintain system security, NA-TDMA performs authentication procedures prior to call setup (see Section 5.6.1). To do so, it uses the three message pairs on the SPACH and RACH: *BASE STATION CHALLENGE*, *SSD UPDATE*, and *UNIQUE CHALLENGE*.

5.6 Network Operations

Relative to AMPS, the principal network management innovations in NA-TDMA are in the categories of authentication and handoff. There are also enhancements to call management procedures and mobility management. The following paragraphs discuss these categories separately and describe the messages that control them.

5.6.1 Authentication and Privacy

NA-TDMA authentication procedures overcome a major AMPS weakness, the vulnerability of analog cellular systems to unauthorized use. First published in IS-54, the secure authentication procedures also appear in the CDMA system specified in IS-95 (see Chapter 6). At the heart of network security in NA-TDMA is a secret key (A-key, Table 5.1) stored in each telephone. The key is also recorded in the subscriber's home system in a secure database referred to as an *authentication center*. Figure 4.3 indicates a possible location for an authentication center (AC). An essential property of the authentication procedure is that the A-key appears nowhere else in the system. The encryption keys used to protect system information and user information are referred to as *SSD* (*shared secret data*). The authentication center and each telephone use the A-key to compute SSD. The authentication center transmits SSD through the network infrastructure to base stations. The remainder of this section describes how the system generates SSD and uses it.

The A-key (Table 5.1) is a 64-bit binary number. The system never transmits the A-key. Instead, the authentication center and the terminal both use the same A-key, in combination with a random number generated at the AC, to compute a 128-bit word, denoted *shared secret data* (SSD). SSD is the cryptographic key used by a terminal and a base station to protect transmitted information. NA-TDMA frequently changes the value of SSD. The frequent changes minimize system vulnerability to an intruder who occasionally succeeds in intercepting a radio message. The system uses half of the SSD for authentication purposes. This is a 64-bit word denoted SSD-A. It uses SSD-B, the other 64 bits of the SSD, as an encryption key for user information.

Figure 5.18 indicates how the authentication center and the telephone compute SSD. CAVE (cellular authentication and voice encryption) is a cryptographic algorithm that operates on 152-bit inputs. The CAVE output is either a 128-bit SSD or an 18-bit authentication code. In Figure 5.18, RANDSSD is a random number generated at the authentication center. As part of the network security system, the base station and terminal obtain a new SSD from time to time. Figure 5.18 shows the procedure that gives a base station and a terminal the same SSD to be used for authentication and encryption. The authentication center in the subscriber's home system uses a random number generator to produce a new value of RANDSSD. Following Figure 5.18, it computes SSD-A and SSD-B and transmits the results, along with RANDSSD, to the base station at the current location of the terminal. The base station sends an SSD UPDATE ORDER message to the terminal. If there is no call in progress, this order travels on a forward control channel or a digital traffic channel. If there is a call in progress on an analog voice channel, the order travels on the forward voice channel. If the terminal is using a digital traffic channel, the update order travels on the fast associated control channel.

As shown in Figure 5.19, the UPDATE SSD ORDER message contains the random number (RANDSSD) produced at the authentication center. On

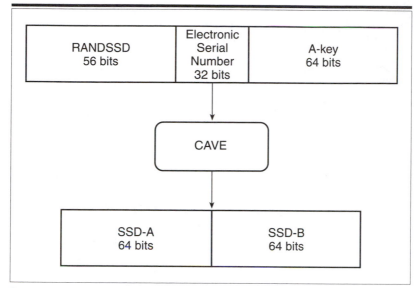

Figure 5.18 Procedure for generating shared secret data at the home system and in a terminal. (Reproduced under written permission from Telecommunications Industry Association.)

receiving RANDSSD, the terminal computes SSD-A and SSD-B as shown in Figure 5.18. It then uses its own random number generator to produce a 32-bit random number, RANDBS. It applies four numbers (152 bits in all) to the CAVE algorithm to compute an 18-bit authentication code, AUTHBS(M), as shown in Figure 5.20. The four inputs to the algorithm are

- RANDBS,
- the new value of SSD-A,
- the electronic serial number (ESN), and
- MIN1, the first 24 bits of the 34-bit mobile identification number.

The terminal then transmits RANDBS to the base station as part of a *BASE STATION CHALLENGE ORDER* message. The base station uses RANDBS and SSD-A to perform the computation shown in Figure 5.20. It sends the result, AUTHBS(B), to the terminal in a *BASE STATION CHALLENGE CONFIRMATION* message. The telephone responds by transmitting an *UPDATE SSD CONFIRMATION* message to the base station. This message indicates whether AUTHBS(M) = AUTHBS(B). If the two numbers match, the system assumes that the

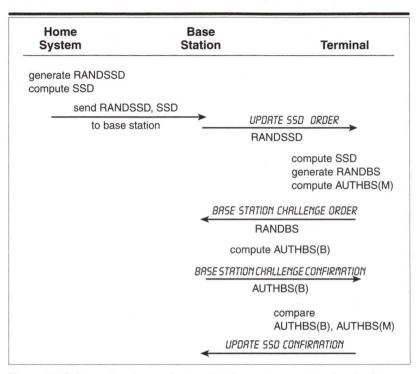

Figure 5.19 Procedure for confirming that the system and the terminal have the same SSD.

base station and the terminal have the correct shared secret data. Note that neither the A-key nor the SSD appears in radio messages.

After computing SSD-A, the telephone uses this data to authenticate its identity each time it originates a call, responds to a paging message, or registers its location. In each of these procedures, the terminal follows a procedure shown in Figure 5.20 to compute an 18-bit authentication code, AUTHR. The 152 input bits to the CAVE algorithm include SSD-A; the electronic serial number; 24 bits of the mobile identity number; and a random number, RAND. This random number is generated by the base station and transmitted, as part of the initialization procedure, either on the FOCC or in an *ACCESS PARAMETERS* message on the digital broadcast control channel. The terminal transmits AUTHR to the base station, which computes its own version of AUTHR. If the two codes match, the system authenticates the identity of the terminal.

The *UNIQUE CHALLENGE ORDER* message provides the system with an additional mechanism for verifying the authenticity of a terminal. At any time,

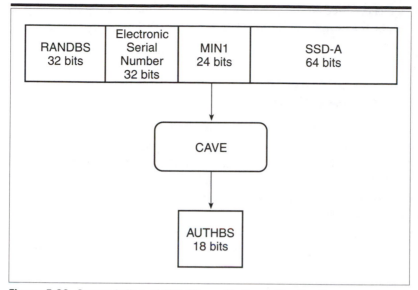

Figure 5.20 Computation of AUTHBS used in the authentication procedure.
(Reproduced under written permission from Telecommunications Industry Association.)

the base station can send this message, which contains a random number RANDU. The terminal then applies RANDU, SSD-A, and other stored parameters to the CAVE algorithm in order to compute AUTHU. It transmits AUTHU to the base station in a *UNIQUE CHALLENGE ORDER CONFIRMATION* message. If the base station fails to receive the correct AUTHU, it can bar the terminal from further access or it can order the terminal to update its SSD and thereby verify its authenticity.

In addition to using the authentication codes AUTHR and AUTHU to verify the identity of a terminal, NA-TDMA contains one more network security mechanism. The base station controls the contents of a memory register in the terminal by means of a *PARAMETER UPDATE* message transmitted on the DCCH when there is no call in progress and on the FACCH during a call. Whenever it receives this message the terminal increments the number, COUNT, stored in an 8-bit *call-history* register. To gain access to the system, the terminal transmits COUNT to the base station, which verifies that the terminal has the correct value of this parameter. This procedure defends the network against unauthorized terminals that find a means to obtain a correct value of SSD. Unless the unauthorized terminal remains close enough to the authorized terminal to intercept all *PARAMETER UPDATE* messages, there will be a discrepancy between the *PARAMETER UPDATE* messages counted at the two terminals. When the unauthorized terminal transmits the wrong value of COUNT, the network can take defensive action.

5.6.2 Radio Resources Management: Mobile-Assisted Handoff (MAHO)

NA-TDMA provides for four types of handoff: from one analog channel to another analog channel, from analog to digital, from digital to analog, and from digital to digital. An important development in NA-TDMA is the transfer of some of the responsibility for handoff control from the switch to mobile stations. A terminal tuned to a digital traffic channel relies on MAHO to move its call to a new base station. With MAHO, a terminal with a call in progress monitors the quality of the signal on the active traffic channel. In addition, during the time intervals in each frame when it is not required to transmit or receive information on the active traffic channel, a terminal measures the strength of signals received from surrounding base stations. Each terminal reports its measurements to its own base station in *CHANNEL QUALITY* messages on the slow associated control channel. The switch and base station control the procedure by transmitting *MEASUREMENT ORDER* and *STOP MEASUREMENT ORDER* messages to terminals with calls in progress. The *MEASUREMENT ORDER* messages identify active channels in surrounding cells. The terminal then tunes to these channels and observes their signal strengths.

The terminal performs two measurements on the active traffic channel. It estimates BER, the binary error rate, and it measures RSSI, received signal strength indication. The *MEASUREMENT ORDER* message identifies either 6 or 12 channels in surrounding cells to be measured. An initial *CHANNEL QUALITY* message contains the BER estimate (3 bits), the RSSI estimate of the active channel (5 bits), and the RSSI measurements of the first six surrounding channels identified in the *MEASUREMENT ORDER* message (5 bits per channel). If the *MEASUREMENT ORDER* message specifies 12 channels, the terminal transmits, in a second *CHANNEL QUALITY* message, the RSSI estimates for the second set of six channels specified in the measurement order.

Channel quality measurements travel from terminals to base stations on the slow associated control channel. Each message causes 132 bits to be transmitted in the manner indicated in Figure 5.13. Thus, when the system requests measurements on six surrounding channels, it obtains one measurement every 240 ms, the time required to transmit 132 bits. With 12 surrounding channels measured, the time between measurement reports is

480 ms. On half-rate channels, the reporting intervals are 480 ms (six measurements of surrounding channels) or 960 ms (12 measurements).

MAHO provides the following advantages over the centralized handoff procedures in AMPS:

- MAHO monitors the quality of the signal received at the terminal as well as at the base station. AMPS monitors signal strength only at the base station. As a consequence, an AMPS system cannot initiate a handoff in response to signal-quality problems at the terminal.

- MAHO responds more promptly to signal-quality problems. An AMPS system begins to search for an acceptable link to a new base station after the signal level at the original base station falls below a programmed threshold. With channel-quality measurements available, NA-TDMA has advance knowledge of the location of a base station that can provide good communications to a terminal. A fast response is particularly helpful in service areas with small, low-power cells. In these environments, moving terminals can encounter abrupt changes in signal strength.

- MAHO provides BER estimates as well as signal-strength estimates. This allows the system to perform handoffs in response to excessive interference on traffic channels. This is not possible with signal-strength measurements alone.

- MAHO moves some of the information processing necessary for network control from switches to base stations and terminals. In a large AMPS system, a single switch handles on the order of 5,000 simultaneous calls. The volume of communications and computing necessary to request and interpret signal-strength measurements can introduce significant bottlenecks. With NA-TDMA, most of this work is performed at terminals and base stations.

Mobile-assisted channel allocation (MACA) is a radio resources management procedure related to MAHO. MACA takes place before a call is set up. When a system implements MACA, the BCCH transmits a *MOBILE-ASSISTED CHANNEL ALLOCATION* message to all of the terminals in a cell. The message contains a list of idle channels that are available to handle new calls. Terminals, when they are not receiving BCCH information, tune to these channels and perform signal-strength measurements. They transmit the measurements to the base station in *MACA REPORT* messages on the RACH. Depending on information in the MACA message, terminals report signal-strength measurements when they originate a call, respond to a *PAGE* message, or when they register. The system uses these signal-strength measurements, as part of a channel allocation algorithm, to assign an appropriate physical channel to a conversation. The aim of the channel assignment is to maximize capacity and minimize mutual interference.

In addition to the MAHO and MACA messages, NA-TDMA performs radio resources management by means of *PHYSICAL CHANNEL CONTROL* messages transmitted on the fast and slow associated control channels. These messages convey power adjustment and time alignment commands to terminals with calls in progress.

5.6.3 Call Management

The call management messages in Tables 5.4, 5.5, and 5.7 control call setup and release. They also coordinate enhanced calling features. *ALERT WITH INFO* and *CONNECT* messages play a role in setting up calls directed to mobile terminals. *ALERT WITH INFO*, like the AMPS *ALERT* message, directs the terminal to produce an audible signal. *ALERT WITH INFO* can also carry the phone number of the calling party to be displayed on the telephone's screen, as a part of a caller identification process. The *ALERT WITH INFO* message also supports telephones capable of producing a variety of audible signals. A variable data field specifies the specific signal to be emitted. This feature is particularly useful in a group calling feature available in NA-TDMA. In group calling, several terminals are provided with the same user group identifier, which is associated with one telephone number. When a call arrives for that number, the system sends *PAGE* messages to all of the terminals with the user group identifier. The system sends an *ALERT WITH INFO* message specifying a special audible ringing tone to all of the telephones that send a *PAGE RESPONSE* message to the system. The call is then taken by the person who responds first. Group calling is the cellular equivalent of a bank of telephones set up to answer calls to one number.

When a subscriber responds to an alerting signal by pressing a button on the telephone, the terminal sends a *CONNECT* message to the base station. The *CONNECT* message in NA-TDMA replaces the on-hook, off-hook indications provided by the AMPS 10 kHz supervisory tone.

NA-TDMA *FLASH* messages indicate to the system that a telephone user wishes to initiate a special action during an ongoing call, for example, accept a waiting call or add a third party to a conversation. With conventional telephones, people initiate these requests by tapping on the switch hook. This provokes a large, detectable voltage change in the telephone switching office. When a cellular phone user presses a button to initiate a special action, the terminal sends a *FLASH WITH INFO* message to the base station. The information included in this message is typically a telephone number. A similar action by the remote party in the conversation can cause the base station to transmit a *FLASH WITH INFO* message to the terminal.

5.6.4 Mobility Management

In addition to the registration mechanism of AMPS, NA-TDMA borrows from GSM technology (see Chapter 7) to provide for registration based on location areas. If a system uses this technique, it defines clusters of contiguous cells as *location areas*. Whenever a terminal that does not have a call in progress enters a new location area, it sends a *REGISTRATION* message to the local base station. When a call arrives for a terminal, the system pages the terminal only in the location area where it last registered. To manage this technique, a NA-TDMA *GLOBAL ACTION* message on the FOCC and a *REGISTRATION PARAMETERS* message on the BCCH transmit LOCAID, the 12-bit location area identifier of the base station. The terminal compares this identifier with the stored identifier of the previous base station. A discrepancy between the two identifiers indicates that the terminal has entered a new location area and is therefore required to send a *REGISTRATION* message to the system.

5.7 Status of NA-TDMA

Dual-mode NA-TDMA entered commercial service in the United States and Canada late in 1992 and the number of subscribers with dual-mode terminals increased gradually in the first few years of service. After the adoption of IS-136, the pace accelerated. Operating companies advertise NA-TDMA service as "digital PCS." Thanks to the technology embodied in IS-136, subscribers obtain, in addition to telephone service, enhanced services delivered to the screens of their portable telephones. These services include short messages (alphanumeric paging), notification of voice mail, and caller identification. The cellular operating companies that adopted NA-TDMA acquired many operating licenses in the 1,900 MHz band in spectrum auctions held in the United States in 1995 to 1997. Their plan is to establish national coverage for NA-TDMA for subscribers with "dual-band" terminals capable of operating at 850 MHz and 1,900 MHz.

With respect to technology, in May 1996 the industry adopted an advanced speech coder, referred to as ACELP (algebraic code excited linear prediction), for operation in NA-TDMA digital traffic channels [TIA, 1996a]. ACELP harnesses advances in digital signal processing to produce higher received voice quality than VSELP at the NA-TDMA transmission rate of 13 kb/s. Other technical innovations are embodied in standards for data transmission and facsimile services delivered over NA-TDMA physical channels [TIA, 1995c; TIA, 1995d].

Review Exercises

1. Why is there a guard time in reverse link NA-TDMA transmissions but not in forward link transmissions? What considerations determine the length of the guard time? What are the disadvantages of making the guard time too short or too long?

2. How does an NA-TDMA system change the time alignment of signals transmitted in reverse digital traffic channels? What operating conditions stimulate changes in time alignment?

3. Why does the reverse DCCH have more synchronization information than the forward DCCH and more synchronization information than the forward and reverse DTCH?

4. What is the purpose of the DL field in forward link NA-TDMA time slots? Why doesn't AMPS transmit a control channel locator?

5. Why does NA-TDMA have both slow and fast associated control channels? What is the disadvantage of sending all control information in the SACCH? What is the disadvantage of sending all control information in the FACCH?

6. How do the slow associated control channel and the fast associated control channel share a physical channel with user information in NA-TDMA? Compare these approaches with the way the AMPS forward voice channel and reverse voice channel share a physical channel with user information.

7. Explain the sleep mode operation in NA-TDMA. What is the advantage of having a large number of paging subchannels (PFN)? What is the disadvantage?

8. In the MACA (mobile-assisted channel allocation) procedure, terminals measure the signal strength on channels identified in a *MOBILE-ASSISTED CHANNEL ALLO-CATION* message. The terminals transmit the measurements to the base station in a *MACA REPORT* message when they originate a call, respond to a *PAGE* message, or register. Why don't they transmit the information as soon as they make the measurements?

9. What is the advantage of using shared secret data as an encryption key relative to using the A-key?

10. How does an NA-TDMA terminal send a *SHORT MESSAGE SERVICES* message to the network: (a) when there is a call in progress? (b) when there is no call in progress?

CHAPTER SIX

North American Cellular System Based on Code Division Multiple Access

6.1 Background and Goals

Code division multiple access is a form of spread spectrum communications. Spread spectrum technology dates from the 1940s. The original applications of this technology were in military communications systems, where a principal appeal of spread spectrum signals is their immunity to interference from other signals. This is important in military settings where forces deliberately create interference in order to jam the communications of their enemies. The 1970s and 1980s saw growing interest in commercial applications of spread spectrum communications. In the United States, the Federal Communications Commission stimulated product development by approving spread spectrum transmissions in unlicensed spectrum bands. This policy has led to the production of several wireless local area networks [LaMaire et al., 1996; Wickelgren, 1996] and cordless telephones that transmit spread spectrum signals. Since the 1980s, Qualcomm has operated a spread spectrum satellite communications system for managing fleets of commercial vehicles. Using this system, each vehicle periodically reports its location to a central site. The fleet manager can also exchange text messages with drivers.

The earliest proposals to apply spread spectrum to cellular systems appeared in the late 1970s. These proposals stimulated theoretical work, which revealed some of the strengths and weaknesses of spread spectrum in cellular applications [Yue, 1983]. However, there were no practical developments until the late 1980s and early 1990s, when Qualcomm proposed a CDMA system and demonstrated, with Pacific Telesis, prototype terminal and base station equipment operating in the AMPS frequency band. In the following years, Qualcomm continued to refine its system and to demonstrate experimental equipment in a variety of operating environments. This led, in July 1993, to the adoption of the Qualcomm system by the Telecommunications Industry Association as Interim Standard 95 [TIA, 1993b].

In common with NA-TDMA, the most important CDMA design goal was high spectrum efficiency. Early Qualcomm projections suggested that CDMA could produce a 20:1 improvement in capacity relative to analog AMPS. These projections were based on idealized assumptions. A variety of other calculations, taking into account the limitations of practical operating environments, suggested capacity increases on the order of 5:1 to 10:1 relative to AMPS. Based on these estimates, as well as operational and economic considerations, several United States operating companies announced plans to adopt CDMA for dual-mode operation. Some of these systems began commercial operation in 1996. Therefore, in 1997, there are two different dual-mode digital cellular systems in operation in the United States, with the industry divided between CDMA systems based on IS-95 and NA-TDMA systems. CDMA cellular systems are also in operation in Hong Kong and South Korea. CDMA has also been adopted by several companies for mobile telephone service in the 1,900 MHz PCS bands (Figure 1.18). Although IS-95 and NA-TDMA address the goal of high capacity differently, they adopt the same approach to furthering the goals of privacy and network security. Thus IS-95 incorporates the cryptographic authentication system presented in Section 5.6.1.

6.2 Architecture

Like NA-TDMA, IS-95 prescribes dual-mode operation. However, the two systems differ substantially in their relationship to the analog AMPS systems in which they operate. Recall that an NA-TDMA signal occupies exactly the same bandwidth as an analog AMPS signal. As a consequence, system operators can replace individual AMPS channel units in analog base stations with TDMA radios that carry three full-rate physical channels. By contrast, IS-95 prescribes spread spectrum signals with a bandwidth of 1.23 MHz in each direction. This is approximately 10 percent of a company's total spectrum allocation. As a consequence, a cellular operating company that adopts CDMA has to convert frequency bands of at least this size, corresponding to 41 contiguous AMPS channels, from analog to digital operation.

IS-95 contains many innovations relative to earlier cellular systems. One of them is a *soft handoff* mechanism, in which a terminal establishes contact with a new base station before giving up its radio link to the original base station. When a call is in a soft handoff condition, the terminal transmits coded speech signals to two base stations simultaneously. Both base stations send their demodulated signals to the switch, which estimates the quality of the two signals and sends one of them to a speech decoder. A complementary process takes place in the forward direction. The switch sends coded speech signals to both base stations, which transmit them simultaneously to the terminal. The terminal combines the signals received from the two base stations and demodulates the result. Thus we have the network architecture illustrated in Figure 6.1, which shows a vocoder in the switch rather than in base stations, their location in many NA-TDMA implementations.

CDMA soft handoff requires base stations to operate in synchronism with one another. In order to achieve the necessary synchronization, all base stations contain global positioning system (GPS) receivers. A network of GPS satellites transmits signals that enable each GPS receiver to calculate its location in coordinates of latitude, longitude, and elevation. The satellite signals also include precise time information, accurate to within one microsecond, relative to universal coordinated time, an international standard.

In common with AMPS and NA-TDMA, CDMA terminals and base stations employ an extensive set of identification codes that help control various network operations. Table 6.1 presents many of the identifiers that are used in the CDMA mode. Note that IS-95 provides for a highly

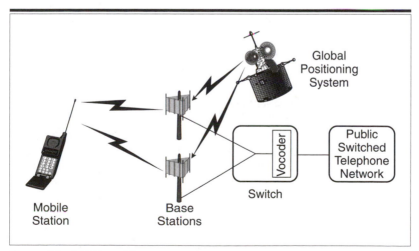

Figure 6.1 CDMA architecture depicting a soft handoff situation.

Table 6.1 CDMA Identification Codes

Notation	Name	Size (bits)	Description
MIN	Mobile identifier	34	Directory number assigned by operating company to a subscriber
ESN	Electronic serial number	32	Assigned by manufacturer to a mobile station
A	A-key	64	Secret key for cryptographic authentication
SCM	Station class mark	8	Indicates capabilities of a mobile station
MOB_MFG_CODE	Mobile manu-facturer code	8	Identifies the brand of the terminal
MOB_MODEL	Mobile model number	8	Model number assigned by manufacturer
MOB_FIRM_REV	Mobile firmware revision	16	Terminal firmware version
MOB_P_REV	Mobile protocol revision	8	The version of IS-95 supported by the terminal
SID	System identifier	15	Assigned by regula-tors to a geographical service area
NID	Network identifier	16	Set of base stations defined by an operat-ing company
BASE_ID	Base station identifier	16	Base station code defined by an operating company
PN_OFFSET	Pseudo-noise code offset	9	Delay applied to a random number sequence at a base station
BASE_CLASS	Base station classification	4	Type of network to which the base station is connected

(continued)

Table 6.1 CDMA Identification Codes *(Continued)*

Notation	Name	Size (bits)	Description
REG_ZONE	Registration zone	12	Location area defined by operating company
BASE_LAT	Base station latitude	22	Geographical location of the base station
BASE_LONG	Base station longitude	23	Geographical location of the base station

detailed indication of the configuration of each terminal. The station class mark of a dual-mode CDMA terminal is an 8-bit identifier. The corresponding identifiers in AMPS and NA-TDMA have lengths of 4 bits and 5 bits, respectively. In addition to the SCM, each terminal stores 40 bits that describe its precise configuration including the manufacturer (MOB_MFG_CODE, 8 bits), the model number assigned by the manufacturer (MOB_MODEL, 8 bits), and the revision number of the firmware running on a particular terminal (MOB_FIRM_REV, 16 bits). The revision number is also specific to each manufacturer. The other configuration code is MOB_P_REV, an 8-bit indicator of the protocol run by the terminal. Initially all terminals operate with MOB_P_REV = 00000001, corresponding to the original version of IS-95. Higher protocol revision numbers will be assigned to future versions of IS-95.

A CDMA base station also contains a rich set of identifiers. Augmenting the 15-bit system identifier (SID) in AMPS and NA-TDMA, CDMA systems specify a 16-bit network identifier (NID). In CDMA, a *network* is a set of base stations contained within a system. Recall that an AMPS system corresponds to a geographical area defined by regulatory authorities. By contrast, CDMA networks can be established by operating companies to meet special requirements. Each base station has its own NID, and each CDMA terminal can be programmed with a SID/NID pair indicating the system and network associated with the terminal's home subscription. Each base station has its own PN_OFFSET. This is a time delay applied to forward direction transmissions that enables the terminals in a cell to decode the desired signal and reject signals from other base stations. The 4-bit BASE_CLASS identifier anticipates terminals that will have access to a variety of wireless services. In the initial issue of IS-95, the only assigned BASE_CLASS is 0000, corresponding to *public macrocellular systems.* Future class numbers could be assigned to other public networks or to various types of private networks such as wireless business systems (PBX) and residential cordless telephones.

The CDMA system anticipates a variety of mobility management schemes including location-area registration, as in NA-TDMA and GSM; timer-based registration; and distance-based registration. To facilitate location-area registration, IS-95 defines a 12-bit REG_ZONE identifier to be assigned to each base station. REG_ZONE plays the same role as the location area identifier, LOCAID, in NA-TDMA (see Table 5.1). The identifiers, BASE_LAT (22 bits) and BASE_LONG (23 bits), specify the exact geographic location of the base station, in latitude-longitude coordinates. Terminals can use this information to perform distance-based registration.

6.3 Radio Transmission

CDMA radio transmission differs significantly from the techniques employed in the other systems examined in this book. Four important differences are:

- CDMA employs different transmission techniques in the forward and reverse directions. In the other systems, forward and reverse transmissions are identical or very similar to each other.

- There are two stages of modulation in a CDMA system, compared with one in the other systems.

- To maximize spectrum efficiency, the CDMA system employs variable-bit-rate traffic channels. The other systems have constant-bit-rate channels.

- The CDMA system has a reuse factor of one. Each radio channel is used in all sectors of all cells. For interference control, the other systems require a spatial separation of transmitters using the same carrier frequency (see Section 9.3.2).

Sections 6.3.1 and 6.3.2 describe the reverse and forward transmission techniques, respectively. Section 6.3.3 discusses reuse. Section 6.4, on logical channels, describes the variable-bit-rate transmission of source information. The remainder of this introductory section describes the two stages of modulation and demodulation in a spread spectrum system. We begin by referring to Figure 6.2, which shows a conventional radio modulator, in which an information signal modulates a radio-frequency carrier. The transmitter amplifies the modulated carrier and sends it to an antenna.

In a spread spectrum system, the digital information signal modulates a digital carrier. In IS-95, the digital carrier is a binary signal with a

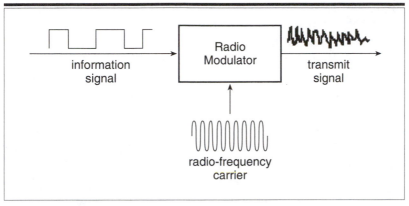

Figure 6.2 Single-stage digital modulation (TDMA and FDMA).

switching rate of $W = 1{,}228{,}800$ chips per second (ch/s). *Chip* is the nomenclature for a binary signal element in a digital spread spectrum carrier. The word *chip* distinguishes these short signal elements from the *source bits* of an information signal and the *channel bits* of an information signal protected by a channel code. The modulated digital spread spectrum signal in turn modulates a radio carrier to produce the transmitted signal. This situation is shown in Figure 6.3. The ratio of the digital carrier chip rate, W ch/s, to the source information rate, R b/s, in a spread spectrum signal is referred to as the *processing gain, G:*

$$G = \frac{W \text{ ch/s}}{R \text{ b/s}} \text{ ch/b.} \qquad (6.1)$$

$$\frac{1{,}228{,}800 \ \frac{ch/s}{}}{9{,}600 \ b/s} = 128 \frac{c}{b}$$

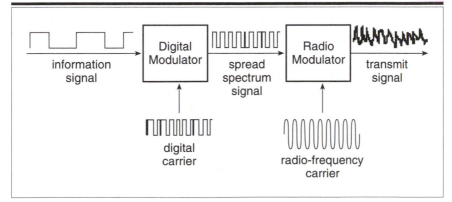

Figure 6.3 Two stages of modulation in a spread spectrum system.

An important information rate in the original implementation of IS-95 is $R = 9,600$ b/s. With a 9,600 b/s source signal, the processing gain in IS-95 is $G = 1,228,800/9,600 = 128$ ch/b. Later implementations employ speech coders operating at 14,400 b/s, with a processing gain of 85.3 ch/b.

The spread spectrum receiver shown in Figure 6.4 performs a pair of inverse operations corresponding to the two modulation stages in Figure 6.3. The radio demodulator produces a lowpass signal with a bandwidth related to the chip rate W ch/s. The digital demodulator, labeled "Correlator" in Figure 6.4, multiplies this signal by the binary digital carrier signal. For each transmitted bit, the correlator obtains G products corresponding to the G chips in the digital carrier. It adds these G products and makes a binary decision about the identity of the transmitted bit on the basis of the polarity of the sum. Section 9.8 describes this procedure mathematically.

In order to function properly, the receiver in Figure 6.4 has to generate a digital carrier that is precisely synchronized with the chip sequence in the received signal. Synchronization is a major task of a spread spectrum receiver [Viterbi, 1995: Chapter 3]. The difficulty of performing this task is magnified in a mobile communications system with signals subject to multipath propagation. The motion of a terminal causes the arrival times of signals at base stations and terminals to change rapidly. Mobility also changes the nature of the multiple signal paths between a terminal and a base station. In this environment, the digital carrier in Figure 6.4 is synchronized with only one of several existing signal paths. The quality of the signal on this and other paths fluctuates as the terminal moves. Location changes cause some paths to disappear and new signal paths to appear.

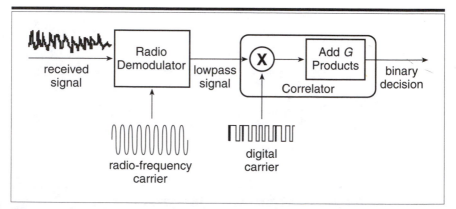

Figure 6.4 Two stages of demodulation in a spread spectrum receiver.

To cope with this situation, IS-95 prescribes multiple correlators in each receiver. The system specification refers to the correlators as *demodulating elements*. At any instant of time the digital carriers in the different correlators are synchronized to signal paths with different propagation times (see Section 9.2.4). Meanwhile, a searching circuit examines the arriving signal in order to detect the appearance of a new path. When a new path appears, the receiver generates a digital carrier synchronized to the signal on that path and assigns a correlator to demodulate the signal. It simultaneously releases a correlator assigned to a weaker signal path. In this manner, IS-95 frequently locks its receiver to strong signal components and stops receiving signal components that have weakened due to changes in terminal location. Figure 6.5 shows a receiver with multiple correlators. Each correlator operates with a digital carrier synchronized to one propagation path. This configuration is referred to as a RAKE receiver. The time differences between the carriers resemble the spacing of prongs on a garden rake. In a process referred to as *multipath diversity combining*, the RAKE receiver adds the correlator outputs before making a binary decision about the polarity of the transmitted bit.

The radio-frequency carrier of an IS-95 signal is identical to an AMPS carrier and it is referred to by the corresponding AMPS channel number

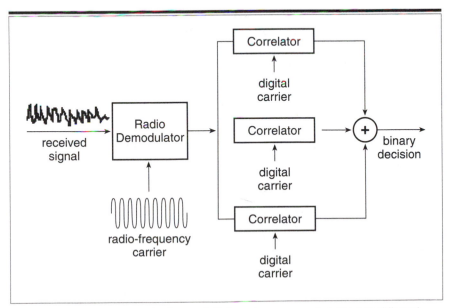

Figure 6.5 RAKE receiver: the three digital carriers are synchronized to different signal paths.

(Equations 3.1 and 3.2). The bandwidth of a CDMA signal is 1.23 MHz. Recall that each physical channel in AMPS occupies 30 kHz. Therefore, the bandwidth of a CDMA signal corresponds to the aggregate bandwidth of 41 AMPS channels. As a consequence, a CDMA signal covers more than 20 AMPS channels on either side of the CDMA radio carrier. Therefore, any possible CDMA radio carrier has to be at least 22 AMPS channels from the edge of an AMPS operating band. In Figure 6.6, the unshaded regions contain AMPS channel numbers corresponding to possible CDMA radio-frequency carriers.

6.3.1 Reverse-Direction Radio Transmission

The digital carrier transmitted by each terminal using a traffic channel is derived from a "long code mask" containing the electronic serial number of the terminal. Closed-loop power control is an important part of the reverse-direction radio transmission technique.

Physical Channels

In the other systems presented in this book, a physical channel corresponds to either a frequency band or a frequency band and a time slot. By contrast, a reverse CDMA physical channel corresponds to a binary code assigned to a terminal. In Figure 6.7, this binary code, labeled channel identifier, is part of the digital carrier. The job of the digital modulator is to combine the information signal with the digital carrier. The digital modulator expands the bandwidth occupied by the information signal. A receiver (Figures 6.4 and 6.5) at a base station uses the digital carrier to

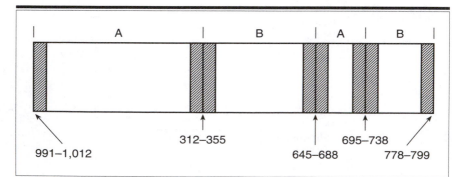

Figure 6.6 CDMA radio-frequency carriers (unshaded regions) expressed as AMPS channel numbers.

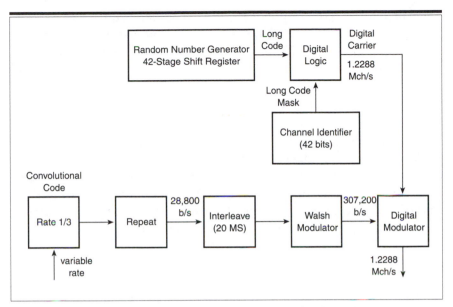

Figure 6.7 Digital modulation on the CDMA reverse traffic channel.

reduce ("de-spread") the bandwidth occupied by the transmitted signal and thereby extract the digital source information. Figure 6.7 indicates that the digital carrier in the reverse channel of IS-95 is a logical combination of a 42-bit channel identifier and the output of a "long code" produced by a binary pseudorandom number generator operating at 1.2288 Mch/s. The IS-95 term for the channel identifier is *long code mask*. The nature of the channel identifier depends on the logical channel (see Section 6.4). In digital traffic channels that carry user information, the long code mask contains the electronic serial number (ESN) of the terminal. The random-number generator in Figure 6.7 is a feedback shift register with 42 stages. The long code repeats itself after $2^{42} - 1$ chips. Thus the repetition rate is

$$\frac{2^{42} - 1 \text{ ch}}{1,228,800 \text{ ch/s}} = 3.6 \times 10^{13} \text{ } s = 41.4 \text{ days.} \qquad (6.2)$$

Figure 6.7 also shows the coding and interleaving operations that take place prior to digital modulation. The digital modulator in the original CDMA implementation processes inputs signals at four different data rates, ranging from 1,200 b/s to 9,600 b/s as indicated in Table 6.2. The

20ms chunks

information signals corresponding to these data rates convey $R_I = 800$ b/s, 2,000 b/s, 4,000 b/s, or 8,600 b/s. The terminal processes these signals in frames, each containing 20 ms of information. Thus, the information content of each frame is

20ms

$$\text{IBPF bits} = R_I \text{ b/s} \times 0.02 \text{ s}, \tag{6.3}$$

which is 16 bits, 40 bits, 80 bits, or 172 bits as shown in Table 6.2.

At the two higher data rates, the transmitter adds either 8 (4,800 b/s data rate) or 12 (9,600 b/s) parity check bits per frame (PBPF) for error detection at the receiver. A convolutional coder of constraint length nine and rate 1/3 processes each frame. It adds 8 tail bits per frame (all 0) prior to computing the convolutional code bits. The coder generates

$$\text{CBPF} = 3(\text{IBPF} + \text{PBPF} + 8) \text{ code bits/frame.} \tag{6.4}$$

Repeating each frame 1, 2, 4, or 8 times, the terminal generates 576 bits of coded information every 20 ms. Thus, regardless of the information rate, the bit rate at this stage is

$$576 \frac{\text{bits}}{\text{frame}} \div 0.020 \frac{\text{sec}}{\text{frame}} = 28,800 \text{ b/s.} \tag{6.5}$$

Table 6.2 Contents of 20 ms Frames, Reverse Channels

Frame = 20ms

Data Rate R b/s	1,200	2,400	4,800	9,600
Information rate R_I b/s	800	2,000	4,000	8,600
Information bits per frame (IBPF)	16	40	80	172
Parity bits per frame (PBPF)	0	0	8	12
Data bits per frame (IBPF + PBPF + 8)	24	48	96	192
Coded bits per frame (CBPF)	72	144	288	576
Repetitions	8	4	2	1
Total bits per frame (BPF)	576	576	576	576

An interleaver (see Section 9.5) permutes the order of the coded bits prior to an operation, referred to in IS-95 documentation as *orthogonal Walsh modulation*. In this operation, each block of 6 input bits produces a sequence of 64 output bits. Each possible output sequence is a row of a 64×64 Walsh Hadamard matrix (see Section 9.9). This matrix contains one row of 0s. Each of the other 63 rows contains 32 1s and 32 0s. The Walsh modulator is, in effect, a (64,6;32) block code with rate 6/64. The signaling rate at the output of the Walsh modulator is

$$64 \, \frac{\text{channel bits}}{\text{code word}} \div 6 \, \frac{\text{source bits}}{\text{code word}} \times 28{,}800 \, \frac{\text{source bits}}{\text{s}}$$

$$= 307{,}200 \text{ b/s}.$$

(6.6)

The output of the Walsh modulator modulates the digital carrier to produce the spread spectrum signal at 1.2288 Mch/s.

Radio Modulation

The radio modulator in a CDMA reverse channel produces a form of four-level modulation referred to as *offset quadrature phase shift keying* (OQPSK). As shown in Figure 6.8, each chip in the spread spectrum signal modulates in-phase and quadrature components of the carrier-frequency sine wave. Prior to the phase modulation, each of the two signals, in-phase and quadrature, modulates its own pseudorandom code sequence with a signaling rate of 1,228,800 ch/s and a periodicity of 2^{15} chips. The numbers are such that each code sequence goes through exactly 75 cycles of operation in 2 seconds. In 2 seconds, there are

$$2 \text{ s} \times 1{,}228{,}800 \text{ ch/s} = 75 \times 2^{15} \text{ch}.$$

(6.7)

To generate the pseudorandom code sequences for the in-phase and quadrature signals, the terminal synchronizes its operation with universal coordinated time. Each base station determines the exact time from the global positioning system and transmits the time in a sync channel message.

Before modulating the in-phase and quadrature (sine and cosine) components of the radio carrier, the system introduces a one-half chip delay to the quadrature signal. (This delay is the *offset* in the OQPSK nomenclature.) At the output of the modulator, the four possible phases of the radio signal are $\pm\pi/4$ and $\pm 3\pi/4$ corresponding to the four possible combinations of the in-phase and quadrature chips. However, the time offset

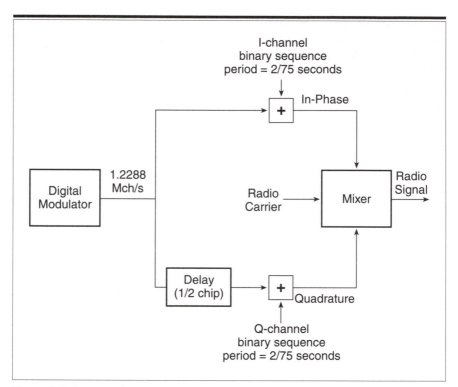

Figure 6.8 Reverse channel radio modulation. (Reproduced under written permission from Telecommunications Industry Association.)

in the two signal components makes it impossible for the in-phase and quadrature components of the output phase angle to change simultaneously. Thus, the possible transitions in the resultant phase are 0 and $\pm\pi/2$, with possible phase transitions occurring at twice the chip rate. There is never a 180° phase shift in the transmitted signal. Baseband filters in the mixer in Figure 6.8 smooth the phase transitions in order to suppress out-of-band energy components in the transmitted signal.

Radiated Power

IS-95 specifies complex procedures for regulating the power transmitted by each terminal. The aim is to make all reverse-direction signals in a cell arrive at the base station with the same strength. Closely matched received power levels are essential for correct CDMA operation. Any significant variation in these power levels has a strong negative impact on system capacity [Viterbi, 1995]. Because the strength of the signal path between a moving terminal and each base station changes continuously, the terminal performs power adjustments 800 times per second under the

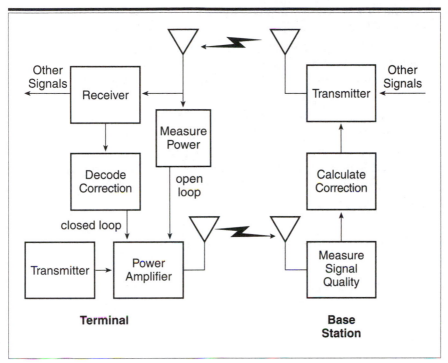

Figure 6.9 Power control at a CDMA terminal.

control of the base station. These adjustments are referred to as *closed-loop power control.* Section 6.4.5 describes the closed-loop power control signal transmitted from the base station to each terminal. In addition, IS-95 specifies *open-loop power control,* which causes the terminal to adjust its transmitter power as a function of the power it measures in the received forward-direction signal. Figure 6.9 shows the elements of the CDMA reverse link power control system. Section 6.6.1 describes the power control operations in detail. Overall, a CDMA terminal is capable of adjusting its transmitter power in steps of 1 dB over a range of 64 dB. The maximum power radiated by any CDMA terminal is 6.3 W.

Variable-Bit-Rate Radio Signals

As described in Section 6.3.3, the principal purpose of variable-bit-rate operation is to reduce interference caused by each transmitter when there are pauses in the input speech. Section 6.3.3 indicates that signal quality depends on the received energy per bit, E_b, at the receiver. At the output of the radio modulator, Figure 6.8, the signal repetition (Table 6.2) in the

[handwritten: 20ms Frame; 20ns; 16 intev; 1.25ms]

digital modulator causes this energy to be inversely proportional to the data rate. Thus, each bit of a signal leaving the radio modulator at 2,400 b/s has four times the energy of each bit of a 9,600 b/s signal. To arrive at equal energy per bit (and equal signal quality) in the transmitted signal, a CDMA terminal transmits reduced-bit-rate signals intermittently with a duty cycle inversely proportional to the data rate. It divides each 20 ms frame into 16 intervals of duration 1.25 ms. A 9,600 b/s signal occupies all of these intervals. A signal at 4,800 b/s occupies one-half of them, and so on. A randomizing function distributes the energy over each frame independently of the operation at other terminals. The purpose of the randomizing function is to distribute uniformly in time the energy received from interfering signals.

6.3.2 Forward Link Radio Transmission

The digital carrier of a forward-direction channel is a row of a 64 × 64 Walsh Hadamard matrix. In contrast to the reverse direction, all digital carriers in the forward direction are mutually orthogonal. To control co-channel interference from different base stations, the modulator in each base station processes a digital signal that is offset in time from the corresponding signals at surrounding base stations.

Physical Channels

The digital carrier corresponding to one physical channel is a binary code sequence containing 64 chips. Each CDMA base station has access to 64 physical channels, each derived from one row of the 64 × 64 Walsh Hadamard matrix, W_{64} (see Section 9.9). The rows of the matrix are numbered from 0 through 63 and the sequence of bits in row n is denoted "Walsh n." Therefore, the 64 digital carriers available for base station transmissions are Walsh 0–Walsh 63. The digital carriers are mutually orthogonal. Mathematically, this means that if any two digital carriers are multiplied bit by bit (with values +1 and −1 assigned to the binary states), the sum of the 64 products is 0. In a system operating with orthogonal carriers, a digital demodulator tuned to one physical channel contains no interference from other physical channels arriving on the same propagation path. All of the interference at a digital demodulator in a CDMA terminal comes from other base stations and from signals arriving on paths other than the one to which the demodulator is synchronized. This is similar to the TDMA and FDMA situation. It is different from the situation at a receiver in a CDMA base station that encounters interference from

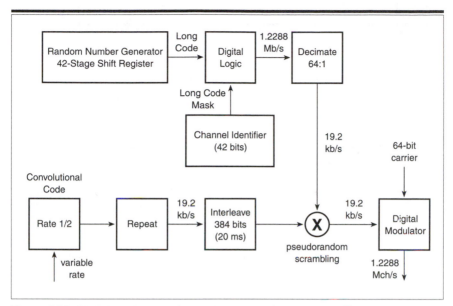

Figure 6.10 Digital modulation for paging channels and forward traffic channels.

signals on other physical channels within its own cell, as well as from terminals in other cells (see Section 9.8).

CDMA base stations transmit information in the four logical channel formats described in Section 6.4: pilot channels, sync channels, paging channels, and traffic channels. Figure 6.10 displays the digital modulation for paging channels and traffic channels. In common with the reverse channel, forward channel signals arrive at data rates ranging from 1,200 b/s to 9,600 b/s. The digital modulator processes these signals in frames of duration 20 ms. Depending on the data rate, the input frames arrive with a variable number of bits. A convolutional code, with constraint length nine, rate 1/2, protects each signal. The resulting channel bit rate can range from 2,400 b/s to 19,200 b/s. When the rate is less than 19,200 b/s, the transmitter repeats code bits to bring the rate up to 19,200 b/s, corresponding to 384 bits per frame. Table 6.3 shows the contents of a frame for each of the four possible data rates. An interleaver permutes the code bits in each frame, which are then scrambled (multiplied by ±1) by a pseudorandom sequence derived from the long code ($2^{42} - 1$ bit sequence) and long code mask (see Section 6.3.1). To match the rate of the pseudorandom sequence to the 19,200 b/s coded information rate, a decimator extracts 1 bit out of every 64 bits from the signal obtained from the long code and the long code mask. The scrambled information signal enters the digital modulator at a rate of 19.2 kb/s. The digital modulator replaces

Table 6.3 Contents of 20 ms Frames, Forward Channels

Data Rate R b/s	1,200	2,400	4,800	9,600
Information rate R_I b/s	800	2,000	4,000	8,600
Information bits per frame (IBPF)	16	40	80	172
Parity bits per frame (PBPF)	0	0	8	12
Data bits per frame (IBPF+ PBPF + 8)	24	48	96	192
Coded bits per frame (CBPF)	48	96	192	384
Repetitions	8	4	2	1
Total bits per frame (BPF)	384	384	384	384

each bit of the information signal with the 64-chip digital carrier or the inverse of the digital carrier, depending on the logical level of the information bit, either +1 or −1. Thus, the output rate of the digital modulator is

$$64 \text{ ch/b} \times 19,200 \text{ b/s} = 1,228,800 \text{ ch/s}. \tag{6.8}$$

Although Figure 6.10 can be used to represent the digital modulation of a CDMA pilot channel, the nature of the input signal and the digital carrier imply that the output of the digital modulator is simply a sequence of 0s. With respect to Figure 6.10, the variable-rate input is simply a sequence of 0s. The digital carrier for the pilot channel is Walsh (0), the first row of the Walsh Hadamard matrix. This row contains 64 logical 0s. As a consequence, the output of the digital modulator of the pilot channel in Figure 6.10 consists of 1,228,800 ch/s, all of which are 0s.

Figure 6.11 shows the digital modulator for the sync channel. The input is a 1,200 b/s information signal. This signal is processed by the rate 1/2 convolutional channel coder. For each output bit of the coder, the system generates 2 identical bits to produce a signal at 4,800 b/s. After interleaving, this signal modulates the digital carrier, which, for the sync channel, is Walsh 32, the thirty-third row of the Walsh Hadamard matrix. This row contains 32 logical 0s, followed by 32 1s. To bring the output rate up to 1.2288 Mch/s, each bit of the interleaver output modulates the 64-bit digital carrier four times.

Radio Modulation

The CDMA forward link employs the four-level modulation technique illustrated in Figure 6.12. It resembles, but is not identical to, the reverse

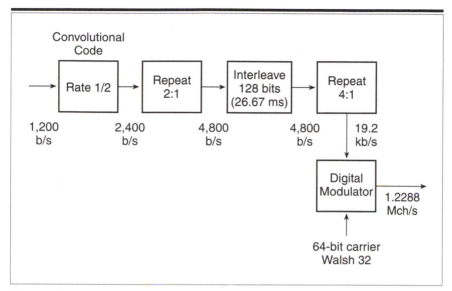

Figure 6.11 Digital modulation for the sync channel.

link modulation in Figure 6.8. One important difference is a time offset in the pseudorandom signals in the in-phase and quadrature signal paths. Each base station has its own time offset, denoted by the parameter PN_OFFSET (Table 6.1), transmitted in network control messages. The offset is synchronized to universal coordinated time. To demodulate a received signal, a terminal synchronizes its receiver with the assigned base station and generates I-channel and Q-channel binary sequences with the value of PN_OFFSET assigned to the local base station. Due to the nature of the binary sequences in the radio modulator, signals received from other base stations, with different values of PN_OFFSET, appear as low-level noise in the receiver of a terminal. Therefore, the time offsets in the demodulators of different base stations play a critical role in suppressing interference from other cells in the receiver of each terminal. There are 512 possible time offsets, with offset i corresponding to a time delay of $64i$ chips. (The cycle time of the binary sequence is 2^{15} chips. There are $2^9 = 512$ possible offsets, separated by multiples of $2^6 = 64$ chips.)

In contrast to the reverse-direction radio modulator (Figure 6.8), the in-phase and quadrature signal components have simultaneous transitions. As a consequence, it is possible for both components of the resultant phase to change simultaneously, producing a 180° phase shift in the output signal.

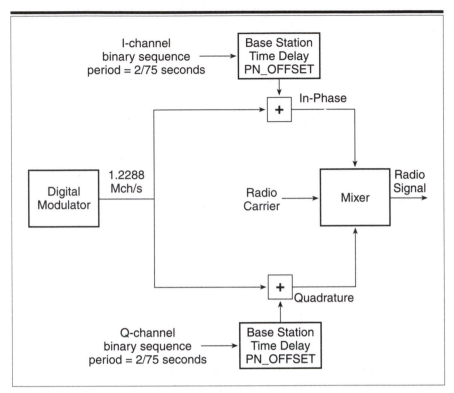

Figure 6.12 Forward-direction radio modulation. (Reproduced under written permission from Telecommunications Industry Association.)

Variable-Bit-Rate Radio Signals

To achieve equal energy per bit, and thereby reduce the interference to other signals, a CDMA base station transmitter varies the radiated power in proportion to the data rate in each frame. Thus at 1,200 b/s, the power is one-eighth of the power at 9,600 b/s. This is in contrast to a CDMA terminal that transmits intermittently to maintain equal energy per bit when the data rate is less than 9,600 b/s (see Section 6.3.1).

6.3.3 Spectrum Efficiency

One of the most striking properties of CDMA is that all cells simultaneously use the same radio bands. This is possible because spread spectrum signals have a high tolerance to interference, provided there is low correlation between the desired signal arriving at a receiver and the signals arriving from adjacent cells. At base station receivers, the low correlation is a property of the reverse-link digital carriers associated with different physical channels. At the receiver in a CDMA terminal, there is low interference from other cells because radio modulators at different base stations operate with pseudorandom sequences (Figure 6.12) that have different time delays (PN_OFFSET) relative to universal coordinated time.

With complete frequency reuse in all cells, the calculation of capacity and spectrum efficiency in a CDMA system differs markedly from the FDMA and TDMA calculations in Sections 3.3.6, 5.3.4, and 7.3.5. The FDMA and TDMA calculations account for reuse by dividing the total number of physical channels by the reuse factor N. In CDMA, Section 9.8 demonstrates that capacity depends on the amount of interference a system can tolerate within the constraint of a signal quality objective, such as binary error rate. The system examined in Section 9.8 is a highly simplified version of IS-95. For a practical system, engineers express the interference limit as a ratio E_b/N_0. E_b is the energy per bit in the desired signal and N_0 is proportional to the sum of the interference power and noise power at the receiver. A system that requires a high value of E_b/N_0 cannot tolerate much interference. It therefore has a low capacity because it has to keep the number of conversations low to limit the interference. A system that can operate with a lower value of E_b/N_0 can tolerate more interference and admit more conversations. The intensive signal-processing operations prescribed by IS-95, including powerful channel coding and a RAKE receiver, serve to increase tolerance to interference and reduce the required E_b/N_0 to the lowest practical value.

CDMA capacity is a statistical phenomenon that depends on a large number of random quantities including speech and silence patterns of subscribers, subscriber locations, and estimates of received power. Capacity calculations are complicated and depend critically on properties of the probability distributions of these random quantities. As a consequence, there is a large body of literature on the subject, based on a combination of intricate mathematical analysis and engineering judgment. The results of all this work are a range of capacity estimates, each depending on system operating conditions and on assumptions that remain to be verified by practical experience. This is in marked contrast to calculations of the capacity of frequency division and time division systems, which depend in a simple way on system design.

The following paragraphs offer a brief summary of the analysis of reverse link CDMA capacity. Viterbi [1995, Sections 1.3 and 6.6] provides a thorough examination of key issues.

For a single-cell system with perfect power control and all terminals transmitting information at R b/s, the number of simultaneous conversations that produce the ratio E_b/N_0 is approximately

$$K \approx \frac{W/R}{E_b/N_0} \text{ conversations/base station.} \qquad (6.9)$$

Here W ch/s is the binary transmission rate and K is the maximum number of simultaneous reverse-direction transmissions in a single cell with an omnidirectional base station antenna. K is a maximum number because with more than K conversations, the average binary error rate in the decoded signals will exceed the system target. Note that in a single-cell FDMA or TDMA system, W/R is the number of physical channels, which is the capacity of a single-cell system. In a cellular system, the capacity per cell is lower by a factor of N, the reuse constant. Thus, in the absence of interference from other cells, CDMA capacity is lower (by the factor E_b/N_0) than TDMA or CDMA capacity. The decrease is due to the interference from other signals in the same cell. The advantage of CDMA lies in its tolerance to interference from other cells.

To move from Equation 6.9 to a formula for the capacity of a CDMA cellular system it is necessary to take into account several phenomena:

- Directional base station antennas increase capacity because they reduce the amount of interference arriving at each receiver.

- Variable-bit-rate speech transmission (see Section 6.4.5) also increases capacity. When a speech coder detects a silent input, it reduces the bit rate of the transmitted signal. This in turn reduces the interference caused by each conversation to other conversations.

- Interference from surrounding cells reduces capacity relative to Equation 6.9. The total interference from all cells has to be low enough to maintain the signal-to-noise-plus-interference ratio above E_b/N_0.

- Imperfect power control also reduces capacity. The CDMA system takes elaborate measures to maintain at each base station equal received powers from all transmitting terminals (see Section 6.6.1). Due to the motion of the terminals, perfect power control is impossible. As a consequence, at any time, some signals arrive at the receiver at excessive power levels and add to the interference.

- An outage margin reduces capacity because it enforces limits on the fraction of time the fluctuating signal quality can be below the system target.

Taken together, these factors appear in a simple capacity formula of the form

$$C = \left(\frac{W/R}{E_b/N_0}\right)\left(\frac{F_{sectors}F_{speech}}{F_{power}F_{cells}F_{outage}}\right) \text{ conversations/base station,} \quad (6.10)$$

where each factor $F_{\text{subscript}} > 1$ takes into account one of the phenomena listed above. Although this formula is simple, the calculation of each F is complex and imprecise. It is beyond the scope of this book to reproduce the calculations. Instead, we cite examples of numerical values obtained for each of the quantities.

In IS-95, $W = 1{,}228{,}800$ ch/s and the nominal information rate of a reverse traffic channel in the original implementation is $R = 9{,}600$ b/s. Thus $W/R = 128$. The bit error rate target is 0.001, and Viterbi [1995] states that the required $E_b/N_0 = 4$ (6 dB). With this limit to E_b/N_0, a single cell with an omnidirectional antenna and constant-bit-rate speech transmission could support

$$K \approx \frac{1{,}228{,}800 \,/\, 9{,}600}{4} = 32 \text{ conversations/base station.} \quad (6.11)$$

This is a modest number for a system with 1.23 MHz bandwidth. The high capacity anticipated for IS-95 depends on the five adjustment constants.

The values given by Viterbi for these constants are:

- $F_{\text{speech}} = 2.67$, based on the assumption that in a conversation people speak 3/8 of the time and are silent the remaining 5/8 of the time. It also assumes that a speech activity detector will distinguish perfectly between speech and silence and that the transmitter is totally inactive during silent periods.

- $F_{\text{sectors}} = 2.4$, based on the properties of base station directional antennas.

- $F_{\text{cells}} = 1.6$, based on a statistical analysis that takes into account radio propagation assumptions and the spatial distribution of terminals in a cluster of cells.

- $1.05 \le F_{\text{power}} \le 3.3$, a range of values appearing in a set of graphs [Viterbi, 1995: Figure 6.5] that represent the effects of operating factors and estimates of the accuracy of some mathematical assumptions. Viterbi suggests a nominal value of $F_{\text{power}} = 1.2$.

- $1.1 \le F_{\text{outage}} \le 1.3$, based on assumptions about admission control policies. Viterbi suggests a nominal value of $F_{\text{outage}} = 1.1$.

Substitution of these estimates into Equation 6.10 produces a range of system-capacity estimates,

$$29.8 \text{ conversations/cell} \le C \le 111 \text{ conversations/cell,} \quad (6.12)$$

with Viterbi's nominal estimate near the high end of the range:

$$C \approx 97.1 \text{ conversations/cell.}$$

This is the number of conversations supported by one CDMA radio carrier that occupies a bandwidth of 1.23 MHz per direction. Therefore, the total bandwidth necessary to maintain 97.1 conversations/cell is 2.46 MHz, which leads to the efficiency estimate:

$$E \approx \frac{97.1}{2.46} = 39.5 \text{ conversations/cell/MHz,} \tag{6.13}$$

within the total range derived from Equation 6.12:

$$12.1 \text{ conversations/cell/MHz} \leq E \leq 45.1 \text{ conversations/cell/MHz.} \tag{6.14}$$

These estimates are based on the speech coding rate of 9,600 b/s in the original CDMA implementation. In 1997, commercial implementations are produced with a vocoder operating at a maximum rate of 14,400 b/s in order to deliver higher speech quality. With this rate, in the denominator of Equation 6.11, all of the numerical estimates in the subsequent equations would be multiplied by 2/3.

The wide range of efficiency estimates that can be extracted from a single reference is evidence of the complex set of phenomena that influence CDMA spectrum efficiency. Starting with Viterbi's analysis, other investigators suggest their own values for the constants in Equation 6.10. One example is the range $1.6 \leq F_{\text{outage}} \leq 2.0$ [Landolsi, Veeravalli, and Jain, 1996].

All of these estimates assume that the entire system bandwidth is available to carry conversations. They would have to be reduced to some extent to account for transmission resources used for network control information (access channels in the reverse direction). Moreover, they pertain only to reverse-direction transmissions. Recall that base station transmitters use modulation and coding schemes that differ substantially from those of terminals, and that the nature of co-channel interference is different in the two directions. Therefore, the forward direction merits its own capacity analysis [Viterbi, 1995: Section 6.7]. This is not examined here because investigators assert that the reverse link of a CDMA system has a lower capacity than the forward link, and therefore determines overall system efficiency.

6.4 Logical Channels

IS-95 specifies the logical channel categories shown in Figure 6.13. Recall that there are 64 physical channels available for CDMA base station transmission. One of these channels is always active as a pilot channel (Section 6.4.1) and another one serves as a sync channel (Section 6.4.2). The remaining physical channels in the forward direction comprise a mixture of paging channels (Section 6.4.3) and forward traffic channels (Section 6.4.5). IS-95 specifies that the base station transmit at most one sync channel and at most seven paging channels. The remainder of the forward-direction physical channels (a minimum of 55 and a maximum of 63) are available to serve as digital traffic channels. In the reverse direction, there are access channels (Section 6.4.4) and reverse traffic channels (Section 6.4.5). Figure 6.13 indicates that the traffic channels carry information in three different formats: signaling messages, a power control signal (in the forward direction), and variable-bit-rate user information, such as speech.

6.4.1 Pilot Channel

As described in Section 6.3.2, the pilot channel transmits a continuous sequence of 0s at a rate of 1.2288 Mch/s. By tuning to the pilot channel, terminals communicating with the base station acquire carrier phase and timing references. Terminals in other cells tune to the pilot channel to obtain signal strength indications as part of a CDMA mobile-initiated handoff procedure (Section 6.6.1).

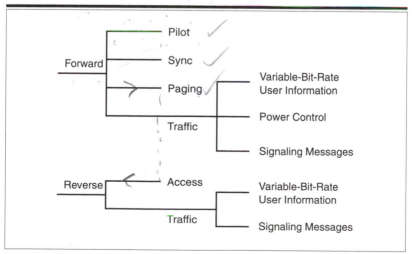

Figure 6.13 CDMA logical channels.

6.4.2 Sync Channel   *Walsh 32*

As indicated in Figure 6.11, the information rate of the sync channel is 1,200 b/s and the digital carrier is Walsh 32. The sync channel repeatedly transmits one message (Table 6.8), which conveys important system information to terminals, including system time (obtained from the global positioning system) and the time delay (determined by PN_OFFSET) introduced to the pseudorandom sequences in the radio modulator. The sync channel also transmits the base station's system identifier and network identifier (SID/NID) and the parameter MIN_P_REV, which indicates the minimum IS-95 protocol revision supported by the base station. Another parameter transmitted by the sync channel is the information rate of the base station paging channels.

6.4.3 Paging Channel   *Walsh 1 − 7 slots $\frac{2048}{0-204}$*

A CDMA signal carries up to seven paging channels that transmit information to terminals that do not have calls in progress. The digital carriers of paging channels are Walsh 1 to Walsh 7 (the second to eighth rows of the Walsh Hadamard matrix). The information rate on a paging channel is either 4,800 b/s or 9,600 b/s. Base stations transmit paging messages in time slots of duration 80 ms (384 bits at a rate of 4,800 b/s and 768 bits at 9,600 b/s). The maximum length of a paging message is 1,184 bits. It follows that a paging message can span from one to four time slots depending on the length of the message body and the transmission rate of the paging channel.

As in NA-TDMA, the CDMA system specifies a sleep mode that organizes paging information to conserve battery energy in portable terminals. To accomplish this, IS-95 defines a mode of paging channel operation that confines messages for each terminal to a specified fraction of the paging channel bit stream. This is achieved by defining a "maximum paging channel slot cycle" consisting of 2,048 paging channel slots, numbered from 0 to 2,047. The total duration of this slot sequence is 2,048 × 80 ms = 163.84 s. Each terminal can be programmed to examine a fraction of the slots in each sequence for paging messages. For example, the terminal can examine one slot in every 16 slots, or 2,048/16 = 128 slots in every cycle. In this example, the terminal in an idle state can operate with reduced power for 15 out of every 16 slots and then "wake up" to examine the sixteenth slot for a paging message. Each terminal operating in the idle mode monitors a fraction of the slots within one paging channel. A terminal determines the specific paging channel and the slots within the channel that it monitors by performing a computation (hash function) based on the terminal's directory number (MIN) and electronic serial number (ESN). The computational procedure aims to assign roughly equal numbers of terminals to different paging channels and to time slots within the channels.

6.4.4 Access Channel

IS-95 access channels perform many of the functions of AMPS reverse control channels. A terminal without a call in progress uses an access channel to send messages to a base station for three principal purposes: to originate a call, to respond to a paging message, and to register its location. Each base station operates with up to 32 access channels. The digital carrier of an access channel depends on the 42-bit channel identifier in Figure 6.7. Referred to in IS-95 as a *long code mask*, the channel identifier contains a 5-bit field that specifies the access channel number. Other components are: the base station identifier (BASE_ID); the number of paging channels monitored by the terminal (up to seven channels); and PN_OFFSET, the relative delay of the signals transmitted by the base station.

The transmission rate of each access channel is 4,800 b/s, with information conveyed in 20 ms frames as indicated in the third column of Table 6.2. Each frame consists of 88 information bits and 8 tail bits for the convolutional coder. The maximum length of an access message body is 880 bits or 10 frames (200 ms). The minimum length of an access message is three frames (60 ms). A preamble consisting of 1–16 frames, each containing 96 0s, precedes each access message. The terminal learns the size of the preamble (PAM_SZ) from an *ACCESS PARAMETERS* message received on the paging channel. Thus, a transmission on an access channel covers a minimum of four frames (80 ms) and a maximum of 26 frames (520 ms).

To send a message on an access channel, a terminal chooses at random an access channel number ranging from 0 to ACC_CHAN. ACC_CHAN is a parameter of an *ACCESS PARAMETERS* message. The number of access channels monitored by the base station is ACC_CHAN + 1. The terminal then performs an access algorithm that determines the permitted time of transmission of an access message. Each transmission is referred to as a *probe*. After transmitting a message, the terminal waits for an acknowledgment to arrive on the paging channel. If the acknowledgment is not received within a timeout period (ACC_TMO) specified by the system (and transmitted in the *ACCESS PARAMETERS* message), the terminal pauses for a random time interval (backoff time) and transmits the message again, at increased power.

A likely reason for not receiving an acknowledgment is that at least one other terminal transmits at the same time on the access channel. The transmissions collide at the base station and the mutual interference prevents the base station from receiving more than one transmission in any slot. The random waiting time reduces the probability that the two terminals will transmit again in the same slot. The variable power levels increase the probability that one message will be received at higher power than interfering messages. If the power difference is sufficiently great, the base station will capture (receive accurately) the strongest message. Capture has a positive effect on access channel throughput [Goodman and Saleh, 1987].

The process continues until the terminal receives an acknowledgment or the number of probes reaches a limit specified by the system. The number of allowed probes is between 1 and 16. Each sequence of unsuccessful probes is an *attempt*. If an access attempt fails, the terminal can begin the process again, repeating the sequence of probes until the number of attempts (probe sequences) reaches a limit, which may be as high as 15. Finally, if all attempts fail, the terminal abandons the access procedure. Figure 6.14 is a flowchart of the IS-95 access protocol.

6.4.5 CDMA Traffic Channels

When a call is in progress, information moves between terminals and base stations over forward and reverse traffic channels. On a reverse traffic channel, the 42-bit channel identifier (see Figure 6.7), corresponding to a CDMA physical channel, is a combination of the 32-bit electronic serial number of the terminal (ESN in Table 6.1) and a fixed 10-bit field. The physical channel carrying a forward traffic channel is the 64-bit carrier in Figure 6.10 (a row of the 64×64 Walsh Hadamard matrix). Thus, a CDMA terminal always uses the same physical channel to carry a reverse traffic channel. The physical channel carrying a forward traffic channel is established by the system at the beginning of a call and it changes as the terminal moves through new cells.

Compared with other systems, IS-95 traffic channels have two distinguishing characteristics:

- They carry speech at variable bit rates ranging from 9,600 b/s, to 1,200 b/s, depending on an analysis of the input speech and on signaling activity.

- They carry a dynamic mixture of user information and network control information.

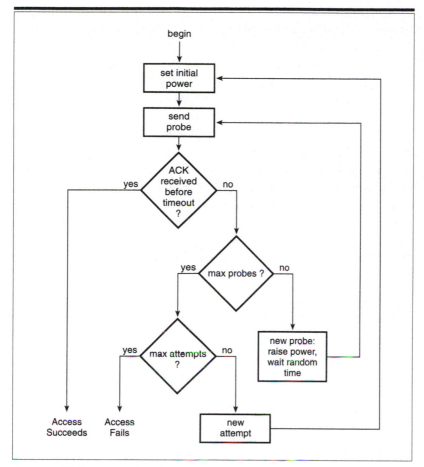

Figure 6.14 Access protocol.

Variable-Bit-Rate Speech Transmission

The CDMA speech coder is a version of CELP (code excited linear prediction), referred to as QCELP, where the Q refers to Qualcomm. The system performs encoding and decoding operations on 20 ms speech *frames*. In each 20 ms time interval, the encoder processes 160 samples of speech, obtained from an analog-to-digital converter operating at 8,000 samples/s at a terminal, or from a digital signal arriving at a switch. The coder represents speech at four bit rates: 8,000 b/s, 4,000 b/s, 2,000 b/s, and 800 b/s, producing 160, 80, 40, or 16 bits per 20 ms frame. This is the speech component of the information bits per frame (IBPF) in Tables 6.2 and 6.3. Table 6.4 describes in detail the composition of IBPF for each code rate.

In each frame, the QCELP speech coder generates ten linear prediction coding filter coefficients (see Section 9.7). It represents each coefficient with an accuracy of 4 bits (8,000 b/s coder), 2 bits (4,000 coder), or 1 bit (2,000 b/s coder and 800 b/s coder). The result is 40, 20, or 10 linear predictor bits per frame, as indicated in Table 6.4. Except at the 800 b/s rate, the speech coder performs a long-term prediction (pitch) analysis that generates two quantities: an estimated pitch period, quantized to 7 bits; and a pitch gain, quantized to 3 bits. At the 2,000 b/s rate, it performs this analysis once for the entire 20 ms frame. At 4,000 b/s it performs the pitch analysis in each of two 10 ms subframes, and at 8,000 b/s it performs the analysis four times per frame. The result is 40, 20, or 10 bits per frame of pitch information. The remaining speech bits in each frame comprise a vector quantizer representation of the excitation. Like the pitch analysis, the excitation analysis has a time resolution that depends on the code rate. At the highest rate, the speech coder introduces

Table 6.4 Contents of Traffic Channel Speech Frames

Data Rate (b/s)	1,200	2,400	4,800	9,600
Speech rate (b/s)	800	2,000	4,000	8,000
Speech content (bits per frame)	16	40	80	160
Filter coefficients (bits per frame)	10	10	20	40
Pitch parameters (bits per frame)	0	10	20	40
Excitation parameters (bits per frame)	6	20	40	80
Error-correcting code (bits per frame)	0	0	0	11
Frame content bit (bits per frame)	0	0	0	1
Information bits per frame (IBPF)	16	40	88	172
Error-detecting code (PBPF)	0	0	8	12

an error-correcting block code to protect the 18 bits in the 20 ms frame that have the strongest influence on speech quality. This block code adds 11 bits per frame to the 8,000 b/s speech coder output. At the two higher rates, the system also adds a cyclic redundancy check error-detecting block code to each speech frame to enable the receiver to monitor transmission quality. The frame content bit at the highest rate indicates whether the contents of the frame are derived entirely from the speech coder or whether the frame contains signaling messages, as described in the next subsection.

Variable-bit-rate speech coding serves two purposes. Relative to continuous, full-rate speech coding, it raises system capacity by reducing the average amount of interference that each transmitter causes to other communications. The encoder examines the contents of each 20 ms speech frame and determines the necessary coding rate. The most important part of this analysis is voice-activity detection. When the encoder determines that no speech is present, it reduces the rate in steps to 16 bits per frame (800 b/s). The effective transmission rate goes from 9,600 b/s to 1,200 b/s and the transmitter produces 1/8 of the interference energy of a full-rate transmission. Unlike the other systems described in this book, the CDMA system cannot interrupt the signal completely when there is no speech present. This is because of the time necessary to synchronize a CDMA receiver when a new signal arrives. In addition to reducing interference, variable-bit-rate speech coding allows the base station to multiplex signaling information with user information on a traffic channel, as described in the next subsection.

Signaling on CDMA Traffic Channels

To exchange network control information while a call is in progress, CDMA terminals and base stations interrupt or reduce the flow of speech information and insert messages into traffic channels. There are five modes of operation, summarized in Table 6.5. In the *blank-and-burst* mode, control messages completely replace the speech. In the three *dim-and-burst* modes, there is a mixture of speech information and control information in each 20 ms frame. With dim-and-burst transmission, speech quality is lower than it is with full-rate transmission, but higher than with blank-and-burst. When control messages are present, the traffic channel always operates at 9,600 b/s. Table 6.5 shows the composition of each frame in each mode of operation.

Table 6.5 Number of Bits per Frame (20 ms) in Full-Rate (9,600 b/s) CDMA Traffic Channels

Transmission Mode	Blank-and-Burst	Dim-and-Burst			Speech Only
Speech	0	16	40	80	171
Control message	168	152	128	88	0
Content indicator	4	4	4	4	1
Parity check	12	12	12	12	12
Coder tail bits	8	8	8	8	8
Information bits	192	192	192	192	192

The first content indicator bit distinguishes speech-only frames from other frames. The other three content indicator bits describe the exact mixture of speech and signaling information. The four possible rates of signaling information transfer are

$$168 \frac{b}{frame} \div 0.020 \frac{s}{frame} = 8,400 \text{ b/s (blank-and-burst)}$$

$$152 \frac{b}{frame} \div 0.020 \frac{s}{frame} = 7,600 \text{ b/s}$$

$$128 \frac{b}{frame} \div 0.020 \frac{s}{frame} = 6,400 \text{ b/s}$$ (6.15)

$$88 \frac{b}{frame} \div 0.020 \frac{s}{frame} = 4,400 \text{ b/s}$$

The corresponding speech rates are 0; 1,200 b/s; 2,400 b/s; and 4,800 b/s.

Power Control Subchannel

CDMA spectrum efficiency depends on precise control of the power radiated by each terminal. With a call in progress, a base station monitors the received power from each terminal and transmits power control commands to the terminal at a data rate of 800 b/s. Each bit transmitted in the power control subchannel commands a terminal either to increase or decrease its transmitter power by 1 dB. The base station multiplexes the power control signal with the mixture of speech and network control information flowing on a forward traffic channel. To do so, the base station inserts 16 power control bits into every frame transmitted on a forward traffic channel. This multiplexing takes place after the convolutional coding, repetition, and interleaving illustrated in Figure 6.10. At this point the signaling rate is 19,200 b/s. Every $1/800$ s $= 1.25$ ms, the multiplexer replaces 2 traffic channel code bits with 1 power control bit. (This causes the energy per bit in the power control subchannel to be twice as high as it would be with only 1 traffic channel code bit replaced.) The process is known as *puncturing* the convolutional code. Its effect is to weaken somewhat the error-correcting capability of the rate 1/2 convolutional code.

6.5 Messages

Like AMPS, IS-95 performs network control operations by exchanging messages between terminals and base stations on four types of logical control channels. The IS-95 paging and access channels correspond to the AMPS forward and reverse control channels, respectively. Corresponding to the forward and reverse voice channels in AMPS, IS-95 transmits signaling messages on forward and reverse traffic channels. These messages share the traffic channels with user information by means of blank-and-burst or dim-and-burst operations.

In addition to paging channels, access channels, and signaling information transmitted on traffic channels, Figure 6.13 indicates a pilot channel and a sync channel. The other information listed in Figure 6.13 is a power control signal. As described in Section 6.4.5, this signal is a continuous 800 b/s stream with each bit directing a terminal to raise or lower its power by 1 dB. The pilot channel has the most simple information format. It carries an unmodulated beacon signal that allows terminals in the local cell to synchronize their code generators. Terminals in nearby cells measure the strength of the pilot as part of CDMA handoff procedures. The sync channel broadcasts one message that conveys system information to all terminals in a cell.

Table 6.6 lists CDMA messages carried on the paging and access channels and Table 6.7 is the corresponding list for the forward and reverse traffic channels.

Table 6.6 Messages Carried on CDMA Paging and Access Channels

Access Channel	Paging Channel
Broadcast Messages	
	SYSTEM PARAMETERS
	ACCESS PARAMETERS
	NEIGHBOR LIST
	CDMA CHANNEL LIST
Call Management Messages	
ORIGINATION	CHANNEL ASSIGNMENT
PAGE RESPONSE	PAGE
	SLOTTED PAGE
	RELEASE*
	FEATURE NOTIFICATION
DATA BURST	DATA BURST
LOCAL CONTROL RESPONSE*	LOCAL CONTROL*
	REORDER*
	INTERCEPT*
	ABBREVIATED ALERT*
Authentication and Privacy Messages	
AUTHENTICATION CHALLENGE RESPONSE	AUTHENTICATION CHALLENGE
SSD UPDATE CONFIRM/REJECT*	SSD UPDATE
BASE STATION CHALLENGE*	BASE STATION CHALLENGE CONFIRM*
Mobility Management Messages	
REGISTRATION	REGISTRATION ACCEPTED/REJECTED*
	REGISTRATION REQUEST*

(continued)

Table 6.6 Messages Carried on CDMA Paging
and Access Channels *(Continued)*

Operations, Administration, and Maintenance Messages	
	LOCK UNTIL POWER CYCLED*
	UNLOCK*
	MAINTENANCE REQUIRED*
	AUDIT*
Other Messages	
MOBILE STATION REJECT *	
MOBILE STATION ACKNOWLEDGMENT	BASE STATION ACKNOWLEDGMENT

* Message classified as an "order" (see Section 6.5.1).

Table 6.7 Messages Carried on CDMA Traffic Channels

Reverse Traffic Channel	Forward Traffic Channel
Call Management Messages	
CONNECT*	ALERT WITH INFORMATION*
ORIGINATION CONTINUATION	
FLASH WITH INFO	FLASH WITH INFO
RELEASE*	RELEASE*
DATA BURST	DATA BURST
SEND BURST DTMF*	SEND BURST DTMF*
CONTINUOUS DTMF TONE*	CONTINUOUS DTMF TONE*
LOCAL CONTROL*	LOCAL CONTROL*
LOCAL CONTROL RESPONSE*	
SERVICE OPTION REQUEST*	SERVICE OPTION REQUEST*
SERVICE OPTION RESPONSE*	SERVICE OPTION RESPONSE*
	SERVICE OPTION CONTROL*

(continued)

Table 6.7 Messages Carried on CDMA Traffic Channels *(Continued)*

Authentication and Privacy Messages	
AUTHENTICATION CHALLENGE RESPONSE	AUTHENTICATION CHALLENGE
SSD UPDATE CONFIRM/REJECT*	SSD UPDATE
BASE STATION CHALLENGE*	BASE STATION CHALLENGE CONFIRM*
PARAMETER UPDATE CONFIRM*	PARAMETER UPDATE*
	MESSAGE ENCRYPTION MODE*
LONG CODE TRANSITION REQUEST*	LONG CODE TRANSITION REQUEST*
LONG CODE TRANSITION RESPONSE*	LONG CODE TRANSITION RESPONSE*

Radio Resources Management Messages	
HANDOFF COMPLETION	HANDOFF DIRECTION
REQUEST ANALOG SERVICE*	ANALOG HANDOFF DIRECTION
POWER MEASUREMENT REPORT	POWER CONTROL PARAMETERS
	NEIGHBOR LIST UPDATE
PILOT STRENGTH MEASUREMENT	PILOT MEASUREMENT REQUEST*

Mobility Management Message	
	MOBILE STATION REGISTERED

Operations, Administration, and Maintenance Messages	
STATUS	STATUS REQUEST*
	AUDIT*
	MAINTENANCE*
	LOCK UNTIL POWER CYCLED*
	MAINTENANCE REQUIRED*

Other Messages	
MOBILE STATION REJECT *	
MOBILE STATION ACKNOWLEDGMENT	BASE STATION ACKNOWLEDGMENT

* Message classified as an "order" (see Section 6.5.1).

6.5.1 Message Structure

CDMA messages have the format illustrated in Figure 6.15. Each message begins with an 8-bit (one octet) message-length field that contains the total message size measured in octets. The following 8 bits indicate the message type. This is followed by a 7-bit acknowledgment field as described in the next subsection. The remainder of the message consists of the message content followed by a cyclic redundancy check (error-detecting code) field, which contains either 30 bits (sync, paging, access channels) or 16 bits (traffic channels).

Layer 2 Acknowledgments

With the exception of the broadcast messages listed in Table 6.6 and the 5YNC CHANNEL message (Table 6.8), the CDMA system inserts into each message at least 7 bits of information for data link layer control at layer 2 of the OSI protocol (see Section 9.10). The purpose of the data link layer is to promote accurate information transfer between terminals and base stations. In CDMA, this goal is advanced by means of acknowledgments and selective-repeat retransmission of messages containing errors. The layer 2 information in every CDMA message consists of a 3-bit message sequence field (MSG_SEQ), a 3-bit acknowledgment sequence field (ACK_SEQ), and 1 acknowledgment required (ACK_REQ) bit. This bit indicates whether the current message requires a layer 2 acknowledgment. When a terminal receives a message that requires an acknowledgment, it inserts the message sequence number of the received message into the ACK_SEQ field of a message transmitted to the base station. It uses the access channel to acknowledge paging channel messages and the reverse traffic channel to acknowledge forward traffic channel messages.

When the base station transmits a message, it expects an acknowledgment to arrive within some specified time interval. If no acknowledgment

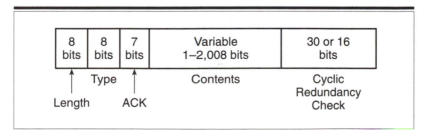

Figure 6.15 CDMA message format.

arrives within this time, the base station assumes that there has been a transmission failure and it retransmits the message, using the same MSG_SEQ number that appeared in the previous transmission of the message. With this selective-repeat procedure, messages arrive at the terminal out of order. The MSG_SEQ numbers enable the terminal to restore the original time sequence of messages. The system operates in a similar manner when messages are transmitted from the terminal to a base station. The paging channel acknowledges messages received on the access channel, and the forward traffic channel acknowledges messages received on the reverse traffic channel.

Orders

As indicated by asterisks (*) in Tables 6.6 and 6.7, about half of the CDMA messages are *orders*. On each logical channel, all orders have the same 8-bit message type. For example, message type 00000111 on the paging channel specifies an order. On the traffic channels, the message type for an order is 00000001. The specific order is indicated by a 6-bit order code and an 8-bit order qualification code appearing in the message content. Orders, for the most part, are simpler than other messages. Many are confirmations of messages. Others are simple directions, such as RELEASE. A minority of the orders carry numerical parameters.

6.5.2 Acknowledgment Messages

In addition to layer 2 acknowledgments and retransmissions described in Section 6.5.1, IS-95, like other systems, provides for signaling acknowledgments at higher protocol layers. These acknowledgments are carried in separate messages listed in Tables 6.6 and 6.7. While a layer 2 acknowledgment confirms the reception of a legitimate bit sequence, an acknowledgment message confirms the logic of the content of an original message and indicates whether a terminal or base station is prepared to perform a function prescribed in the signaling message. Some of the IS-95 acknowledgment messages are labeled "acknowledgments." Tables 6.6 and 6.7 label others as "confirmations" or "responses." The contents of one of these messages, and the actions taken when there is a signaling system problem that is not detected at layer 2, depend on the specific control functions that the message supports.

6.5.3 Message Content

The sync channel carries one message with the principal purpose of synchronizing the random-number generator in Figure 6.7. Derived from a 42-stage shift register, this random-number generator produces the digital carrier to be used for reverse traffic channel transmissions. Table 6.8 displays the contents of the 162-bit sync channel message. The PN_OFFSET, long code state, and system time fields contain the information required by the terminal to synchronize its random-number generator. Other information in the message include the system and network identifiers (Table 6.1) and the bit rate of paging channels at the present base station. The minimum protocol revision field informs a terminal of whether it is properly configured to use this base station. Terminals with configurations characterized by lower protocol revision numbers are unable to obtain service.

Tables 6.6 and 6.7 classify the paging channel, access channel, and traffic channel messages according to network control categories. Many of them are counterparts of NA-TDMA messages, especially in the categories of call management, authentication, and mobility management. Reflecting the unique characteristics of the CDMA physical layer, there are broadcast messages, privacy messages, radio resources management

Table 6.8 Contents of a Sync Channel Message

Bit Position	Information
1–8	message type 00000001
9–16	protocol revision
17–24	minimum protocol revision for a terminal
25–39	system identifier (SID)
40–55	network identifier (NID)
56–64	PN_OFFSET
65–106	long code state
107–142	system time obtained from global positioning system
143–157	relationship of local time to system time
158–159	paging channel bit rate 4,800 b/s or 9,600 b/s
160–162	not used for information (000)

messages, and maintenance messages that have no counterparts in earlier systems. Among the messages carried on paging channels, the *SYSTEM PARAMETERS* message carries information on mobility management procedures to be used in the current cell (Section 6.6.2), and information on the maintenance of channel sets used in soft handoff procedures (Section 6.6.1). The *ACCESS PARAMETERS* message is similar to an AMPS *GLOBAL ACTION* message. It broadcasts the number of access channels in use in the current cell, parameters of the access protocol (Section 6.4.4), and an indication of whether terminals are required to respond to an authentication challenge when transmitting a message on an access channel. The *CDMA CHANNEL LIST* message broadcasts the radio-frequency carriers (AMPS channel numbers) of all CDMA signals in the present cell.

The *NEIGHBOR LIST* message broadcasts to all terminals in a cell numerical values of the 9-bit parameter, PN_OFFSET, associated with nearby base stations. This parameter controls the timing (see Figure 6.12) of the signals transmitted by each base station. By introducing different time delays to different base stations in an area, CDMA makes it possible for a terminal to separate the signal arriving from the local base station from signals arriving from interfering base stations. The information in a *NEIGHBOR LIST* message helps terminals that do not have calls in progress to move from one cell to another and to promptly lock their receivers to a sync channel and then to a paging channel. For terminals with calls in progress, the *NEIGHBOR LIST* message plays an important role in CDMA soft handoff procedures. Base stations on the neighbor list are likely to become candidates for a handoff (see Section 6.6.1). After a terminal moves to a traffic channel, the system can use a *NEIGHBOR LIST UPDATE* message on the forward traffic channel to modify the list of likely candidate base stations.

A *LOCK UNTIL POWER CYCLED* order instructs a specific terminal to turn off its transmitter until it receives an *UNLOCK* order or the user turns off the power in the terminal and turns it on again. The presence of this order reflects the sensitivity of a CDMA system to terminals that transmit too much power. Such terminals can cause harmful interference to a large number of other transmissions. When the system detects a problem at a terminal it can also send a *MAINTENANCE REQUIRED* order, which indicates to the user, by means of a displayed message, that the terminal requires service.

Other paging channel messages with nomenclature unique to CDMA include:

- an *ABBREVIATED ALERT* order, which causes the terminal to emit an audible tone before assigning a voice channel;

- a *FEATURE NOTIFICATION* message, which can deliver special services such as caller identification, or a voice-mail status indication;

- a *DATA BURST* message (also available on the access and traffic channels), which moves a sequence of alphanumeric characters to and from terminals; and

- a *SLOTTED PAGE* message, which carries more than one paging notification in each 80 ms time slot (see Section 6.4.3) of a paging channel.

On the traffic channels, the *LONG CODE TRANSITION* privacy orders are unique to CDMA. They allow a terminal and a base station to move between a public long code and a private code. The public code is produced by a published version of the random-number generator in Figure 6.7. The private code is produced by a random-number generator configuration that is stored in a terminal and known to the system operator. By using the private code, the system inhibits eavesdropping on the call in progress. The *MESSAGE ENCRYPTION MODE* order specifies whether certain network control messages will be transmitted in encrypted form.

The radio resources management messages transmitted on the traffic channels coordinate the CDMA soft handoff and forward link power control operations described in Section 6.6.1. The *NEIGHBOR LIST UPDATE* order and the *PILOT MEASUREMENT REQUEST* and *PILOT STRENGTH MEASUREMENT* messages prepare the system for a handoff. The *POWER CONTROL PARAMETERS* and the *POWER MEASUREMENT REPORT* messages provide a base station with information used to establish the power of signals transmitted on forward traffic channels.

6.6 Network Operations

Although many CDMA network control procedures resemble those of other systems, the radio resources management operations of IS-95 include innovations uniquely matched to the requirements of the CDMA transmission technology. Section 6.6.1 describes the most important of these innovations, transmitter power control and soft handoff. Sections 6.6.2 through 6.6.4 cover the other categories of network operations.

6.6.1 Radio Resources Management

Theoretical studies reveal that in order to achieve high capacity [Holtzman, 1991; Viterbi, 1995], a CDMA system has to maintain a narrow range of received power levels among the signals arriving at a base station. When there is a disparity in received powers, the strong signals interfere excessively with the weaker ones and unduly limit the total number

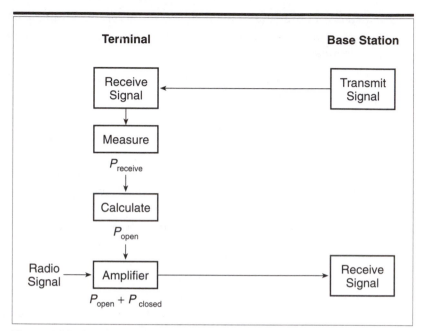

Figure 6.16 Open-loop power control.

of simultaneous transmissions consistent with a signal-quality objective (Section 6.3.3). In order to confine the received signal powers to a narrow range, IS-95 performs a combination of *open-loop* and *closed-loop* operations to control the power of the transmitters at mobile terminals (Figure 6.9).

Power Control at Terminals

To perform open-loop power control, a terminal measures the strength of the received pilot signal associated with the forward traffic channel. A strong received signal, suggesting that the terminal and base station are in close proximity, causes the terminal to transmit at low power. Conversely, a weak pilot signal indicates a long distance between terminal and base station and causes the terminal to transmit at high power. Specifically, the open-loop component of the transmitter power is given by:

$$P_{\text{open}} \text{ (dBm)} = -P_{\text{receive}} \text{ (dBm)} + P_{\text{target}} \text{ (dB)}. \qquad (6.16)$$

P_{receive} is the measured power level of the received signal at the terminal and P_{target} is the difference between the transmitted power level at the base station and the desired received signal level at the base station. The terminal calculates P_{target} on the basis of information received from the base station in an *ACCESS PARAMETERS* message:

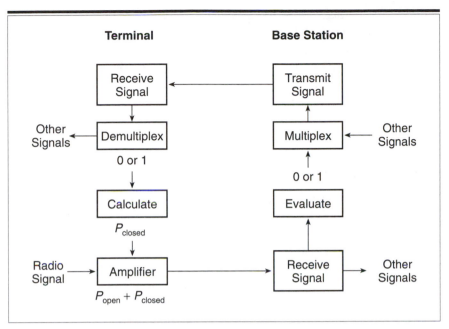

Figure 6.17 Closed-loop power control.

$$P_{target} \text{ (dB)} = -73 \text{ dB} + P_{control}(\text{dB}), \qquad (6.17)$$

where $P_{control}$ is derived from the *ACCESS PARAMETERS* message. With $P_{control} = 0$, the received power at the base station is 73 dB lower than the transmitted power at the base station, provided forward link and reverse link path attenuations are identical. Figure 6.16 indicates the operations performed at the base station and terminal to perform open-loop power control.

Using information obtained from the base station, a terminal combines this open-loop power control with closed-loop control. The closed-loop control produces a power adjustment P_{closed} (dB). The transmitted power at the terminal is

$$P_{transmit} \text{ (dBm)} = P_{open} \text{ (dBm)} + P_{closed} \text{ (dB)}. \qquad (6.18)$$

To control the closed-loop power adjustment, the base station decides, every 1.25 ms, whether the power received from the terminal is too high or too low. It transmits this decision as 1 bit in the power control subchannel (see Section 6.4.5). A 0 indicates that the received power is too low, and a 1 indicates that the received power is too high. On receiving this signal, the terminal modifies P_{closed} by increasing P_{closed} by 1 dB

(0 received) or decreasing P_{closed} by 1 dB (1 received). Thus, every 1.25 ms, the terminal calculates

$$P_{closed} \text{ (dB)} = P_{closed} \text{ (dB)} + 1 \text{ dB}; \ 0 \text{ received, or}$$

(6.19)

$$P_{closed} \text{ (dB)} = P_{closed} \text{ (dB)} - 1 \text{ dB}; \ 1 \text{ received.}$$

Figure 6.17 illustrates the closed-loop power control.

Power Control at Base Stations

IS-95 specifies an exchange of information between a base station and a terminal that enables the base station to control the power of forward traffic channel transmissions. However, the procedures for using this information to establish transmitter power settings are outside the scope of IS-95. Each base station contains its own algorithms for regulating forward link power. The base station uses a *POWER CONTROL PARAMETERS* message to request information from the terminal about the quality of the forward traffic channel signal. The terminal responds with a *POWER MEASUREMENT REPORT* message that contains the strength of the received forward traffic channel signal. A terminal in a soft handoff state receives two or more forward traffic channel signals simultaneously and reports the strength of each one. The *POWER MEASUREMENT REPORT* message also reports on the number of detected errors in the forward traffic channel signal. The base station uses the data in this message to determine the power of the forward traffic channel signal.

Soft Handoff

Soft handoff is the most novel IS-95 network control operation. As a terminal moves from one cell to another, it communicates simultaneously with the base stations in both cells, as indicated in Figure 6.1. The terminal plays an active role in establishing this communication. In common with NA-TDMA (Chapter 5) and GSM (Chapter 7), CDMA handoff is *mobile assisted*, with each terminal performing measurements that influence handoff decisions. CDMA handoff is exceptional in being *mobile initiated*. Terminals in the other systems simply report signal measurements to their base stations. A CDMA terminal analyzes the measurements and informs the system when it detects that a handoff might be necessary. As in NA-TDMA and GSM, CDMA handoff is *switch controlled*. The switch makes handoff decisions and assigns new physical channels. In a soft handoff state, a terminal receives signals on two different physical channels, one associated with each of the participating base stations. Recall that each terminal has its own unique physical channel, determined by the 32-bit electronic serial number (ESN). The terminal uses this physical channel for all reverse traffic channel transmissions. During soft handoff, two different base stations assign correlators to receive signals on this physical channel.

Soft handoff requires two sets of signal-processing functions: measurement and diversity reception. It also requires an exchange of messages on the forward and reverse traffic channels. Terminals and base stations use the collection of correlators in each IS-95 receiver (Figure 6.5) to perform the signal-processing functions. In common with the handoff procedures in NA-TDMA and GSM, CDMA soft handoff is mobile assisted. Using instructions received in NEIGHBOR LIST and NEIGHBOR LIST UPDATE messages, the terminal examines pilot signals transmitted in neighboring cells and reports the received power of these pilots.

A terminal dedicates at least one correlator, referred to as a *searcher*, to performing the measurements. The other correlators participate in diversity reception. When a receiver is tuned to only one base station, the correlators comprise a RAKE receiver. They process signals from different signal paths and combine the signals to form a single output bit stream, as described in Section 6.3. During soft handoff, at least one correlator receives the forward traffic channel signal from the original base station and at least one correlator receives the same signal from the new base station. At the conclusion of the soft handoff, all active correlators receive the digital traffic channel signal from the new base station. With respect to the reverse traffic channel, two base stations receive the same signal during soft handoff. Each base station demodulates the signal, using multiple correlators, and sends the result as a sequence of digital frames to the switch. For each digital frame, a channel decoder at the switch determines which arriving signal has the lower error rate and sends this signal to the source decoder.

The use of dedicated searching hardware in each terminal for examining signals in neighboring cells marks one departure from NA-TDMA mobile-assisted handoff. Another difference is the way in which a terminal reports signal strength measurements in IS-95. Instead of reporting these measurements continuously, a CDMA terminal sends measurement reports in response to observed changes in signal strengths. To perform its measurement and reporting tasks, a terminal maintains four lists of base stations, with each base station characterized by a 9-bit PN_OFFSET.

Recall that the radio modulators (Table 6.3) in all CDMA base stations generate the same pair of binary sequences (I-channel sequence and Q-channel sequence) with a 2/75 second repetition period. However, each base station transmits these sequences with its own time offset relative to a system reference. This PN_OFFSET identifies each base station in a system. To measure the strength of the pilot signal from a base station, the terminal programs the searching correlator with the PN_OFFSET of that base station.

To participate in the soft handoff procedure, each terminal segregates the set of PN_OFFSET values (and implicitly the set of base stations) in a system into four categories:

- The *active list* contains base stations currently used for traffic channel transmissions. In a soft handoff condition, there is more than one base station on this list.

- The *candidate list* consists of base stations classified by the terminal, on the basis of measured signal quality, as available for traffic channel transmissions.

- The *neighbor list* is a set of nearby base stations that could soon be available for handoff.

- The *remaining list* contains the base stations that are not in any of the other categories.

The searching correlators in the terminal monitor the signal strengths of all pilots in the active, candidate, and neighbor lists. When these measurements suggest that a handoff might be necessary, the terminal initiates the handoff by sending a PILOT STRENGTH MEASUREMENT message. This message reports signal quality calculated from measurements of the output of the searcher element. Several events, each characterized by a change in signal measurements, can stimulate the terminal to change the status of a base station. Examples are:

A. The strength of an active set signal drops below a threshold.

B. The strength of a neighbor set signal rises above a threshold.

C. The strength of a candidate set signal exceeds the strength of an active set signal by some margin.

D. The strength of a candidate set signal drops below a threshold.

Event B causes the terminal to move the relevant base station from the neighbor set to the candidate set. Event D causes the terminal to move the base station from the candidate set to the neighbor set.

On receiving a *PILOT STRENGTH MEASUREMENT* message, the system decides whether to begin a handoff or complete a soft handoff in progress. To control either of these operations, a base station commands a terminal to change the list of pilots in the active set. The command appears in a *HANDOFF DIRECTION* message that conveys the new members of the active set. When a soft handoff begins, the new active set contains the PN_OFFSET of the original base station and the PN_OFFSET of the new base station. The message that conveys these parameters also contains the forward traffic channel number to be used at the new base station. On receiving this message, the terminal then tunes one or more correlators to this forward traffic channel. When soft handoff ends, the *HANDOFF DIRECTION* message specifies an active set with only one base station. On receiving this message, the terminal ends its reception of forward traffic channel information from the base station that has been removed from the active set.

Figure 6.18 shows the operations that take place as a terminal moves from one cell to another. In order to simplify the message flow lines, this figure contains two columns (at opposite ends) for the same switch. The sequence of operations begins with the terminal in communication with the old base station. In processing its measurements, the terminal recognizes that a signal from a neighbor base station is sufficiently strong to support reliable communications. It notifies the old base station in a *PILOT STRENGTH MEASUREMENT* message. The old base station relays this information to the switch, which responds by setting up a soft handoff. To do so, it instructs the old base station to send a *HANDOFF DIRECTION* message to the terminal. This message contains the forward traffic channel number to be used at the new base station. The switch also directs the new base station to transmit user information on this forward traffic channel and to receive the reverse traffic channel signal from the terminal. After the terminal acknowledges the *HANDOFF DIRECTION* message with a *HANDOFF COMPLETE* message, the terminal is in communication with both base stations. Both of them receive the same user information from the switch and send it, on their assigned physical channels, to the terminal. The terminal programs correlators to receive both signals and combines the results prior to digital demodulation. Each base station demodulates the reverse traffic channel signal separately and sends the results to the switch, which selects one of the two digital signals.

After a while, the terminal recognizes that the strength of the signal from the old base station has dropped below a threshold. It informs the system by transmitting a *PILOT STRENGTH MEASUREMENT* message. When the signal strength information reaches the switch, the switch decides to remove the old base station from the call. It does so by sending messages to both base stations, which transmit the necessary information to the terminal in a *HANDOFF DIRECTION* message. The terminal responds by releasing

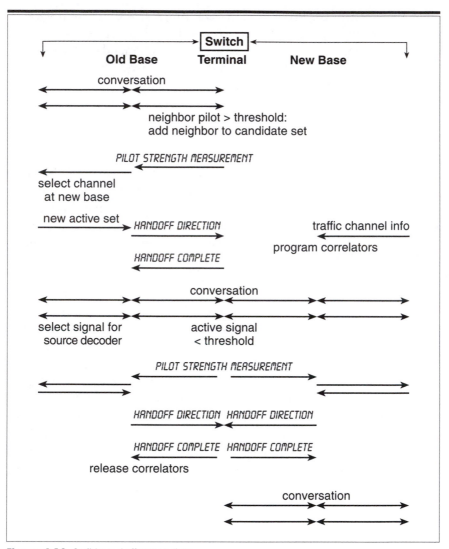

Figure 6.18 Soft handoff procedure.

the correlators tuned to the signal at the old base station. The old base station similarly releases the correlators tuned to the reverse traffic channel. After the terminal transmits a *HANDOFF COMPLETE* message, the call travels only through the new base station.

This discussion has referred to soft handoff procedures that involve two base stations. Although this is the most common form of soft handoff, IS-95 admits the possibility of a terminal communicating with up to six base stations simultaneously. While the call is in a soft handoff state, all base stations perform closed-loop power control. Each transmits its own signal on a power control subchannel. The terminal responds at each instant by examining all of the power control signals. If all of them command a 1 dB power increase, the terminal obeys this command. If at least one base station commands a 1 dB decrease, the terminal decreases its power by 1 dB.

Other Types of Handoff

As a dual-mode system, IS-95 is capable of transferring a call from a CDMA traffic channel to an analog voice channel. This type of handoff is required when a terminal moves to a cell that is not yet equipped for CDMA operation. To perform this handoff, a base station transmits an *ANALOG HANDOFF DIRECTION* message to the terminal. The system is not capable of transferring a call in progress from an analog voice channel to a CDMA digital traffic channel. In addition to the soft handoff described in the previous subsection, IS-95 also anticipates the need for *hard handoff*. This is necessary when the system directs a call from one CDMA frequency band to another or the terminal moves to a cell that is not controlled by the same switch as the previous cell. In a hard handoff, the terminal breaks its connection with the previous base station and then synchronizes to the new base station before establishing voice communications with the new base station.

6.6.2 Mobility Management

In common with other cellular systems, IS-95 provides for a terminal to send *REGISTRATION* messages to the system in order to facilitate efficient paging when the system receives a call directed to the terminal. IS-95 specifies five different types of autonomous registration, each characterized by the event that causes a terminal to transmit a *REGISTRATION* message on the access channel. The *SYSTEM PARAMETERS* message on the paging channel indicates to terminals which types of registration are in effect. It also contains numerical registration parameters, such as the local registration zone identifier, the time limit for timer-based registration, and the distance limit for distance-based registration. Table 6.9 lists the five types of autonomous registration and the events that cause a terminal to register.

Each *REGISTRATION* message contains, in addition to the identity of the terminal, information about the paging channel time slots monitored by a

Table 6.9 IS-95 Registration Modes

Registration Type	Event Triggering Registration
power up	subscriber turns on the terminal
power down	subscriber turns off the terminal
timer	elapsed time since previous registration exceeds a limit
distance	distance between present base station and base station that received previous registration exceeds a limit
zone	terminal enters a new registration zone

terminal operating in sleep mode (see Section 6.4.3) in order to conserve battery power.

6.6.3 Authentication and Privacy

IS-95 incorporates the authentication technology specified for NA-TDMA (see Section 5.6.1). It also specifies the same methods of encrypting user information that NA-TDMA specifies. In addition, IS-95 includes a privacy technique unique to a CDMA system. This technique allows each user to operate with a private long code mask (Figures 6.7 and 6.10). Like the encryption A-key, the private long code mask is stored in the memory of a CDMA telephone and in a secure location managed by the network operator. The private long code mask is more secure than the public long code mask described in Section 6.3.1. The public long code mask contains the electronic serial number, which is transmitted without protection when the terminal operates in the analog (AMPS) mode. After a call is set up, either the terminal or the system can initiate operation with a private long code mask by transmitting a *LONG CODE TRANSITION* order.

6.6.4 Operations, Administration, and Maintenance

With signals in all cells covering a bandwidth of 1.23 MHz, CDMA systems are especially vulnerable to malfunctions in individual terminals. For example, a terminal radiating excessive power would cause harmful interference to dozens of conversations. To protect a system against this possibility, IS-95 specifies messages that a system can send on paging channels and forward traffic channels in order to stimulate corrective action at a terminal. A *LOCK UNTIL POWER CYCLED* message disables a terminal's transmitter until the user turns off the terminal power and turns it on again. A *MAINTENANCE REQUIRED* message causes the terminal to inform the user that there is a problem that requires attention.

6.7 Status of CDMA

After several years of product development and field testing, CDMA systems began commercial service in the United States in 1996. The first commercial systems were dual-mode cellular systems in the 850 MHz AMPS band. At the end of 1996, CDMA systems began commercial operations in the 1,900 MHz PCS bands. Cellular CDMA systems also operate in Hong Kong and Korea.

Technology adoption decisions announced by network operators make it clear that pace of CDMA service expansion will accelerate in 1997 and subsequent years. Although the initial implementations used the 9,600 b/s variable-rate speech coder described in this chapter, the preference in the United States in 1997 is for implementations that include a coder operating at 14,400 b/s [TIA, 1996c]. This coder delivers improved voice quality at the expense of lower spectrum efficiency (Section 6.3.3).

In 1997, work in progress on technology enhancement includes development of equipment that conforms to a standard for digital data transmission over CDMA channels [TIA, 1995b], and standards creation for short message services [TIA, 1996b] and packet data communications [Garg, Smolik, and Wilkes, 1997: Section 10.6]. There is also activity in the research community on transmitting signals at data rates up to 64 kb/s over IS-95 channels and multiplexing on the same radio carrier signals from a variety of sources with different bit rates and different error requirements.

Review Exercises

1. Describe some differences between the CDMA physical channels used for forward-direction transmissions and physical channels used in the reverse direction.

2. What is the purpose of the timing offset in the pseudo-noise sequences of base station radio modulators?

3. How do CDMA base stations use signals received from the global positioning system? Why is it important that all base stations operate in synchronism?

4. Why does the CDMA system employ closed-loop power control in reverse-direction transmissions?

5. When terminal T is in a soft handoff condition communicating simultaneously with base station A and base station B, describe the nature of the four physical channels in use (T to A, T to B, A to T, and B to T). Describe how these physical channels are selected and how the network elements learn their identity.

6. How does variable-bit-rate speech transmission help the performance of the CDMA system?

7. The highest bit rate of the QCELP coder is 9,600 b/s. In what situations does the system transmit speech at lower bit rates?

8. Why is it useful for a CDMA terminal to acquire "neighbor list" information? How does a terminal use the information acquired in NEIGHBOR LIST and NEIGH-BOR LIST UPDATE messages?

9. What is an advantage of sending an ABBREVIATED ALERT message on a paging channel relative to sending an ALERT message on a forward traffic channel?

10. What is an advantage of open-loop power control over closed-loop power control? Why is open-loop power control unable to maintain a narrow range of received power levels at a base station?

CHAPTER SEVEN

GSM: Pan-European Digital Cellular System

7.1 Background and Goals

The Pan-European digital cellular system traces its origins to 1982, when analog cellular services were in their earliest stages of commercial deployment. At that early date, European authorities anticipated the long-term potential of mobile communications and stimulated CEPT, the Conference of European Postal and Telecommunications administrations, to study the creation of a mobile telephone standard to be adopted throughout Western Europe. CEPT responded by forming the *Groupe Special Mobile* (Special Mobile Group). Group members used the initials *GSM* to refer to their project. Eventually, GSM produced a compatibility specification that was adopted as a European standard. The system embodied in the standard acquired the name *Global System for Mobile Communications*. During the GSM development period in the 1980s, European economic integration proceeded rapidly; the idea of a telecommunications system that could be used conveniently by travelers throughout Europe was an exciting symbol of this integration. A continental standard for mobile telephones stands in strong contrast to fixed telephone networks in Europe, which have different dialing procedures and charging systems in each country. Similar disparities exist in European analog cellular systems, which have five incompatible analog air interfaces scattered around the continent [Mouly and Pautet, 1992].

In this telecommunications environment of incompatible interfaces, the most important technical goal of GSM was full roaming in all countries. Another goal was to accommodate diverse service plans to conform to the separate needs and policies of participating countries. These characteristics of GSM make it possible for a subscriber to carry one telephone throughout Europe, initiating and receiving phone calls in all locations, without the burden of learning new dialing codes every time he crosses a national boundary. Calling features and charges reflect the service plan of the subscriber's home service provider. As an incentive for the diverse participants in GSM to reach an agreement, authorities made new frequency bands available for Pan-European operation with the requirement that all transmissions in these bands conform to a single standard. GSM reached a major milestone in 1987 when all participants agreed on the framework of a compatibility specification. Two principal technologies in the specification are an air interface based on hybrid frequency-division/ time-division multiple access (Section 9.1) and infrastructure communications based on Signaling System Number 7 (Section 9.12). Over the subsequent three years, GSM produced a complete standard that became the responsibility of ETSI, the European Telecommunications Standards Institute. In addition, network operators signed a memorandum of understanding that specifies how GSM systems are brought into commercial service. The memorandum of understanding also governs the business arrangements between different operators of GSM networks.

GSM is a comprehensive standard. Like the other standards covered in this book, it specifies the air interface that links terminals and base stations. However, it is unique in also prescribing open interfaces between infrastructure network elements, most notably between base stations and switches. The development of GSM reflects a remarkable cooperative effort undertaken over many years by dozens of people from fifteen countries. Many GSM innovations were later embodied in other systems. Two examples of these innovations are location-based mobility management and mobile assisted handover.[1]

Although there are several similarities between GSM and the North American digital cellular systems described in Chapters 5 and 6, GSM stands apart in two respects. It is a purely digital system. There is no provision for dual-mode operation with an analog cellular system. This difference reflects the contrast between the diversity of analog cellular systems in Europe and the ubiquitous deployment of AMPS systems in North America. Another difference is the large number of network interfaces specified by GSM, in contrast with the CDMA standard and the NA-TDMA standard, which specify only the air interface. The GSM open interfaces reflect the major influence of network operators in the development of GSM. Open interfaces favor network operators by giving them flexibility in procurement. By contrast, equipment vendors take the lead

---

[1] The term *handover*, used in all GSM documentation, is a synonym for *handoff*, used in North American cellular standards. This chapter follows the GSM convention of referring to *handover*.

in the creation of North American standards. Compared with service providers, equipment vendors have a stronger tendency to favor proprietary interfaces.

Although the principal goal of GSM is international roaming, the project formally adopted a broad set of aims, which included

- full international roaming,
- provision for national variations in charging and rates,
- efficient interoperation with ISDN systems,
- signal quality better than or equal to that of existing mobile systems,
- traffic capacity higher than or equal to that of present systems,
- subscriber costs lower than or equal to those of existing systems,
- accommodation of non-voice services, and
- accommodation of portable terminals.

GSM adopted this ambitious list of objectives in 1985. As the standardization proceeded, it became clear that achieving them fully would not be consistent with the service introduction date of 1991 approved by the initial GSM network operators. To reconcile the performance and cost objectives with early deployment, GSM decided that the standard would evolve through a set of "phases." This decision implicitly added the goals of early deployment and adaptability (Section 2.3.21) to the list. The Phase 1 GSM specifications were divided into more than 100 sections, with a total length of 5,320 pages. Phase 2 specifications have been developed section by section in the mid-1990s. The main goal of Phase 2 is to enrich the set of information services available to GSM subscribers [Mouly and Pautet, 1995].

The remainder of this chapter is a description of the principal properties of GSM, as defined in Phase 1 of the specifications. The services specified in Phase 1 include

- telephony with some special features,
- emergency calls,
- data transmission at rates up to 9,600 b/s, and
- a short message service for transmitting up to 160 alphanumeric characters between terminals and a network.

Phase 2 adds additional non-voice services and enriched telephony features.

To present the salient features of GSM, this chapter follows the framework established in Chapter 2 and used throughout this book. The focus is on communications across the air interface. For more details, readers can refer to the excellent book by Mouly and Pautet, *The GSM System for Mobile Communications,* which is a 700-page tutorial on GSM [1992]. Of course, the ultimate authority is the GSM standard, but that consists of more than one hundred documents published by ETSI, with a total length of more than 5,000 pages.

7.2 Architecture

The GSM network architecture reflects a strong influence of ISDN (Section 9.11). Figure 7.1 displays a large set of standard network elements with interfaces between the elements designated by letters. The nomenclature U_m for the air interface is taken from the ISDN U interface within the subscript "m" appended to denote "mobile."

The GSM terminology for the three essential network elements is *mobile station* (terminals), *base station,* and *mobile switching center* (switches). In

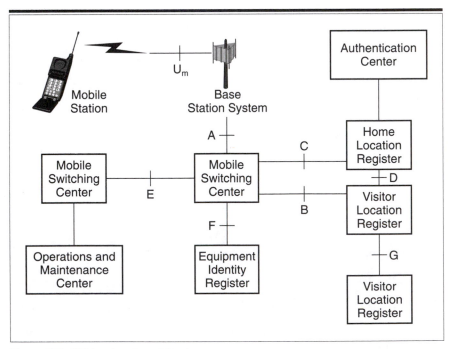

Figure 7.1 GSM network architecture.

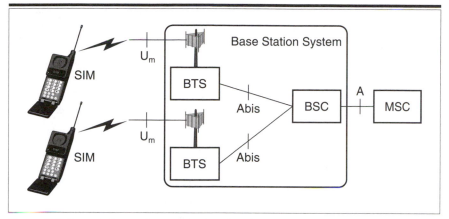

Figure 7.2 Base station system and subscriber identity module installed in mobile station.

addition, GSM specifies three databases: *home location registers* (HLR), *visitor location registers* (VLR), and *equipment identity registers* (EIR). The HLR and VLR are innovations essential to fulfilling the principal aim of GSM: full roaming in all service areas. Although this architecture appears earlier in this book (in Chapter 4 on North American intersystem operations [TIA, 1991]), it originated with GSM. In keeping with the preference of the GSM service industry for a large number of open interfaces, GSM specifies cellular base stations in greater detail than any other system. Thus, Figure 7.1 indicates a base station system, which is shown in Figure 7.2. Figure 7.2 contains two elements: a *base transceiver station* (BTS) and a *base station controller* (BSC), connected by a standard interface, designated *Abis,* suggesting an addition to the A interface. This specification reflects trends in the design of cellular hardware to serve small cells. In contrast to the original cellular configuration, with high-power transmitters connected to antennas 50–60 m above the ground, microcells transmit at low power with antennas on the order of 10 m above the ground, mounted on the sides of buildings or on lampposts. Because it is desirable to reduce to the greatest extent possible the cost and size of these installations, equipment manufacturers separate the essential radio equipment from the network control components of a base station. This partition of functions into separate pieces of equipment is reflected in the GSM designation of distinct *base transceiver stations* (BTS), consisting primarily of radio equipment, and *base station controllers* (BSC), that perform network control operations and signal processing functions. Typically, one BSC controls several BTS.

Another important GSM innovation appears in every mobile station. GSM specifies that every terminal contain a *subscriber identity module* (SIM). The SIM is a removable card that stores essential subscriber information, including identification numbers, details of the subscriber's service plan, and abbreviated dialing codes selected by the subscriber. The SIM is the subscriber's link to a cellular system. By removing the SIM, the subscriber disables the telephone, with the exception of an ability to place emergency calls. To change telephones—for example, from a portable phone to a vehicle-mounted phone—the subscriber simply moves the SIM from one telephone to the other. Using the new phone, the subscriber retains her own telephone number, her special calling features, and the telephone directory she has programmed into the SIM.

This situation differs substantially from the other cellular systems, which store subscriber information in fixed hardware within a terminal. Thus in AMPS, CDMA, and NA-TDMA, the telephone unit is part of the subscription. When a person changes telephone equipment, the service provider gets involved, changing the subscription to reflect the identity of the new telephone. A GSM subscription, like a fixed telephone subscription, is unaffected by the telephone instrument used by a subscriber. GSM specifies two types of SIM distinguished by their physical characteristics. One is like a credit card and is easily inserted into or removed from a terminal. The other is much smaller—comparable in size to a postage stamp—and better suited to compact portable telephones. It is also harder to insert and remove than the credit card type.

As in the other systems presented in this book, GSM base stations and telephones store and transmit a variety of identification codes that participate in network operations. Table 7.1 is a partial list of these codes. Some of the codes, including the IMEI and the classmark, are properties of the telephone equipment and are stored in the terminal itself. Other codes, including the international mobile subscriber identifier (IMSI) and the secret encryption key (Ki) belong to the subscriber. These codes are stored in the SIM and can be moved from one telephone to another. The temporary mobile subscriber number is a GSM innovation. After the network assigns a TMSI to a terminal, the terminal and the network transmit this number in call management and mobility management procedures. This adds to privacy and network security because it avoids transmitting over the air the IMSI, which identifies the subscriber. It also economizes on transmission bandwidth resources because it is shorter than the IMSI. The secret authentication key (Ki) is at the heart of GSM security and privacy operations. The system operator determines the length of this key. The maximum length is 128 bits. This key is stored on the SIM and in the

Table 7.1 GSM Identifiers

Notation	Name	Size	Description
IMSI	International mobile subscriber identity	15 digits	Directory number assigned by operating company to a subscriber
TMSI	Temporary mobile subscriber identity	32 bits	Assigned by visitor location register to a subscriber
IMEI	International mobile equipment identifier	15 digits	Unique serial number assigned by manufacturer to a terminal
Ki	Authentication key		Secret key assigned by operating company to a subscriber
Kc	Cipher key	64 bits	Computed by network and by mobile station
—	Mobile station classmark	32 bits	Indicates properties of a mobile station
BSIC	Base station identity code	6 bits	Assigned by operating company to a base transceiver station
—	Training sequence	26 bits	Assigned by operating company to a base transceiver station
LAI	Location area identity	40 bits	Assigned by operating company to a base transceiver station

subscriber's home system. Terminals and the network use Ki to compute the cipher key, Kc, which protects user information and network control information from unauthorized interception.

The 32-bit classmark describes the capabilities of a terminal. It has several components, three of which are essential: the revision level is the version of the GSM standard to which the terminal conforms; the RF power capability indicates the power levels available to the mobile transmitter (see Section 7.3.4); and the encryption algorithm indicates the manner in which the terminal encrypts user information and network control information. These three components comprise 8 bits of the classmark. In many network control procedures, only this part of the classmark is transmitted.

The remainder of the classmark indicates the frequency capability of the terminal and whether the terminal is capable of operating a short message service. The base station identity code and the training sequence serve the same purposes as the supervisory audio tone in AMPS and the digital verification color code in NA-TDMA. They help a terminal verify that it receives information from the correct base station rather than another base station using the same physical channel. The base station uses these codes in the same manner to verify that the received signal comes from the correct terminal. The location area identity (LAI) has three components. A mobile country code and a mobile network code are like the system identifier in North America. Together they identify the network to which a cell belongs. The third component of the LAI is the location area code, which controls mobility management operations.

7.3 Radio Transmission

Figure 7.3 shows the GSM spectrum allocation. As in AMPS (see Figure 3.2), there are two 25 MHz bands separated by 45 MHz, with the lower band used for transmissions from terminals to base stations and the upper band for transmissions from base stations to terminals. In some countries, analog cellular systems occupy the lower 15 MHz of each band. In these countries, initial GSM systems operate in the upper 10 MHz. As the demand for GSM grows, GSM channels gradually displace analog channels at the lower carrier frequencies. Eventually, analog operations will be discontinued and GSM systems will completely occupy the two bands in Figure 7.3. The GSM technology described in this chapter is also used in the European 1,800 MHz personal communications bands and the North American 1,900 MHz personal communications bands (see Figure 1.18). The systems operating in these bands are designated DCS 1800 and DCS 1900, respectively.

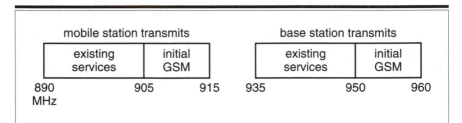

Figure 7.3 GSM frequency bands.

7.3.1 Physical Channels

As a hybrid frequency-division/time-division system (see Figure 9.3), GSM organizes radio transmission by assigning carriers and time slots to logical channels. Figure 7.4 shows that each GSM band has carriers spaced at 200 kHz. The frame duration in GSM is 4.62 (= 120/26) ms, derived from the definition of a 120 ms traffic multiframe, divided into 26 frames. Each frame contains eight time slots. In order to make it unnecessary for a terminal to transmit and receive signals simultaneously, the time reference for a reverse-direction frame is retarded by three time slots relative to the time reference for a forward-direction frame. Thus we have Figure 7.4, which shows the relative timing of forward and reverse transmissions.

With the 200 kHz carrier spacing, the frequency allocation of 25 MHz per direction in Figure 7.3 admits the possibility of

$$\frac{25 \text{ MHz}}{200 \text{ kHz/carrier}} = 125 \text{ carriers per direction.} \qquad (7.1)$$

However, GSM specifies only 124 carriers, leaving unoccupied guard bands at the edges of the GSM spectrum allocation in Figure 7.3. As indicated in

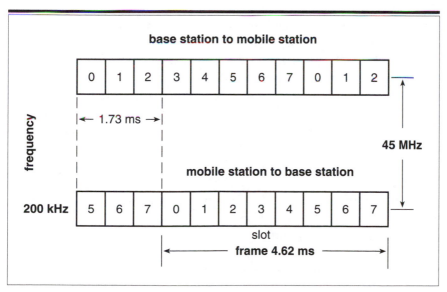

Figure 7.4 GSM frames and slots.

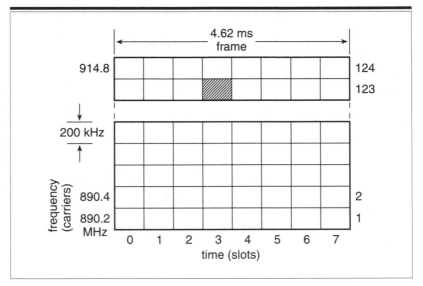

Figure 7.5 GSM physical channel.

Figure 7.5, the carrier numbers are $C = 1 - 124$, which correspond to center frequencies

$$f(C) = 890 + 0.2\ C\ \text{MHz} \tag{7.2}$$

for mobile station transmissions. The corresponding base station transmission frequencies are $f(C) + 45$ MHz.

Although a rectangle in Figure 7.5 is necessary to specify a GSM physical channel, it is not sufficient. In addition to a time slot and carrier, a physical channel consists of a repetitive frame pattern that depends on the logical channel carrying specific information. Among personal communications systems, GSM has the most elaborate timing structure, with definitions of time intervals ranging from 900 ns (one-quarter of a bit) to 3 h 28 m 53.76 s (encryption hyperframe). Figure 7.6 displays the most important time intervals defined by GSM.

To gain an understanding of GSM timing, the best place to begin is at the traffic multiframe in Figure 7.6. It has a duration of 120 ms, which can be synchronized to the timing of other networks. For example, in ISDN, an important time interval is 125 µs, corresponding to the 8 kHz sampling rate of telephone speech. The duration of a GSM traffic multiframe spans 960 ISDN speech samples. Figure 7.7 shows the 26 frames in a traffic multiframe. The logical channel that carries telephone speech in GSM is a *full-rate*

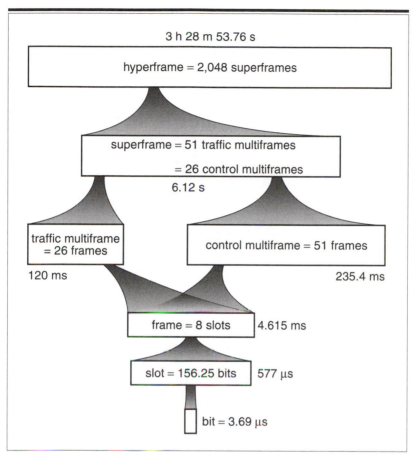

Figure 7.6 GSM time intervals.

traffic channel (TCH/F), which occupies one time slot in 24 of the 26 frames in every multiframe. Traffic channel information travels in frames 0–11 and frames 13–24. As in NA-TDMA, a slow associated control channel (SACCH) accompanies every GSM traffic channel. The SACCH occupies one frame in every traffic multiframe. A SACCH associated with a full-rate traffic channel alternatively occupies one slot in frame 12 and one slot in frame 25. Each GSM carrier can convey eight full-rate traffic channels together with their associated control channels.

Like NA-TDMA, GSM also specifies half-rate traffic channels (TCH/H). In GSM, a half-rate traffic channel occupies a specific time slot in 12 of the 26 frames in every multiframe. Another TCH/H occupies the other 12 frames available for user information. It follows that each carrier can

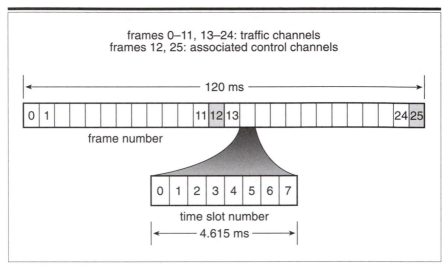

Figure 7.7 Traffic multiframe.

carry up to 16 half-rate traffic channels, which together fill all time slots in 24 frames per multiframe. Eight of these traffic channels have a SACCH in frame 12 in Figure 7.7, and the other eight half-rate channels have a SACCH in frame 25.

Other physical channels in GSM correspond to frame patterns with repetition periods related to the control multiframe, containing 51 frames, in Figure 7.6. A complete cycle of traffic multiframe transmissions and control multiframe transmissions constitutes a superframe with a duration of $51 \times 26 = 1,326$ frames, or 6.12 s.

7.3.2 GSM Bit Stream

To examine GSM radio transmission in detail it is necessary to look inside each time slot. GSM documentation refers to the signal transmitted in one time slot as a *burst*. Figure 7.8 shows the composition of time slots used by the majority of the GSM logical channels. Section 7.4 describes each of the logical channels and presents the time slot structures of the ones that deviate from Figure 7.8.

The 26-bit TRAINING sequence in the middle of the time slot in Figure 7.8 serves a purpose similar to that of the SYNC field in NA-TDMA. A receiver has advance knowledge of the training sequence and uses this information to estimate the characteristics of the time-varying radio channel. This estimate trains an adaptive equalizer, which compensates for the effects of multipath propagation (see Section 9.6). GSM specifies eight

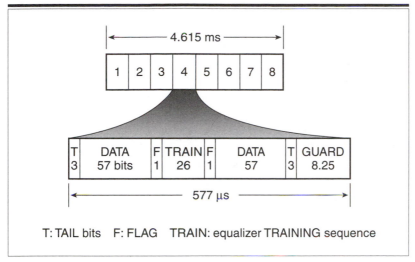

Figure 7.8 Contents of a GSM time slot.

different training sequences with low mutual cross-correlation. Network operators assign different training sequences to nearby cells that use the same carrier. The GSM training sequence therefore performs the function of the AMPS Supervisory Audio Tone (SAT) and the NA-TDMA digital verification color code (DVCC). It enables terminals and base stations to confirm that the received signal comes from the correct transmitter and not a strong interfering transmitter.

The two DATA fields carry either user information or network control information. Each of these fields is accompanied by a 1-bit FLAG and 3 TAIL bits. The FLAG indicates whether the DATA field contains user information or network control information. The TAIL bits, all set to 0, can be used to enhance equalizer performance. There is also a guard time of 30.5 μs (corresponding to 8.25 bits) when no information is transmitted. The guard time includes ramp time for the transmitter to turn off at the end of one time slot and turn on at the beginning of the next slot. It also prevents signals assigned to adjacent time slots from arriving simultaneously at a base station receiver.

Figure 7.8 contains a total of 156.25 bits, which implies that the GSM transmission rate is

$$\frac{26 \text{ frames/multiframe}}{120 \text{ ms/multiframe}} \times 8 \text{ slots/frame} \times 156.25 \text{ b/slot}$$

$$= 270.833 \tfrac{1}{3} \text{ kb/s}. \tag{7.3}$$

This transmission rate corresponds to a bit duration of $1/0.270833 = 3.69$ μs. GSM specifies that each receiver be capable of equalizing signals that arrive over multiple propagation paths with delay differences as high as 16 μs. This delay spread corresponds to more than four bit periods. To unscramble the intersymbol interference caused by this large spread, an adaptive equalizer (see Section 9.6) is an essential component of every GSM receiver.

The modulation scheme in GSM is Gaussian minimum shift keying (GMSK), a form of frequency shift keying (FSK). A GMSK modulator performs signal processing operations to reduce the bandwidth occupied by an FSK signal. The principal operation is linear filtering, with a Gaussian transfer function. The filter confines the modulated signal to the 200 kHz band allocated to each carrier. Thus the modulation efficiency of GSM is

$$\frac{270.833 \text{ kb/s}}{200 \text{ kHz}} = 1.35 \text{ b/s/Hz}, \qquad (7.4)$$

significantly higher than AMPS frequency shift keying with modulation efficiency 0.33 b/s/Hz (10 kb/s ÷ 30 kHz). However, it is lower than the modulation efficiency of NA-TDMA (1.62 b/s/Hz). In exchange for this lower modulation efficiency, GMSK has the advantage of a constant signal envelope, which reduces the drain on the battery of a portable telephone, relative to the NA-TDMA modulation scheme. The GMSK signal is also more robust in the presence of channel impairments than its NA-TDMA counterpart.

Note that in contrast to NA-TDMA (see Figure 5.6), GSM has only one time-slot configuration for transmission of user information. Both base stations and mobile stations use this configuration. Thus, a GSM base station turns off its transmitter at the end of each time slot. When it has information to send to another terminal in the next time slot, the base station resumes transmitting after a pause of 30.5 μs. Recall that NA-TDMA base stations transmit continuously even if only a fraction of the time slots per frame are assigned to conversations or digital control channels. In GSM, the base station turns off its transmitter in unassigned time slots. This has the effect of reducing interference to signals in nearby cells using the same carrier. It is also essential when the system employs slow frequency hopping.

7.3.3 Slow Frequency Hopping

GSM has two definitions of radio carriers. One is the conventional definition of a sine wave at a single frequency (among the 124 carriers in the

GSM band). The other definition of a radio carrier is a frequency hopping pattern, consisting of a repetitive sequence of frequencies occupied by a signal. When the radio carrier is a frequency hopping pattern, the signal moves from one frequency to another in every frame. The purpose of frequency hopping is to reduce the vulnerability of GSM signals to transmission impairments. Without frequency hopping, the entire signal is subject to distortion whenever the assigned carrier is impaired. When the distortion is severe and sustained, an error-correcting code is incapable of recovering the transmitted bit stream. Many impairments are frequency-dependent. When a transmitter employs frequency hopping, it is likely that the signal will encounter these impairments for only a fraction of the time (when it hops to a frequency with a poor propagation path or high interference). In this situation, it is possible that error-correcting codes applied to GSM signals will mitigate the sporadic effects of the transmission impairments. Figure 7.9 shows, as a function of time, the frequency bands occupied by two conventional carriers and two frequency hopping carriers.

Frequency hopping can also reduce harmful effects of co-channel interference between signals in nearby cells. The interference in a conversation depends on the location of a mobile phone in another cell using the same carrier. If a network operator assigns different hopping patterns to different cells, two mobile phones that are in vulnerable positions with respect to one another will use the same carrier frequency for only a fraction of the time. With only part of a signal subject to interference, error-correcting codes have a chance of overcoming the effects of the interference.

All GSM terminals are capable of frequency hopping. Network operators decide whether to introduce frequency hopping, and if so, which patterns to use. To avoid interference with one another, all of the signals in a cell have to hop in a coordinated manner, so that two of them do not use the same frequency simultaneously. Moreover, to reduce the effects of interference from other cells, as described in the previous paragraph, the hopping patterns in a group of cells have to be coordinated with one another. Frequency hopping thus adds a new dimension of complexity to cellular reuse planning.

7.3.4 Radiated Power

GSM specifies five classes of mobile stations distinguished by maximum transmitter power, ranging from 20 W (43 dBm) to 0.8 W (29 dBm). When a terminal transmits in a full-rate channel, the transmitter is active during only one time slot per frame (one-eighth of the time). This implies that the average radiated power is lower than the maximum by a factor of eight

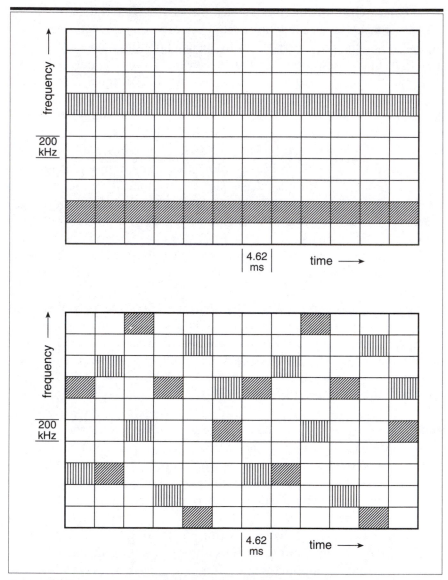

Figure 7.9 Two conventional carriers (upper drawing). Each carrier uses the same frequency all the time. Two frequency hopping carriers (lower drawing). The hopping patterns repeat every six frames.

(9 dB). Typically, the maximum power capability of vehicle-mounted terminals is 8 W (1 W average). Portable terminals typically have 2 W maximum transmitter power (250 mW average). In common with other cellular systems, GSM employs power control. Terminals can adjust their power to any of 16 power levels that range over 30 dB in steps of 2 dB.

7.3.5 Spectrum Efficiency

The GMSK modulation technique, combined with error-correcting codes and adaptive equalization, makes GSM less vulnerable than NA-TDMA to interference. The system can meet signal-quality objectives with a signal-to-interference ratio as low as 7 dB [Mouly and Pautet, 1992]. This allows networks to operate with a reuse factor of $N = 3$ or $N = 4$, depending on the environment. The entire GSM spectrum allocation contains

$$124 \text{ carriers} \times 8 \text{ channels/carrier} = 992 \text{ physical channels.} \quad (7.5)$$

Without taking into account the overhead imposed by the need for common control channels, the efficiency of GSM is

$$E = \frac{992 \text{ channels}}{4 \text{ cells/cluster} \times 50 \text{ MHz}}$$
$$= 4.96 \text{ conversations/cell/MHz } (N = 4), \text{ or}$$

$$(7.6)$$

$$E = \frac{992 \text{ channels}}{3 \text{ cells/cluster} \times 50 \text{ MHz}}$$
$$= 6.61 \text{ conversations/cell/MHz } (N = 3).$$

These numbers are slightly lower than the efficiency of NA-TDMA. Even though NA-TDMA has many more physical channels (up to 2,500 channels in 50 MHz) than GSM, its efficiency is not substantially higher because of its greater vulnerability to interference.

7.4 Logical Channels

Figure 7.10 displays the logical channels defined in GSM. The traffic channels are both two-way channels with identical transmission formats in the two directions. In this respect, GSM differs from NA-TDMA and CDMA. In both of those systems, the multiplexing scheme on the forward traffic channel differs from the multiplexing scheme on the reverse traffic channel. There are three categories of control channels (in GSM terminology, they are together referred to as *signaling channels*). A base station uses *broadcast channels* to transmit the same information to all terminals in a cell. The *common control channels* carry information to and from specific terminals. However, they use physical channels that are available to all of the terminals in a cell. The *dedicated control channels* use physical channels that are assigned to specific terminals. The following paragraphs describe the GSM logical channels individually.

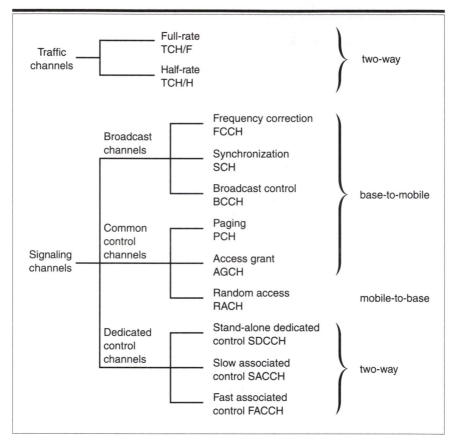

Figure 7.10 GSM logical channels.

7.4.1 Broadcast Channels and Common Control Channels

Together the broadcast and common control channels serve the same purposes as the digital control channel in NA-TDMA and the pilot, sync, paging, and access channels in CDMA. They make it possible for a terminal without a call in progress to synchronize its operation with a base station, to gain essential information about system operation, and to set up calls. In each cell, GSM multiplexes the broadcast and common control channels on the same carrier (either a single frequency or a frequency hopping sequence). The broadcast channels always occupy time slot 0 in repetitive frame patterns on the carrier. The common control channels also occupy time slot 0, and if they need more capacity than time slot 0 can provide, they can occupy time slots 2, 4, or 6 of the same carrier.

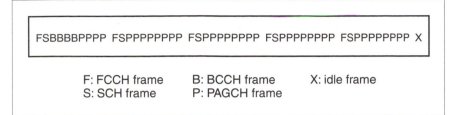

Figure 7.11 Control multiframe.

The frames occupied by each channel are specified with respect to their positions within the 51-frame control multiframe in Figure 7.6. Figure 7.11 shows the contents of time slot 0 in each of the 51 frames. In each multiframe there are five groups of frames, each containing ten frames beginning with a frequency-correction frame and a synchronization frame. At the beginning of the multiframe, four broadcast control frames follow the FCCH and SCH. With the exception of one idle frame at the end of the multiframe, all of the remaining frames carry paging and access grant information, referred to together as PAGCH [Mouly and Pautet, 1992] in Figure 7.11.

The pattern illustrated in Figure 7.11 applies to time slot 0 in one carrier in the forward direction. In the reverse direction, time slot 0 of the corresponding carrier is assigned to a random access channel in all 51 frames of the multiframe. All of the terminals in a cell without a call in progress share this channel on a contention basis. The other seven time slots on this carrier are independent of the one dedicated to control channels. Typically, they carry traffic channels or stand-alone dedicated control channels. However, the even-numbered slots can also be used for common control channels if the number of control messages in a cell exceeds the capacity of a single physical channel.

Frequency Correction Channel (FCCH)

On beginning its operation in a cell, a terminal without a call in progress searches for a frequency-correction channel. The FCCH is one of the logical channels with a time-slot structure that deviates from that shown in Figure 7.8. Instead of the DATA fields, TRAINING fields, FLAG bits, and TAIL bits of Figure 7.8, the FCCH simply transmits 148 0s. This causes the GMSK modulator to emit a constant sine wave for the entire duration of a time slot. After detecting this sine wave, each terminal adjusts its frequency reference to match that of the base station. The FCCH always occupies time slot 0 in a frame of eight time slots. After a terminal detects

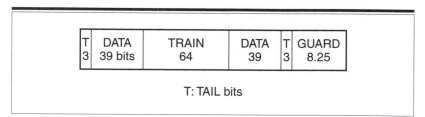

Figure 7.12 Time-slot structure for the synchronization channel.

the distinctive sine wave of an FCCH, it can keep track of the number (between 1 and 7) of each successive time slot. After finding an FCCH, a terminal obtains timing information from a synchronization channel that arrives eight slots after the arrival of the FCCH sine wave.

Synchronization Channel (SCH)

A base station transmits SCH information in time slot 0 of every frame that follows a frame containing an FCCH. The SCH also has its own slot structure, shown in Figure 7.12, that deviates from the one shown in Figure 7.8. To help terminals synchronize their operation to a new base station, the SCH contains a long TRAINING sequence (64 bits) that is the same in all cells. The DATA fields in the SCH contain the base station identity code (BSIC) (see Table 7.1) and the present frame number. The frame number is the position of the current frame within the 3.5-hour GSM hyperframe (Figure 7.6). The hyperframe is a sequence of $2,048 \times 26 \times 51 = 2,715,648$ frames.

Each SCH transmission consists of one message containing 25 bits. It is protected with an error-detecting code that adds 10 parity bits and by a rate 1/2 convolutional code (see Section 9.4.2). Figure 7.13 shows the coding operations on the SCH.

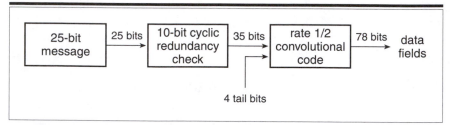

Figure 7.13 Coding on the SCH.

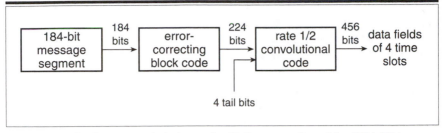

Figure 7.14 Coding on control channels with the exception of the FCH, SCH, and RACH.

Broadcast Control Channel (BCCH)

Base stations use the broadcast control channel to transmit the information that terminals need to set up a call, including the control channel configuration of the local cell and details of the access protocol. The BCCH transmits one message segment,[1] of length 184 bits, in every control multiframe. This message segment is protected by an error-correcting block code (referred to as a *fire code*) that adds 40 parity check bits and by a rate 1/2 convolutional code to produce 456 bits, the DATA content of the four BCCH frames in Figure 7.11. Thus the BCCH sends one message segment every 235 ms, the duration of a 51-frame control multiframe. Figure 7.14 shows the coding operations on the BCCH.

Paging Channel (PCH) and Access Grant Channel (AGCH)

As its name implies, the purpose of the PCH is to notify terminals of arriving calls. The purpose of the AGCH is to direct a terminal to a stand-alone dedicated control channel (SDCCH). Together the PCH and AGCH share time slot 0 in each of the frames designated "P" in the control multiframe in Figure 7.11. Mouly and Pautet [1992] use the designation PAGCH to refer to the aggregate of these two channels. Both channels use the same coding scheme as the BCCH (Figure 7.14). Together they occupy 36 frames per multiframe. With each message occupying four frames, one time slot has a capacity to send nine messages in every 235-ms multiframe.

As in NA-TDMA and CDMA, GSM terminals without a call in progress are capable of sleep-mode operation, turning on their receivers

---

[1] GSM documents, following ISDN, use the term *frame* to refer to a 184-bit block of information transmitted on GSM control channels. However, for the purposes of this presentation, we introduce the term *message segment* to avoid confusion with time frames (Figures 7.5 through 7.8) of the GSM physical layer. Most GSM messages consist of only one segment.

for only a fraction of the time in order to monitor paging messages. To coordinate sleep-mode operation, a base station assigns each block of four P frames shown in Figure 7.11 to either PCH operation or AGCH operation. It then divides the PCH blocks into a number of paging groups ranging from 4 to 81. It uses the BCCH to communicate this allocation of signaling resources to AGCH operation and paging groups. On receiving this information, an idle terminal determines its paging group and monitors only the time slots occupied by that paging group, conserving its battery power the remainder of the time.

Random Access Channel (RACH)

Terminals without a call in progress use this channel to initiate signaling dialogs with the remainder of the system. GSM terminals send messages on the random access channel to originate phone calls, initiate transmissions of short messages, respond to paging messages, and register their locations. As in the counterparts to the RACH found in the other systems described in this book (AMPS reverse control channel, NA-TDMA random access channel, and CDMA access channel), dispersed terminals contend in an uncoordinated manner for access to the RACH. However, in GSM the contention is simpler than in other systems and the information transmitted on the RACH is far more restricted. A RACH occupies all of the reverse direction time slots of a common control channel (one slot in each frame of the 51-frame control multiframe). Terminals with information to transmit use the slotted ALOHA protocol [Tannenbaum, 1988, Section 3.2] to gain access to these time slots. A terminal with information to transmit simply chooses a time slot, transmits a message, and waits for an acknowledgment. An acknowledgment contains, in place of an address, the slot number in which the uplink message arrived, and a 5-bit code word transmitted in the RACH message. The acknowledgment directs the terminal to a stand-alone dedicated control channel (SDCCH) to be used for further signaling messages transmitted between the terminal and the base station. A terminal, after transmitting a RACH message, waits for a fixed time interval for an acknowledgment. If no acknowledgment arrives, the terminal transmits another RACH message, and repeats the procedure until it reaches a maximum number of attempts as specified by a message on the BCCH.

Transmissions on the RACH use shortened bursts, of a duration of 87-bit periods, to ensure that they are confined to the boundaries of a single time slot when they arrive at the base station. Figure 7.15, which shows the time-slot structure of the RACH, indicates that the guard time is 69.25 bits

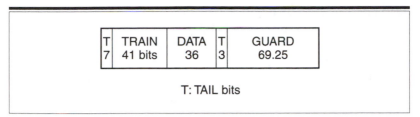

Figure 7.15 Time-slot structure of the RACH.

(256 μs), sufficient to allow transmissions from all parts of a cell to arrive within the 577 μs duration (156.25 bit periods) of a time slot. On observing the time of arrival of the RACH message, a base station determines the correct timing for subsequent transmissions from the terminal and sends this "timing advance" information to the terminal in the channel assignment message.

The 36-bit field labeled DATA in Figure 7.15 carries a simple 8-bit message protected with an error-detecting code and an error-correcting code, as shown in Figure 7.16. Three of the 8 message bits indicate the purpose of the access attempt. The other 5 bits are part of the RACH access protocol. These 5 bits are produced by a random number generator for the purpose of distinguishing messages from two terminals that transmit in the same time slot. When the base station receives a message on the RACH, it transmits this 5-bit number and the time slot number in its acknowledgment. If only one terminal transmits a message in a time slot, the long training sequence and the error-correcting code make it likely that the base station will decode this message accurately.

When two terminals contend for the same time slot, it is likely that their mutual interference will make it impossible for the base station to detect either message. However, if one signal is considerably stronger than the other, this signal could capture the base station receiver and be detected accurately. In this event, the 5-bit random code is likely (with

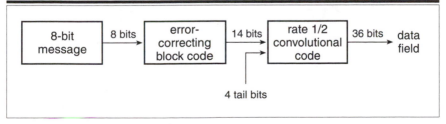

Figure 7.16 Coding on the RACH.

probability 31/32) to distinguish the successful terminal from the other one. However, it is possible (probability 1/32) that both terminals have generated the same 5-bit random code. When this happens, both of them will receive positive acknowledgments and tune to the same traffic channel. In this event, a call management procedure resolves the conflict.

The flowchart in Figure 7.17 is a summary of the access protocol. Compared with other systems, GSM transmits very little information on the RACH. In GSM, stand-alone dedicated control channels carry information that other systems send over the counterparts of the RACH.

7.4.2 Stand-Alone Dedicated Control Channel (SDCCH)

The stand-alone dedicated control channel (SDCCH) is a two-way channel assigned to a specific terminal. The physical channel used by an SDCCH is a set of four time slots in each 51-frame control multiframe (Figure 7.6).

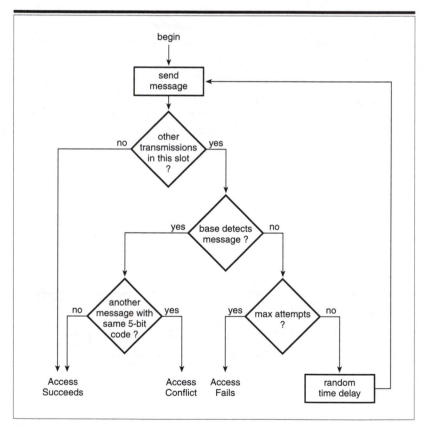

Figure 7.17 Access protocol of the random access channel.

With 114 data bits per time slot (Figure 7.8), the data rate of the SDCCH can be calculated with respect to the duration of a superframe (26×51 frames or 6.12 seconds):

$$4 \, \frac{\text{slots}}{\text{multiframe}} \times 114 \, \frac{\text{bits}}{\text{slot}} \times 26 \, \frac{\text{multiframes}}{\text{superframe}} \div 6.12 \, \frac{\text{sec}}{\text{superframe}} \quad (7.7)$$
$$= 1937.25 \text{ b/s}.$$

This is less than 10 percent of the data rate of a full-rate traffic channel. GSM uses an SDCCH to economize on transmission resources in performing network control procedures, including mobility management and call management, that do not require a high average data rate. The SDCCH is an efficient alternative to using a RACH or a traffic channel to perform network control. The RACH is inefficient due to the contention that takes place in the access protocol. A traffic channel has a data rate that is higher than necessary for the control procedures. We have observed that in contrast to the other systems presented in this book, not much network control information moves on the RACH in GSM. To transfer all the information necessary to set up a call, GSM assigns a terminal to a SDCCH. After performing the necessary transfer of network control information, the system commands the terminal to move to a traffic channel.

Like traffic channels, each SDCCH has a slow associated control channel (SACCH, see Section 7.4.3). In the case of the SDCCH, the SACCH occupies an average of two time slots per control multiframe. Therefore, its bit rate is one-half that of the SDCCH (which occupies four time slots per multiframe), or approximately 969 b/s, which is about 2 percent higher than the bit rate of a SACCH associated with a traffic channel (which is 950 b/s). Channel coding on the SDCCH conforms to Figure 7.14.

7.4.3 Traffic Channels (TCH)

GSM defines two traffic channels. As described in Section 7.3.1, a full-rate channel (TCH/F) occupies 24 time slots in every 26-frame traffic multiframe. A half-rate channel (TCH/H) occupies 12 time slots in every multiframe. Both traffic channels use the time-slot structure of Figure 7.8 with 114 data bits. Therefore the bit rate of a full-rate traffic channel is

$$24 \, \frac{\text{slots}}{\text{multiframe}} \times 114 \, \frac{\text{bits}}{\text{slot}} \div 0.120 \, \frac{\text{sec}}{\text{multiframe}} = 22{,}800 \text{ b/s}. \quad (7.8)$$

The bit rate of a half-rate traffic channel is, as the name implies, half of this, or 11,400 b/s.

Speech Coding and Interleaving

The principal purpose of GSM traffic channels is to carry conversational speech. Initial implementations of GSM use only full-rate traffic channels with the speech coding technique described in this section. In later developments, GSM has adopted standards for two new speech coders [Mouly and Pautet, 1995]. One new speech coder, which performs *enhanced full-rate* (EFR) coding, is used in full-rate traffic channels. Like the advanced coder developed for NA-TDMA, it uses the ACELP technique to achieve higher voice quality than the original GSM speech coder achieves. The other coder, operating at a lower bit rate, can be used to transmit speech in half-rate traffic channels.

The original speech coding technique of GSM is referred to as *linear prediction coding with regular pulse excitation* (LPC-RPE). As indicated in Figure 7.18, the LPC-RPE coder uses $36 + 188 + 36 = 260$ bits to represent each block of 20 ms of speech (160 samples at the 8 kHz sampling rate). Therefore the speech coding rate is

$$260 \text{ bits/block} \div 0.02 \text{ sec/block} = 13,000 \text{ b/s}. \qquad (7.9)$$

This is higher than the coding rates of NA-TDMA and CDMA. The higher bit rate reflects the early date at which the GSM coder was developed. For a given speech quality, the required code rate predictably decreases with time, reflecting advances in signal processing hardware technology. In the LPC-RPE coder, 36 bits per block carry information about eight linear pre-

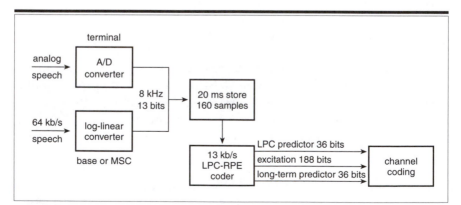

Figure 7.18 Linear prediction coding with regular pulse excitation.

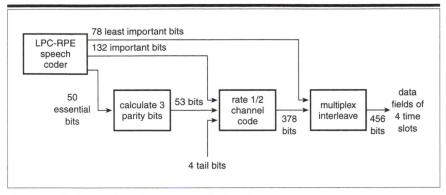

Figure 7.19 Channel coding for speech signals.

diction coefficients (see Section 9.7), another 36 bits represent the long-term predictor, and the remaining 188 bits carry excitation information. As in NA-TDMA, GSM applies different amounts of error correction to the 260 bits in each speech coding block.

Figure 7.19 is a summary of the error protection applied to each 20 ms speech block. It identifies 50 bits as "essential," in the sense that errors in these bits have a severe effect on speech quality. The speech coder adds 3 error-detecting parity bits to these 50 bits. If the corresponding parity checks fail at a receiver, the receiver does not send this block of data to the speech decoder. Instead the receiver performs an operation similar to the bad frame masking specified in NA-TDMA (see Figure 5.9). GSM specifies that on receiving a single block with parity errors, the receiver will repeat the previous block. If subsequent blocks also contain parity errors, the receiver continues to repeat the previous block, gradually decreasing the speech amplitude. Eventually (after 320 ms without a valid received block), the receiver sends silence blocks to the decoder. A group of 132 bits, identified as "important," also have a strong effect on speech quality when it contains errors. A rate 1/2 convolutional coder adds forward error correction to the combination of these 132 important bits, the 50 "essential" bits, and the 3 parity bits that detect errors in the essential bits. The remaining 78 bits generated by the speech coder are transmitted without error protection. The channel coding process generates a total of 456 bits every 20 ms, corresponding to a speech transmission rate of

$$456 \frac{\text{bits}}{\text{block}} \div 0.020 \frac{\text{ms}}{\text{block}} = 22{,}800 \text{ b/s}, \qquad (7.10)$$

which is, of course, the information rate of a full-rate traffic channel.

Note that with 114 DATA bits per time slot (Figure 7.8), the 456 bits produced for each speech block correspond to the information content of four time slots. Rather than fill four time slots sequentially, a GSM transmitter performs the interleaving operation (see Section 9.5) in Figure 7.20 over the contents of two speech blocks, corresponding to 40 ms of speech or $2 \times 456 = 912$ coded speech bits. It distributes these bits over eight frames.

Slow Associated Control Channel

When GSM assigns a traffic channel or a stand-alone dedicated control channel (SDCCH, see Section 7.4.2) to a terminal, it also allocates re-

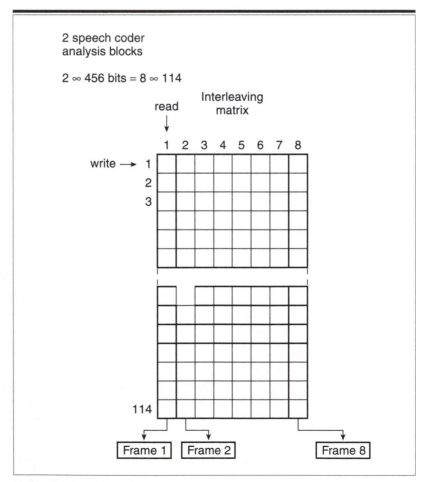

Figure 7.20 GSM interleaving.

sources for an SACCH. Although NA-TDMA performs a corresponding allocation, the multiplexing of SACCH information in the transmitted bit stream is different in the two systems. NA-TDMA places SACCH information in each traffic time slot; GSM establishes separate time slots that contain only SACCH information. These slots are in frames 12 and 25 of each 26-frame traffic multiframe (see Figure 7.7). The SACCH associated with each traffic channel occupies one slot per traffic multiframe. With 114 information bits per time slot, the transmission rate of a traffic SACCH is

$$114 \frac{\text{bits}}{\text{slot}} \times 1 \frac{\text{slot}}{\text{multiframe}} \div 0.120 \frac{\text{sec}}{\text{multiframe}} = 950 \, \text{b/s}. \quad (7.11)$$

Channel coding on the SACCH corresponds to the coding shown in Figure 7.14. With 456 bits transmitted per message, a message spans four traffic multiframes, a time interval of 480 ms.

Fast Associated Control Channel

The 480-ms transmission time of a message on the SACCH is too slow for some network operations. GSM transmits the messages that control these operations on an FACCH, which is an in-band signaling channel created by interrupting user information on a traffic channel or an SDCCH. When a mobile station or a base station transmits an FACCH message, it indicates that user information (on a TCH) or signaling information (on an SDCCH) has been interrupted by altering the polarity of the FLAG bit assigned to each 57-bit DATA field (Figure 7.8) occupied by the FACCH. Like the SACCH, the channel coding on the FACCH corresponds to Figure 7.14. Each FACCH message is multiplexed with user information and interleaved over eight frames in the manner indicated in Figure 7.20. Therefore, for a traffic channel, the transmission of an FACCH message spans eight frames, approximately 40 ms.

7.5 Messages

GSM specifies the communications protocols employed on all of the labeled network interfaces in Figures 7.1 and 7.2. By contrast, the systems described in other chapters have a limited set of open interfaces. The wireless systems in Chapters 3, 5, and 6 specify only air interface protocols (corresponding to the U_m interface in Figures 7.1 and 7.2). IS-41, described in Chapter 4, deals mainly with communications between mobile switching centers and databases (corresponding to interfaces B, C,

D, E, F, and G in Figure 7.1). A substantial fraction of the GSM specification covers the A interface between a base station controller and a mobile switching center and the Abis interface between a base station controller and a base transceiver station. Figure 7.21 is a summary of the protocols on these interfaces as well as on the GSM air interface. The figure indicates that the A interface uses Signaling System 7 (see Section 9.12) protocols and that the Abis interface uses LAPD, the ISDN data link layer

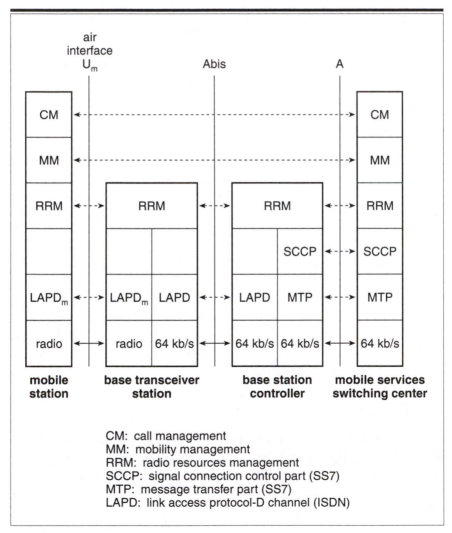

Figure 7.21 GSM protocol layers.

protocol. On the air interface, the corresponding protocol is $LAPD_m$. Like the U_m nomenclature, this terminology appends "m," denoting "mobile," to the name of an ISDN protocol.

In this section, we follow the practice of Chapters 3, 5, 6, and 8 by focusing attention on information transfer over the air interface. Earlier sections of this chapter describe the physical layer (labeled "radio" in Figure 7.21). Section 7.5.1 describes the GSM message structure specified in the $LAPD_m$ protocol. Section 7.5.2 then examines message content, with messages classified according to the network management operations they perform: radio resources management, mobility management, or call management. Figure 7.21 indicates that the base station system participates in radio resources management. By contrast, the mobile switching center and the terminal coordinate call management and mobility management functions. For these two categories of system operations, the base station system simply relays messages between terminals and switching centers.

7.5.1 Message Structure

All of the signaling channels listed in Figure 7.10, with the exception of the frequency correction channel (FCCH), the synchronization channel (SCH), and the random access channel (RACH), transmit information in the $LAPD_m$ format. The physical layer carries these messages in *segments* of 184 bits, as shown in Figure 7.14. Most messages fit into a single segment that spans four physical layer time slots. The exceptions are a few call management messages that require multiple segments.

Figure 7.22 shows the five information fields that appear in $LAPD_m$ messages. Although every message contains a length indicator field, the

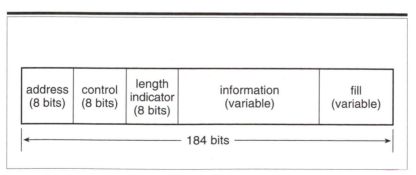

Figure 7.22 Data fields in a GSM message segment.

presence of the other fields depends on the message type and the channel carrying the message. For example, when there are no paging or access grant messages to send in frames labeled "P" in Figure 7.11, the base station simply transmits a length indicator field (indicating 0-length information) and 176 fill bits (set to logical value 1). The message structure of Figure 7.22 is similar to LAPD [Ramteke, 1994: 311–314], with a few notable differences. For example, LAPD messages begin and end with 8-bit synchronization flags: 01111110. In GSM, these flags are unnecessary due to the organization of messages within the DATA fields of time slots (see Figures 7.8, 7.12, and 7.15). Each LAPD message segment also contains a 16-bit frame-check sequence for error detection. By contrast, GSM provides error control at the physical layer by means of powerful channel coding schemes (see Figures 7.13, 7.14, and 7.16). Although the GSM standard allows for the possibility of longer address fields and length-indicator fields, these fields have a length of 8 bits in all messages in the current specification.

Following LAPD, $LAPD_m$ classifies each message as either a *command* (C) or a *response* (R). One *C/R bit* in the address field indicates the nature of the message. The remainder of the address consists of a 3-bit protocol discriminator, a 1-bit extended address indicator (always set to 0 in the initial version of GSM), and a 3-bit *service access point identifier* (SAPI). Anticipating revised versions of the GSM standard, the protocol discriminator indicates to which revision the current message conforms. The SAPI allows a network element to engage simultaneously in more than one type of communication. SAPI = 0 corresponds to a network management message (call control, mobility management, or radio resources management) and SAPI = 3 corresponds to a short message service message.

Following the ISDN convention [Ramteke, 1994], GSM specifies three message types: *information* (I), *supervisory* (S), and *unnumbered* (U). Information messages perform the main tasks of GSM network management. S messages and U messages control the flow of I messages between terminals and base stations. Message flow is also controlled by two message sequence numbers in the control field of each information message. N(S) is the 3-bit sequence number of the current message, and N(R) is the sequence number of a message received by the network element that is sending the current message. For example, when a terminal transmits an I message, the value of N(S) in the control field is the sequence number of that message. The value of N(R) is the sequence number of the last message received from the base station. When the base station observes N(R), it may infer that a message transmitted to the terminal did not arrive successfully. This would cause the base station to retransmit the lost message.

The length-indicator field contains a 6-bit field that specifies the number of octets in the information field. Length = 0 indicates that the current message does not carry an information field. There is 1 bit in the length-indicator field that indicates whether or not the present message segment is the final one in the current message. When the other fields occupy less than 184 bits, the remainder of the message contains fill bits that are all logical 1s.

7.5.2 Message Content

Table 7.2 is a list of the unnumbered (U) messages and supervisory (S) messages that control the flow of information (I) messages. The unnumbered messages initiate and terminate exchanges of information messages. The SET ASYNCHRONOUS BALANCED MODE (SABM) command begins a flow of I messages. This message causes the base station and terminal to set the message numbers N(R) and N(S) to 0 at the beginning of a sequence

Table 7.2 Data Link Control Messages

Message Name	Function	Type	Purpose
SET ASYNCHRONOUS BALANCED MODE (SABM)	command	Unnumbered	initiate transfer of information messages
DISCONNECT	command	Unnumbered	terminate transfer of information messages
UNNUMBERED ACKNOWLEDGMENT (UA)	response	Unnumbered	confirm a command
RECEIVE READY	command or response	Supervisory	request transmission of information message
RECEIVE NOT READY	command or response	Supervisory	request retransmission of information message
REJECT	command or response	Supervisory	suspend transmission of information messages

of I messages. The *SABM* message can be "piggy-backed" in the same transmission as the first information message. In this case, the *SABM* message occupies the address and control fields of the message segment (Figure 7.22) and the information message occupies the information field. The *SABM* message is confirmed by an *UNNUMBERED ACKNOWLEDGMENT (UA)* response message. A network element terminates the flow of messages by means of a *DISCONNECT* command that also stimulates a *UA* response. Figure 7.23 illustrates the flow of data link control messages (layer 2) and network management messages (layer 3) in a network control procedure. In this example, the procedure originates at the terminal, and the base station concludes the procedure with a *DISCONNECT* command. In general, either network element can begin a procedure by sending an *SABM* message and either network element can conclude the procedure by sending a *DISCONNECT* message.

The three S messages are *RECEIVE READY, REJECT,* and *RECEIVE NOT READY.* Each of these messages contains a 3-bit message sequence number, N(R), corresponding to the sequence number of an I message received by the

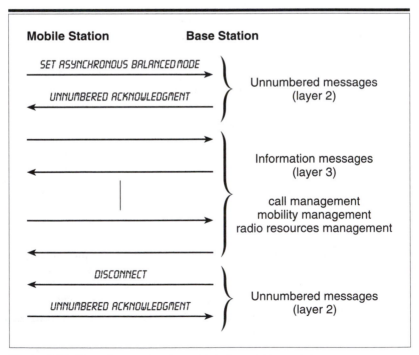

Figure 7.23 Flow of messages in a network control procedure.

network element sending the S message. *RECEIVE READY* requests transmission of the I message numbered N(R). A *REJECT* message requests retransmission of the I message numbered N(R). *RECEIVE NOT READY* indicates that a network element is not able to receive an I message. When this condition changes, the network element sends a *RECEIVE READY* message requesting transmission of message number N(R).

Only I (information) messages contain information fields. With the exception of messages carried on the random access channel and messages carried on the synchronization channel, the first 16 bits of the information field of each I message contain a protocol discriminator, a transaction identifier, and a message type indicator. The protocol discriminator indicates the category of the network operation controlled by the message: either radio resources management, mobility management, or call management. The transaction identifier is a code selected by either the terminal or the network at the beginning of a network control operation. All messages pertaining to that operation carry the same transaction identifier. This allows the system to perform more than one network-control operation at any given time. The messages pertaining to different operations are distinguished by their transaction identifiers. The message type indicator specifies the purpose of each message. The contents of the remainder of the I field depend on the message type. In some simple messages, such as acknowledgments, the information field ends with the message type. In most messages, the message type is followed by "mandatory" data that appears in every message of that type. The mandatory data is then followed by "optional" data that appears in some messages of a given type, but may be omitted from other messages depending on current conditions.

As an example, Table 7.3 shows the structure of an *ASSIGNMENT COMMAND* message (Table 7.4) that appears in handover procedures. The protocol discriminator 0110 identifies a radio resources management message. The network selects a transaction identifier when it initiates the handover procedure. The contents (00101110) of the message type field indicate that this is an *ASSIGNMENT COMMAND* message. The following 24 bits identify, by means of a radio frequency carrier and time slot number, the new physical channel that the call will use at the end of the handover procedure. The power command data indicates the initial mobile station transmitter power on the new channel. An example of optional data is frequency hopping information if it is different on the new physical channel from the frequency hopping employed on the present channel.

Table 7.3 Information Field of an *ASSIGNMENT COMMAND* Message

Bit Positions	Information Elements
1–4	Protocol discriminator 0110
5–8	Transaction identifier
9–16	Message type 00101110
17–40	Channel description
41–48	Power command
variable	Optional data

Radio Resources Management Messages

GSM formally classifies the information messages in the categories indicated in Figure 7.21: call management; mobility management, which includes authentication; and radio resources management. Table 7.4 is a list of radio resources management messages.

On powering up or entering a new cell, a terminal first receives a *SYNC CHANNEL INFORMATION* message on the synchronization channel (SCH). This is one of two messages with a format that does not conform to Figure 7.22. The total message length is 25 bits (Figure 7.13). The message contains two numbers: a base station identifier and a frame number that allows the terminal to synchronize its operation with respect to the hyperframe, superframe, and multiframes of Figure 7.6.

After acquiring synchronism, the terminal tunes to the broadcast control channel, which transmits a variety of *SYSTEM INFORMATION* messages to all of the terminals in a cell. These messages contain information necessary to operate in the current cell, including the location area identifier, information on the physical channels that carry signaling information, parameters of the random access protocol (see Section 7.4.1), and radio frequency carriers active in neighboring cells. There are five types of *SYSTEM INFORMATION* messages defined for broadcast control channels in the initial GSM standard. Phase 2 of the standard adds two broadcast messages, *SYSTEM INFORMATION TYPE 7* and *SYSTEM INFORMATION TYPE 8*. The message *SYSTEM INFORMATION TYPE 6* travels on a slow associated control channel to a terminal with a call in progress. GSM uses this message to transmit local system information to active terminals that move away from the cell in which the call originated.

Table 7.4 Radio Resources Management Messages

Message Name	Logical Channel	Transmitted by
SYNC CHANNEL INFORMATION	SCH	Base
SYSTEM INFORMATION (TYPE 1, 2, 3, 4, 5)	BCCH	Base
SYSTEM INFORMATION (TYPE 6)	SACCH	Base
CHANNEL REQUEST	RACH	Mobile
PAGING REQUEST (TYPE 1, 2, 3)	PCH	Base
IMMEDIATE ASSIGNMENT	AGCH	Base
IMMEDIATE ASSIGNMENT EXTENDED	AGCH	Base
IMMEDIATE ASSIGNMENT REJECT	AGCH	Base
ASSIGNMENT COMMAND*	FACCH	Base
ADDITIONAL ASSIGNMENT	FACCH	Base
PAGING RESPONSE	SDCCH	Mobile
MEASUREMENT REPORT	SACCH	Mobile
HANDOVER COMMAND*	FACCH	Base
HANDOVER ACCESS	TCH	Mobile
PHYSICAL INFORMATION	FACCH	Base
HANDOVER COMPLETE	FACCH	Mobile
CIPHERING MODE*	FACCH	Base
CHANNEL RELEASE	FACCH	Base
PARTIAL RELEASE*	FACCH	Base
FREQUENCY REDEFINITION	SACCH/ FACCH	Base
CLASSMARK CHANGE	SACCH/ FACCH	Mobile
CHANNEL MODE MODIFY*	FACCH	Base
RR STATUS	FACCH/ SACCH	Mobile/Base

* There is also an acknowledgment message from the mobile station corresponding to this message.

To move to a dedicated control channel, a terminal first sends a CHANNEL REQUEST message on the random access channel (RACH). Like the SYNC CHANNEL INFORMATION message, the CHANNEL REQUEST message deviates from

Figure 7.22. As described in Section 7.4.1, the length of the CHANNEL REQUEST message is only 8 bits. Three bits indicate the purpose of the request (such as page response or emergency call). The other 5 bits are a randomly generated code that helps the base station resolve conflicts when two or more terminals transmit CHANNEL REQUEST messages in the same random access channel time slot.

To set up a call to a terminal, the network sends a PAGING REQUEST message on a paging channel (PCH). There are three types of PAGING REQUEST messages, distinguished by the number of terminals paged in a single message and the way in which the paged terminals are identified in the PAGING REQUEST message. After receiving a PAGING REQUEST message, a terminal transmits a CHANNEL REQUEST message. On receiving this message, the base station directs the terminal to a stand-alone dedicated control channel (SDCCH) by means of a message transmitted on an access grant channel. This message can be an IMMEDIATE ASSIGNMENT message, an IMMEDIATE ASSIGNMENT EXTENDED message, or an IMMEDIATE ASSIGNMENT REJECT message. An IMMEDIATE ASSIGNMENT message directs one terminal to a stand-alone dedicated control channel (SDCCH). An IMMEDIATE ASSIGNMENT EXTENDED message conserves transmission resources on the access grant channel by assigning two terminals to two different physical channels. An IMMEDIATE ASSIGNMENT REJECT message contains negative responses to CHANNEL REQUEST messages from up to five terminals. It informs these terminals that the system is not able to assign dedicated channels to them.

After moving to an SDCCH, a terminal that received a PAGING REQUEST message transmits a PAGING RESPONSE message to the system. This message identifies the terminal and stimulates the system to initiate an authentication procedure.

After exchanging mobility management messages and call management messages on an SDCCH with a terminal setting up a call, the network moves the terminal to a traffic channel by means of an ASSIGNMENT COMMAND message. An ADDITIONAL ASSIGNMENT message allocates an additional traffic channel to a terminal that is already operating on a traffic channel. This message is carried on the fast associated control channel (FACCH) of the existing traffic channel.

The MEASUREMENT REPORT message plays a key role in mobile-assisted handover. With a call in progress, a terminal measures signal quality and signal level on the active physical channel. During the time intervals in which it is not transmitting or receiving information on this channel, the terminal performs signal-level and signal-quality measurements on signals received from surrounding cells. The network uses these channel-quality measurements to determine when it is necessary to transfer the call to another cell or to another channel in the same cell. When this is necessary, the *ASSIGNMENT COMMAND* message and the *HANDOVER COMMAND* message serve the same purpose. They move a call from one physical channel to another. When there is no need to adjust the timing advance of the terminal, the network transmits an *ASSIGNMENT COMMAND* message (see Table 7.3) to identify the new physical channel. When it is necessary to adjust the timing of the terminal, the network transmits a *HANDOVER COMMAND* message. When the terminal receives this message, it tunes to the new physical channel and transmits a repetitive sequence of *access bursts*, each with a duration of 321 µs, in the format of Figure 7.15. These bursts comprise a *HANDOVER ACCESS* message, containing one 8-bit reference number, coded in the manner indicated in Figure 7.16. The base station, on receiving information in these shortened time slots, measures the timing adjustment that the terminal requires. The base station transmits this information to the terminal in a *PHYSICAL INFORMATION* message.

The *CIPHERING MODE* message indicates whether or not user information is to be encrypted on the traffic channel. There are two messages that command a channel to stop using a traffic channel. The system can send a *PARTIAL RELEASE* message to a terminal using two or more traffic channels. This message commands the terminal to stop using one of the assigned channels. When there is only one active traffic channel, the system sends a *CHANNEL RELEASE* message to command the terminal to stop using this channel. The base station sends a *FREQUENCY REDEFINITION* message to inform a terminal with a call in progress of a change in the frequency hopping pattern. This message contains a starting-time indication to inform the terminal of the first frame in which the new hopping pattern will be used. The *CLASSMARK CHANGE* message is complementary to the *FREQUENCY REDEFINITION* message. It allows a mobile station to inform the network that its available transmitter power has changed. This is useful when a portable telephone is plugged into or removed from the power supply and external antenna of a vehicle during a call.

The channel mode indicates the type of information transmitted on a traffic channel. For speech, the channel mode defines a specific source coder. For data, it identifies the data speed. The *CHANNEL MODE MODIFY* message allows the system to command a terminal to change from one channel mode to another.

The *RR STATUS* message is the only radio resources management message that can be transmitted by both terminals and base stations. It allows either network element to inform the other one of certain error conditions.

Mobility Management Messages

The mobility management messages in Table 7.5 travel on an SDCCH. In addition to the mobility management function described in Section 2.4.2, the messages classified formally by GSM as mobility management messages also perform authentication functions (Section 2.4.3). In common with many other GSM innovations, many properties of the authentication scheme have been adopted in North American systems, as described in Section 5.6.1. The authentication procedure begins with an *AUTHENTICATION REQUEST* message transmitted from a base station to a terminal. This message contains a 128-bit random number. Using a GSM encryption algorithm, the terminal computes a 32-bit number that depends on the number received from the base station and on a secret key (Ki) stored on the subscriber identity module (SIM). The terminal transmits this number to the network in an *AUTHENTICATION RESPONSE* message. If this 32-bit number is not identical to the response anticipated by the network, the base station transmits an *AUTHENTICATION REJECT* message informing the terminal that all communications with the network have been aborted.

Table 7.5 Mobility Management Messages

Message Name	Transmitted by
AUTHENTICATION REQUEST	Base
AUTHENTICATION RESPONSE	Mobile
AUTHENTICATION REJECT	Base
IDENTITY REQUEST	Base
IDENTITY RESPONSE	Mobile
*TMSI REALLOCATION COMMAND**	Base
LOCATION UPDATING REQUEST	Mobile
LOCATION UPDATING ACCEPT	Base
LOCATION UPDATING REJECT	Base
IMSI DETACH INDICATION	Mobile
*CM SERVICE REQUEST**	Mobile
*CM RE-ESTABLISHMENT REQUEST**	Mobile
MM-STATUS	Mobile/Base

* There is also an acknowledgment message from the receiving network element corresponding to this message.

The *IDENTITY REQUEST* message and the corresponding *IDENTITY RESPONSE* message enable the network to obtain from a terminal, as part of the authentication procedure, any of three identifiers: the international mobile subscriber identity (IMSI) stored on the SIM, the international mobile equipment identity (IMEI) permanently stored in the terminal, and a temporary mobile subscriber identity (TMSI) assigned by a network to a visiting terminal. The network assigns a new TMSI to a terminal by means of a *TMSI REALLOCATION COMMAND* message.

The three *LOCATION UPDATING* messages correspond to mobility management procedures, as described in Section 2.4.2. A terminal registers its location by means of a *LOCATION UPDATING REQUEST* message. The network reports the action it takes in response to this message in either a *LOCATION UPDATING ACCEPT* or a *LOCATION UPDATING REJECT* message. The other mobility management message (in the sense of Section 2.4.2) is the *IMSI DETACH INDICA-TION* message. A terminal sends this message to cancel its registration, typically because the user has turned off the terminal's power.

The final three messages in Table 7.5 coordinate the flow of mobility management messages. At the beginning of a mobility management operation, a terminal sends a *CM SERVICE REQUEST* message to the network. Here CM refers to connection management, in this case the connection established to conduct mobility management operations. This message contains information about the mobile station and information that enables the network and the terminal to encrypt subsequent messages. The *CM RE-ESTABLISHMENT REQUEST* message serves a similar purpose. The terminal transmits this message in order to resume a mobility management operation that has been interrupted. Like the *RR-STATUS* message in Table 7.4, the *MM-STATUS* message allows terminals and base stations to report error conditions to one another.

Call Management Messages

Call management procedures in GSM conform closely to ISDN procedures specified in International Telecommunications Union standard Q.931 [Ramteke, 1994: 288]. Table 7.6 lists, in four categories, the messages that control these procedures. At the beginning of a call, messages travel on a stand-alone dedicated control channel (SDCCH). GSM assigns a traffic channel when the called party accepts the call. Subsequently, messages travel on the fast associated control channel (FACCH) of the assigned traffic channel. A terminal initiates a call by transmitting a *SETUP* message or an *EMERGENCY SETUP* message to the base station. The network responds with a *CALL PROCEEDING* message. To initiate a call to a GSM

Table 7.6 Call Management Messages

Message Name	Transmitted by
Starting a Call	
SETUP	Mobile/Base
EMERGENCY SETUP	Mobile
CALL PROCEEDING	Base
PROGRESS‡	Base
CALL CONFIRMED	Mobile
ALERTING‡	Mobile/Base
CONNECT‡*	Mobile/Base
During a Call	
START DTMF*	Mobile
STOP DTMF*	Mobile
MODIFY*	Mobile/Base
USER INFORMATION‡	Mobile/Base
Ending a Call	
DISCONNECT‡	Mobile/Base
RELEASE‡	Mobile/Base
RELEASE COMPLETE‡	Mobile/Base
Abnormal Conditions	
STATUS	Mobile/Base
STATUS ENQUIRY	Mobile/Base
CONGESTION CONTROL	Mobile/Base

* There is also an acknowledgment message from the receiving network element corresponding to this message.

‡ This message contains an optional data field that carries user-to-user information as part of a GSM supplementary service.

telephone, a base station transmits a *SETUP* message to the terminal. The terminal responds with a *CALL CONFIRMED* message. An *ALERTING* message transmitted by the terminal indicates that the terminal is attempting to inform the user that a call has arrived. When the user accepts the call, the terminal transmits a *CONNECT* message. In the case of a call initiated by the mobile subscriber, the network transmits an *ALERTING* message to the terminal to indicate that the called party is being alerted. This is followed by a *CONNECT* message when the remote party responds. In setting up some calls, the network sends a *PROGRESS* message to the terminal. This message can stimulate an audible indication to the caller that the call is being transferred to a different network; for example, from a public network to a private network.

At the end of a call, GSM transmits three call-clearing messages across the air interface. If the mobile subscriber concludes the call, the terminal transmits a *DISCONNECT* message to the base station. After this message reaches the remote subscriber, the network sends a *RELEASE* message to the terminal. The terminal responds with a *RELEASE COMPLETE* message. In the case of the remote subscriber concluding the call, the same three messages flow in the opposite direction. The terminal receives a *DISCONNECT* message. It responds with a *RELEASE* message, and eventually receives a *RELEASE COMPLETE* message, indicating that the call is over.

During a call, a terminal can transmit a *START DTMF* message, indicating that the subscriber is pressing a button on the numeric keypad of the phone. This causes the network to transmit a dual-tone multiple frequency signal to the remote terminal. A *STOP DTMF* message turns off the signal. The purpose of a *MODIFY* message is to indicate a change in the nature of the transmission path between the two ends of the connection. For example, a terminal can transmit a *MODIFY* message when the user discontinues speech communication and begins a facsimile session [Mouly and Pautet, 1992: 547–548]. As indicated in a footnote to Table 7.6, some GSM messages are capable of carrying user-to-user information between the two end points of a call. This information can be inserted into certain call-establishment messages and call-clearing messages, as indicated in Table 7.6. In these messages, user-to-user information consists of a string of up to 35 characters. There is also a *USER INFORMATION* message that can be transmitted during a call. This message can carry up to 131 characters.

Three messages in Table 7.6 are used in abnormal conditions. As in radio resources management and mobility management procedures, either base stations or terminals can transmit *STATUS* messages to describe error conditions. Either network element can use a *STATUS REQUEST* message to ask the other one for a *STATUS* message. A terminal or a base station transmits a *CONGESTION CONTROL* message to initiate a flow control procedure. This retards the flow of call management messages during certain overload conditions.

7.6 Network Operations

Many of the network operations described in earlier chapters originated with GSM and were later adopted, in modified form, in other networks. Examples include cryptographic authentication, location-based mobility management, and mobile-assisted handover. This section provides examples of how GSM performs these and other procedures, using the messages listed in Section 7.5 and the logical channels described in Section 7.4.

7.6.1 Call to a GSM Terminal

Figure 7.24 shows the messages transmitted in the course of setting up and clearing a call from the network to a mobile telephone. This figure indicates how GSM uses its repertory of logical channels and performs the three categories of network operations (call management, mobility management, and radio resources management) in the course of a single telephone call. (Figure 7.24 omits the four LAPD$_m$ messages shown in Figure 7.23.) The sequence of operations covered in Figure 7.24 begins when the terminal arrives in a cell and tunes to a carrier that contains a broadcast control channel. It first uses the frequency correction channel (FCCH, see Section 7.4.1) to synchronize its local oscillator with the frequency reference of the base transceiver station. It then gains timing information from the synchronization channel in order to operate properly within the complex timing structure of GSM (Figure 7.6). The terminal then obtains other important system information from *SYSTEM INFORMATION* messages transmitted on the broadcast control channel.

After this initialization procedure is complete, the terminal monitors a paging channel (PCH), where eventually it detects a *PAGING REQUEST* message that contains the temporary mobile subscriber number (TMSI) previously assigned by the network. This message causes the terminal to transmit a *CHANNEL REQUEST* message on the random access channel (RACH). The network responds to this request by transmitting an *IMMEDIATE ASSIGNMENT* message on an access grant channel (AGCH). This message establishes a stand-alone dedicated control channel (SDCCH) to be used for the exchange of mobility management messages and call management messages. When it moves to the SDCCH, the terminal transmits a *PAGING RESPONSE* message to the base station.

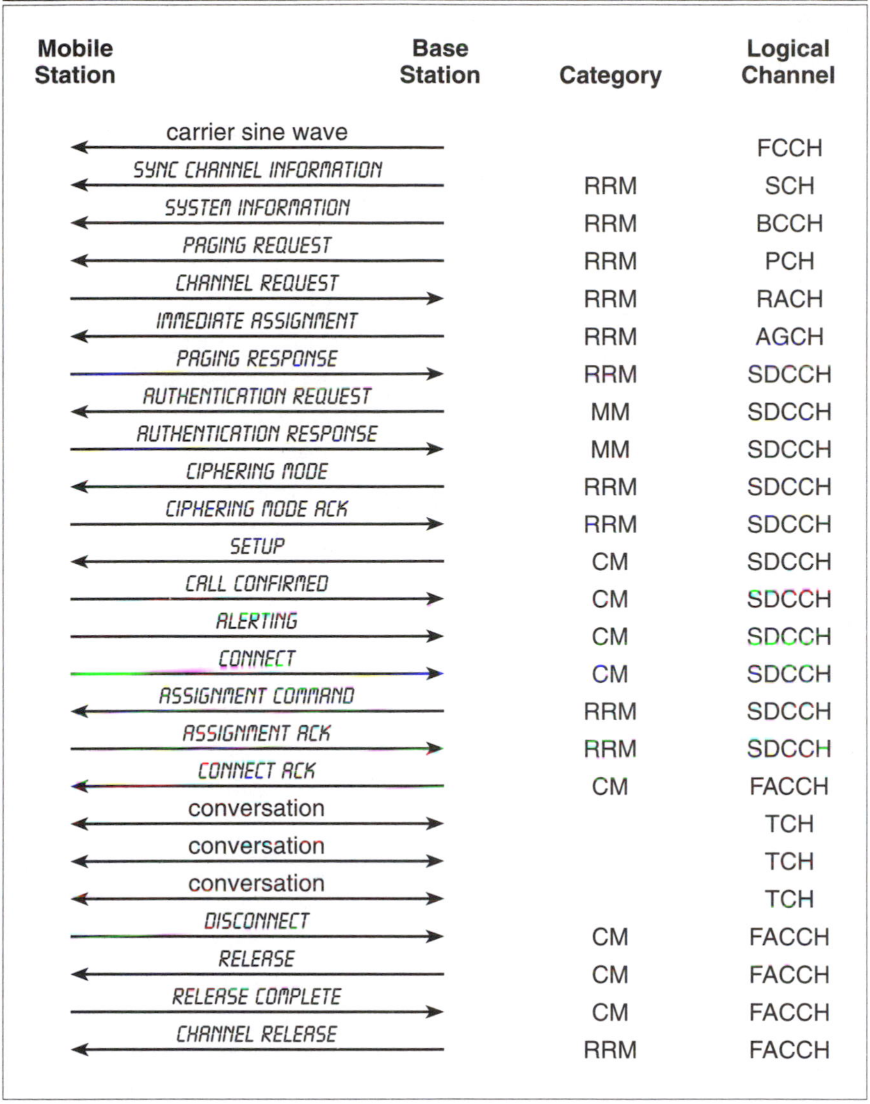

Mobile Station	Base Station	Category	Logical Channel
carrier sine wave			FCCH
SYNC CHANNEL INFORMATION		RRM	SCH
SYSTEM INFORMATION		RRM	BCCH
PAGING REQUEST		RRM	PCH
CHANNEL REQUEST		RRM	RACH
IMMEDIATE ASSIGNMENT		RRM	AGCH
PAGING RESPONSE		RRM	SDCCH
AUTHENTICATION REQUEST		MM	SDCCH
AUTHENTICATION RESPONSE		MM	SDCCH
CIPHERING MODE		RRM	SDCCH
CIPHERING MODE ACK		RRM	SDCCH
SETUP		CM	SDCCH
CALL CONFIRMED		CM	SDCCH
ALERTING		CM	SDCCH
CONNECT		CM	SDCCH
ASSIGNMENT COMMAND		RRM	SDCCH
ASSIGNMENT ACK		RRM	SDCCH
CONNECT ACK		CM	FACCH
conversation			TCH
conversation			TCH
conversation			TCH
DISCONNECT		CM	FACCH
RELEASE		CM	FACCH
RELEASE COMPLETE		CM	FACCH
CHANNEL RELEASE		RRM	FACCH

Figure 7.24 Message sequence in a call to a GSM telephone.

The base station then initiates the GSM authentication procedure by sending an *AUTHENTICATION REQUEST* message to the terminal. This message contains a 128-bit random number (RAND). The terminal applies a GSM *encryption algorithm,* referred to as "A3," to compute a 32-bit *signed response,* SRES. As shown in Figure 7.25, the inputs to A3 are RAND and the secret key Ki stored in the subscriber information module (SIM, Section 7.2). The terminal applies another encryption algorithm, A8, to compute a 64-bit ciphering key, Kc, from SRES and Ki. The terminal transmits

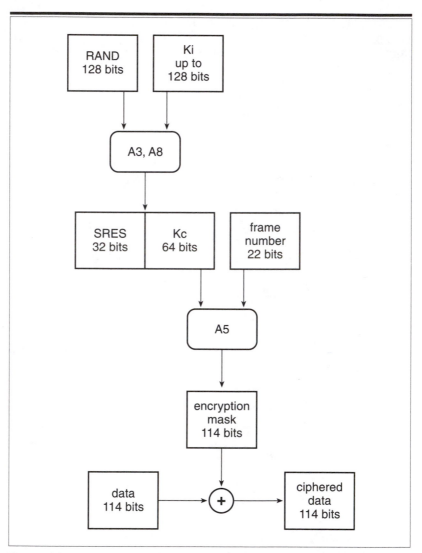

Figure 7.25 GSM security procedures: calculation of signed response (SRES), ciphering key (Kc), and encryption mask.

SRES to the network in an *AUTHENTICATION RESPONSE* message. The network also uses A3 to compute SRES from RAND and Ki. If the two values of SRES are identical, the network accepts the user as an authorized subscriber. It then transmits a *CIPHERING MODE* message to the terminal in order to establish the mode of encryption to be used in the remainder of the call. To encrypt user information and network control information, the base station and network derive a 114-bit mask to be added (modulo 2) to the two DATA fields (Figure 7.8) in each time slot. This mask is obtained from another GSM encryption algorithm, A5. The inputs to A5 are the 64-bit ciphering key, Kc, and the current 22-bit frame number. The base station computes the same mask to decipher the encrypted data. Because A5 uses the frame number to compute the ciphering mask, the mask changes from frame to frame. This property contributes to the security of the communication.

With the authenticity of the subscriber established and the ciphering arrangement in place, the network proceeds to set up the call. To do so, it transmits a *SETUP* message to the terminal. The terminal acknowledges this message with a *CALL CONFIRMED* message and then proceeds to alert the subscriber. When the subscriber accepts the call, the terminal sends a *CONNECT* message to the network. In response, the network moves the call to a traffic channel by means of an *ASSIGNMENT COMMAND* message. At this point, the network establishes a connection between the two parties to the conversation. In Figure 7.24, the mobile subscriber concludes the conversation. When this happens, the terminal sends a *DISCONNECT* message to the network. The network responds with a *RELEASE* command. After the terminal acknowledges this command, the network releases the traffic channel by sending a *CHANNEL RELEASE* message to the terminal. The terminal then returns to monitoring its paging channel.

This example contains a procedure referred to as *off-air call setup* (OACSU). With OACSU, GSM assigns a traffic channel after the mobile subscriber accepts the call. Compared to an earlier assignment of a traffic channel, this conserves transmission resources while the network is setting up a call. OACSU is optional in GSM. It is also possible to move the call to a traffic channel prior to alerting the subscriber. OACSU can also be used when the mobile party initiates the call. In this situation, the network assigns a traffic channel after the remote party accepts the call.

7.6.2 Location-Based Registration

Mouly and Pautet use the phrase *mobility and security management* to refer to the procedures defined by GSM as *mobility management.* These procedures encompass location registration, authentication, and distribution of temporary mobile subscriber identities (TMSI). The system performs all of these procedures in a location-updating procedure that takes place when a terminal finds itself in a new location area. Figure 7.26 shows the air interface messages that control these operations. The sequence begins when a terminal enters a cell and detects a location area identifier in a *SYSTEM INFORMATION TYPE 3* message on the broadcast control channel. If the

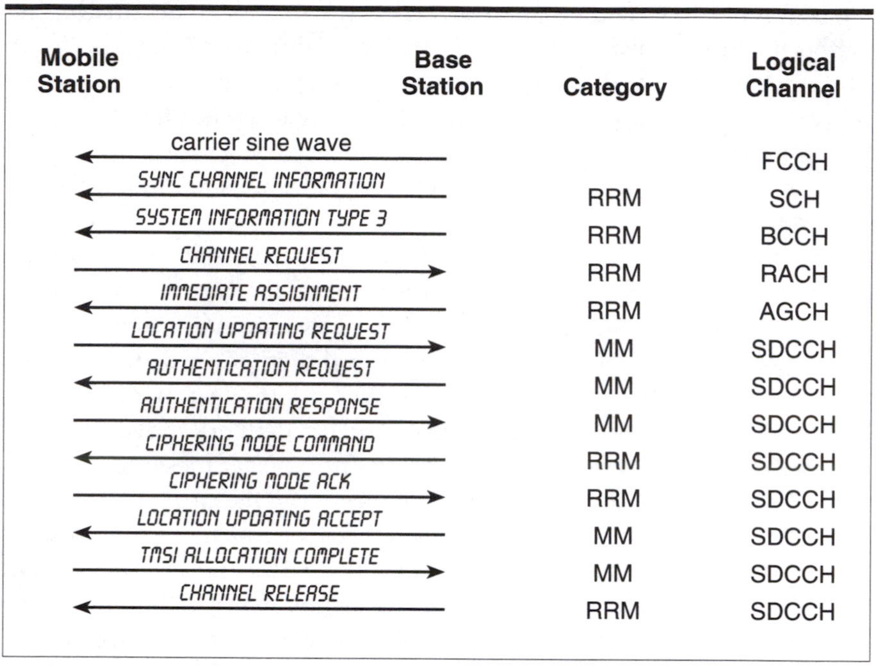

Mobile Station	Base Station	Category	Logical Channel
← carrier sine wave			FCCH
← SYNC CHANNEL INFORMATION		RRM	SCH
← SYSTEM INFORMATION TYPE 3		RRM	BCCH
CHANNEL REQUEST →		RRM	RACH
← IMMEDIATE ASSIGNMENT		RRM	AGCH
LOCATION UPDATING REQUEST →		MM	SDCCH
← AUTHENTICATION REQUEST		MM	SDCCH
AUTHENTICATION RESPONSE →		MM	SDCCH
← CIPHERING MODE COMMAND		RRM	SDCCH
CIPHERING MODE ACK →		RRM	SDCCH
← LOCATION UPDATING ACCEPT		MM	SDCCH
TMSI ALLOCATION COMPLETE →		MM	SDCCH
← CHANNEL RELEASE		RRM	SDCCH

Figure 7.26 Location-updating procedure.

location area of the new cell is different from the location area of the previous cell, the terminal registers its location. To do so, it requests access to a stand-alone dedicated control channel (SDCCH) by means of a *CHANNEL REQUEST* message on the random access channel. The network sends an *IMMEDIATE ASSIGNMENT* message to move the terminal to an SDCCH. The terminal uses this channel to transmit a *LOCATION UPDATING REQUEST* message. In this message, the terminal has two ways to identify the subscriber. It can transmit the subscriber's IMSI (international mobile subscriber identity, stored on the SIM in the terminal). The preferred alternative is to transmit the temporary mobile subscriber identity (TMSI). The network frequently changes the TMSI assigned to a terminal. This makes it difficult for an eavesdropper monitoring the network to identify the subscriber even if the eavesdropper detects the TMSI.

The network then verifies the identity of the terminal in the authentication procedure described in the call setup example (see Section 7.6.1). After authenticating the terminal, the network establishes message ciphering and then assigns a new TMSI to the terminal in a *LOCATION UPDATING ACCEPT* message. The terminal acknowledges receipt of the new identity with a *TMSI ALLOCATION COMPLETE* message, which concludes the mobility management procedure. This causes the base station to transmit a *CHANNEL RELEASE* message to the terminal. The terminal then returns to monitoring the paging channel.

7.6.3 Mobile-Assisted Handover

Figure 7.27 shows the exchange of information that takes place in a mobile-assisted handover procedure. While using a traffic channel, mobile terminals perform channel quality measurements on the active channel, on other channels in different sectors of the same cell, and on channels in surrounding cells. They report the results of these measurements to the network in *MEASUREMENT REPORT* messages on the slow associated control channel. When the network determines that the call should be moved to another channel, it transmits a *HANDOVER COMMAND* message to the terminal. This message identifies the new channel. It also contains an 8-bit handover reference identifier. On receiving this message, the mobile station tunes to the new traffic channel and transmits a sequence of *HANDOVER ACCESS* messages. These messages have the same form as the *CHANNEL REQUEST* message transmitted on the random access channel. They arrive in shortened bursts (see Figure 7.15), with long guard times that allow the base transceiver station to synchronize its receiver to the arrival times of the messages. When the base station acquires this synchronism, it transmits a *PHYSICAL INFORMATION* message. This message contains a timing advance parameter that causes the terminal to adjust its transmission time within the assigned time slot. After performing this adjustment, the terminal sends a *HANDOVER COMPLETE* message in a normal format (Figure 7.8) on the fast associated control channel of the new physical channel. The terminal and base station then resume the exchange of user information and measurement reports.

7.7 Status of GSM

At the beginning of 1997, there were tens of millions of subscribers to GSM systems in more than a hundred countries. In addition to the original 900 MHz band, GSM systems operate in 1,800 MHz Personal Communication Network bands in Europe and Thailand, and in 1,900 MHz Personal Communications Services bands in the United States and Canada. With GSM established as the digital mobile telephone system with the greatest market share, the technical community concentrates on enhancements to the specifications. ETSI defines three phases of the evolution of GSM technology [Mouly and Pautet, 1995]. Phase 1 consisted of

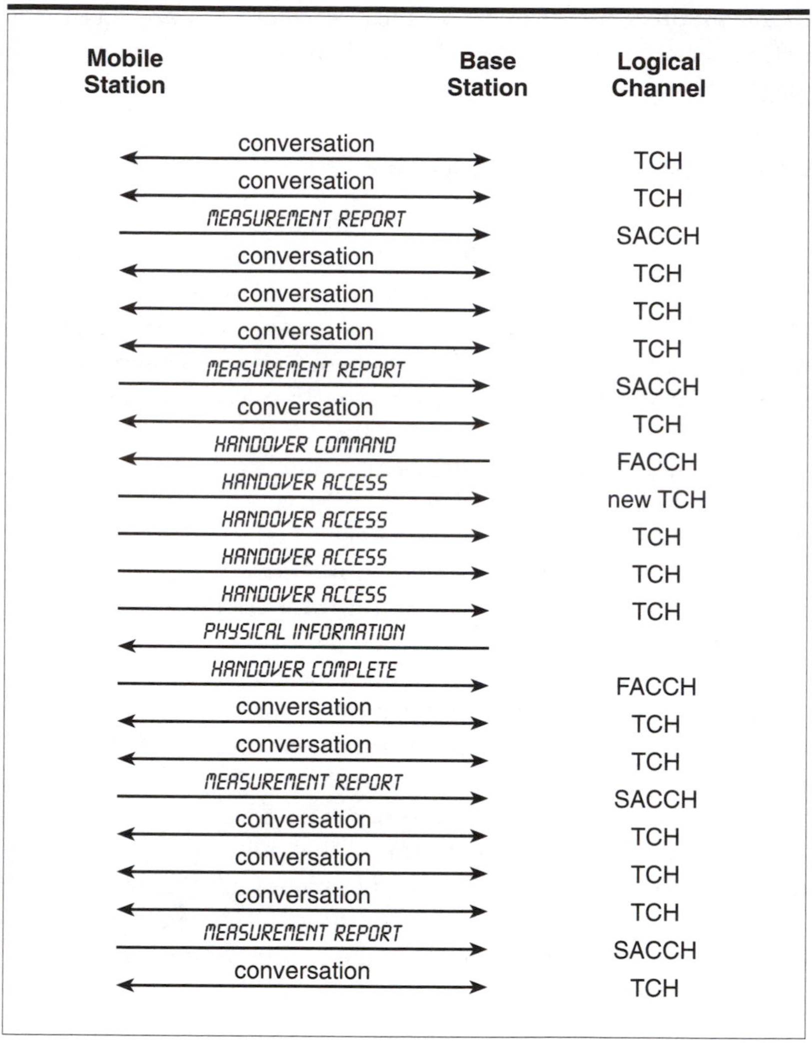

Mobile Station	Base Station	Logical Channel
←—— conversation ——→		TCH
←—— conversation ——→		TCH
—— MEASUREMENT REPORT ——→		SACCH
←—— conversation ——→		TCH
←—— conversation ——→		TCH
←—— conversation ——→		TCH
—— MEASUREMENT REPORT ——→		SACCH
←—— conversation ——→		TCH
←—— HANDOVER COMMAND ——		FACCH
—— HANDOVER ACCESS ——→		new TCH
—— HANDOVER ACCESS ——→		TCH
—— HANDOVER ACCESS ——→		TCH
—— HANDOVER ACCESS ——→		TCH
←—— PHYSICAL INFORMATION ——		
—— HANDOVER COMPLETE ——→		FACCH
←—— conversation ——→		TCH
←—— conversation ——→		TCH
—— MEASUREMENT REPORT ——→		SACCH
←—— conversation ——→		TCH
←—— conversation ——→		TCH
←—— conversation ——→		TCH
—— MEASUREMENT REPORT ——→		SACCH
←—— conversation ——→		TCH

Figure 7.27 Mobile-assisted handover.

sufficient technology development to introduce commercial services in 1992. Phase 2 developments consisted mainly of ironing out some wrinkles in the initial version of GSM and introducing mechanisms for gracefully absorbing new technology. Since 1995, the GSM community has anticipated a future of ongoing modifications to the standard in a process referred to as Phase 2+. In 1997, the existing and anticipated enhancements to GSM fall into two main categories: improved voice-coding techniques and new services.

In addition to the original GSM speech coder based on linear prediction coding with regular pulse excitation, there are two enhanced coders now available in GSM. A enhanced full-rate (EFR) coder delivers improved voice quality relative to LPC-RPE (see Section 7.4.3) within the 22,800 b/s transmission rate of full-rate traffic channels. A half-rate coder offers higher bandwidth efficiency, with somewhat higher voice quality, at the 11,400 b/s rate of half-rate traffic channels. These new speech coders reflect advances that have been made in signal processing hardware since the adoption of the LPC-RPE coder in 1988. In the late 1990s, it is possible to incorporate in portable telephones complex signal processing algorithms that offer improved tradeoffs between bit rate and speech quality.

New GSM services include a packet data transmission protocol referred to as GPRS (generalized packet radio service) and multiple-full-rate circuit switched services that assign more than one time slot per frame to a terminal. These services will be appropriate for advanced applications involving high-throughput data transfers and multimedia communications.

Review Exercises

1. Explain the function of the SIM in a GSM system. How is this function performed in other cellular systems and in a fixed telephone system?

2. In contrast to NA-TDMA, GSM turns the base station transmitter off at the end of each time slot. What are some benefits of this approach? What are the benefits of a base station transmitting continuously on any carrier with an active physical channel (as in NA-TDMA)?

3. What is the advantage of transmitting the training sequence in the middle of a time slot?

4. What are the reasons for the different time slot structures for traffic channels (Figure 7.8), the synchronization channel (Figure 7.12), and the random access channel (Figure 7.15)?

5. How does slow frequency hopping improve the performance of a GSM system?

6. When does the GSM system use a stand-alone dedicated control channel (SDCCH)? Other systems do not have this type of logical channel. Give an example of how another system performs the functions of an SDCCH.

7. Describe the physical channel used by a slow associated control channel in GSM. What are some differences between this physical channel and the physical channel used by the slow associated control channel in NA-TDMA?

8. How does GSM use a temporary mobile subscriber identity (TMSI)? How does the TMSI improve system operation?

9. What is the benefit of using the frame number to calculate encryption masks (Figure 7.25)?

10. What are the benefits of off-air call setup (OACSU) in GSM?

CHAPTER EIGHT

Low-Tier Personal Communications Systems

Although the cellular communications systems described in detail in Chapters 3 through 7 differ from one another in many respects, they also have many common attributes. All are designed to serve a highly mobile population by means of sophisticated high-power radios. Cell diameters in sparsely populated areas can exceed 10 km. Terminals are capable of radiating on the order of 1 watt. Base station antennas can be higher than 50 meters and radiate tens of watts. Systems with these characteristics are referred to as *high-tier wireless systems* [Cox, 1995].

By contrast, *low-tier systems* are designed primarily to serve subscribers moving at pedestrian speeds. They operate with base stations separated by less than 500 meters outdoors and less than 30 meters indoors. Transmitter powers are one to two orders of magnitude lower than those of the high-tier systems. Base station antennas in many cases are indoors. Outdoor antennas are typically 10 meters above the ground. In 1997, four low-tier systems are in production. Two of them, CT2 and DECT, are digital cordless telephone systems, primarily used as wireless extensions of other telephone systems, such as PBXs (private branch exchanges). The other two operate in public networks offering wide-area coverage. The Personal Handyphone System (PHS) in Japan has met with strong commercial success since its introduction in 1995. PACS (Personal Access Communications System) is a United States relative of PHS. It has been adopted by a few network operators for application in the 1,900 MHz PCS band.

The principal goals of all of the low-tier systems include low service price; low terminal price; small, lightweight terminals with long battery life; and speech quality to that of wired telephones. With respect to radio transmission details, all four systems represent speech by means of 32 kb/s adaptive differential pulse code modulation (ADPCM). Compared with the voice coders of the high-tier systems, the voice quality of ADPCM is at least as high, the bit rate is considerably higher, but the equipment complexity and delay are considerably lower. In contrast with the linear prediction speech coders used in NA-TDMA, CDMA, and GSM, ADPCM can accommodate many non-voice signals, such as transmissions from computer modems and facsimile machines. Three of the low-tier systems use hybrid frequency division/time division multiple access. CT2 uses pure FDMA. The remainder of this chapter summarizes the salient characteristics of the four low-tier systems.

8.1 Cordless Telephone, Second Generation (CT2)

CT2 was the first commercial digital wireless telephone system. In addition to the goals listed in the preceding paragraph, important goals for CT2 designers were early deployment and low infrastructure costs. CT2 was embodied in British Standards published in 1987 [Dept. of Trade and Industry, 1987] and 1989 [Dept. of Trade and Industry, 1989] and later adopted as a European standard by ETSI, the European Telecommunications Standards Institute. Subsequently, Northern Telecom developed an enhanced version of CT2, referred to as CT2Plus [Radio Advisory Board, 1990].

8.1.1 Architecture

CT2 provides telephone services in three environments: residential, business, and telepoint. Telepoints are wireless public telephones. People near a telepoint can initiate phone calls. In many applications, telepoints are unable to deliver calls to wireless terminals. One of the enhancements added in CT2Plus is a procedure that enables terminals to register at telepoints and subsequently receive incoming calls. In telepoint and residential environments, the CT2 network architecture conforms to the cordless telephone model of Figure 2.6. Figure 8.1 shows the architecture of a business telephone application. In CT2 terminology, a terminal is a *cordless portable part* (CPP) and a base station is a *cordless fixed part* (CFP) of the system.

8.1.2 Radio Transmission

The original operating band for CT2 is 864–868 MHz. The access technology is FDMA, and within the 4 MHz operating band there are 40 physical channels, each spanning a bandwidth of 100 kHz. CT2 uses time division

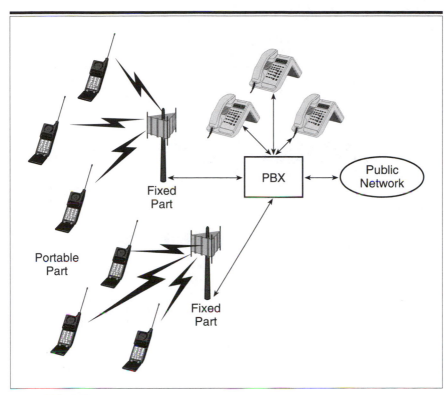

Figure 8.1 CT2 architecture of a business telephone system.

duplex with a frame duration of 2 ms. Figure 8.2 shows the physical channels of CT2. Each frame contains two time slots of 1 ms duration. The fixed part transmits in one time slot and the portable part transmits in the other time slot.

CT2 time slots, shown in Figure 8.3, are the simplest that are described in this book. Within each slot, there are 64 bits for user information and 4 bits for system control information. (The specification also admits an alternate multiplex arrangement with only 2 bits of control information per slot and correspondingly longer guard times between slots.) This comes to 136 bits per frame, to which CT2 adds two guard times of a total duration of 8 bits, for an aggregate of 144 bits per 2 ms frame, or 72 kb/s. The CT2 modulation technique is binary frequency-shift keying. With a channel spacing of 100 kHz, the modulation efficiency of CT2 is

$$\frac{72 \text{ kb/s}}{100 \text{ kHz}} = 0.72 \text{ b/s/Hz,} \tag{8.1}$$

or approximately half of that of GSM.

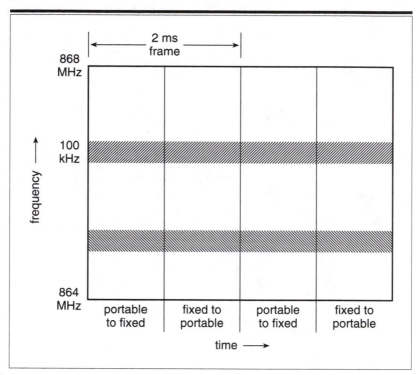

Figure 8.2 CT2 physical channels.

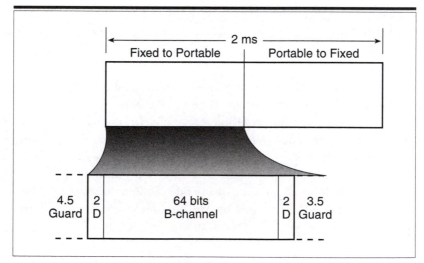

Figure 8.3 CT2 time slots and frame.

8.1.3 Logical Channels

There are only two logical channels in CT2. Their names, B-channel and D-channel, shown in Figure 8.3, are borrowed from ISDN (see Section 9.11). The B-channel carries user information at a rate of 32 kb/s, and the D-channel carries network control information at a rate of 2 kb/s. Figure 8.4 indicates that to provide accurate delivery of a control message, CT2 organizes the D-channel bit stream in code words (each with 64 bits) and packets (one to six code words per packet). One bit in each packet indicates whether the packet is the last one in the message. In a packet, the first code word transmits an address, and subsequent code words (if the packet contains two or more words) transmit data. The final 16 bits of each code word contain an error-detecting code. This 16-bit code begins with a 15-bit cyclic redundancy check operating on the 48 content bits of the code word. The code word ends with a single bit that provides even parity to the entire code word. The CT2 signaling system provides for acknowledged transfer of some messages and unacknowledged transfer of others. In the case of acknowledged information transfer, the 1-bit parity check and the cyclic redundancy check must succeed for every code word in a packet. Otherwise, a negative acknowledgment stimulates retransmission of the entire packet. A binary sequence number in each

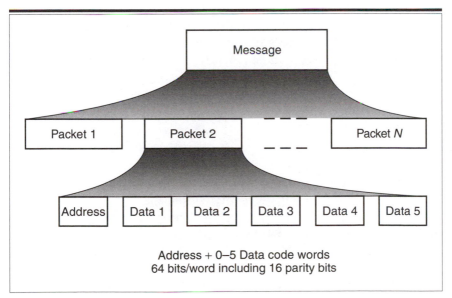

Figure 8.4 CT2 message structure.

packet and each acknowledgment provides a "1-bit sliding window" that controls packet transmission and retransmission.

8.1.4 Network Operations

CT2 differs radically from the cellular systems of Chapters 3 through 7 in the procedures used for radio resources management. In CT2, the most important of these procedures are referred to as link establishment and link re-establishment, corresponding roughly to cellular channel assignment and handoff procedures, respectively. In cellular systems, the network infrastructure controls these operations by exchanging signaling messages with terminals. By contrast, in CT2, channel selection is part of the radio transmission technique. A portable phone that requires a radio link acts autonomously to select a physical channel and make contact with a base station. For network-originated calls, a base station, in a similar manner, acts independently to select a physical channel and then use it to signal to one or more portables.

In the idle state, a base station or a portable phone scans all 40 physical channels, making signal-strength measurements and attempting to detect signaling messages. When it has to establish a link, a CPP or CFP chooses a physical channel at random from the physical channels that have a received signal strength below a certain threshold. It then transmits a message for up to 1.4 seconds. If it receives no response within 1.4 seconds, it selects another physical channel and repeats the process. This decentralized link establishment procedure is vulnerable to failures that do not occur in cellular systems, where network controllers assign channels to base stations and terminals. One failure, referred to as *call deadlock*, occurs when a CFP and CPP simultaneously try to establish a link. In this case, the CFP can transmit on one physical channel waiting for a response from the CPP, while the CPP is transmitting on another channel waiting for a response from the CFP. In this event, it is likely that no link will be established.

In contrast to the elaborate techniques for measuring channel quality and for stimulating handoffs in cellular systems, CT2 relies on a simple means—referred to as "handshaking"—of monitoring the quality of the transmission link between terminal and base station. Each terminal is manufactured with a 19-bit identification code. At the beginning of a call, the base station learns this code. During the call, the CPP and the CFP independently transmit the code on the D-channel at least once per second. When there is a problem with the radio link, transmission errors cause one or both receivers to fail to detect the handshaking code. If either handshake disappears for 10 seconds, CT2 discontinues the communication. Rather than passively wait for a clear handshake message when the flow of handshakes is interrupted, the CPP or CFP may attempt to re-establish the lost link on the same physical channel. If after 3 seconds of interruption, the handshake is not received properly, the CPP may attempt to re-establish the link on another physical channel, selected on the basis of received signal strength. Meanwhile, the 10-second timer, initiated when the handshake failed, continues to run. The communication ceases when this timer expires before a valid handshake is detected.

CT2Plus, devised for personal communications in Canada in the band 944–952 MHz [Radio Advisory Board, 1990], is an enhanced version of CT2. In addition to telephone services, CT2Plus contains provisions for data and facsimile communications. To provide stronger network control than CT2 provides, CT2Plus assigns four physical channels to function as common control channels. It specifies mobility management techniques based on local registration, authentication procedures, faster handoff, and more robust channel allocation than CT2 can provide.

8.2 Digital European Cordless Telecommunications (DECT)

This system [European Telecommunications Standards Institute, 1991] is considerably more ambitious than CT2. Although CT2 is a step forward for cordless telephones, enabling people to make and receive phone calls at a large number of public and private locations, DECT aims to go beyond telephony and deliver a wide range of communications services to wireless terminals. A principal goal of DECT is to bring to wireless terminals all of the facilities of advanced communications networks with no sacrifice in quality.

8.2.1 Architecture

In keeping with its ambitious goals, the DECT system description displays a large set of possible interconnections between DECT and other networks. In contrast to GSM, which defines many standard devices specific to that system (see Figure 7.1), there are only three network elements unique to DECT. As shown in Figure 8.5, they are fixed radio terminations (FT), portable radio terminations (PT), and interworking units (IWU). The PT corresponds to the radio part of a wireless terminal, and the FT corresponds to the radio part of a base station. Rather than a complete network such as AMPS or GSM, DECT is an access system. Its purpose is

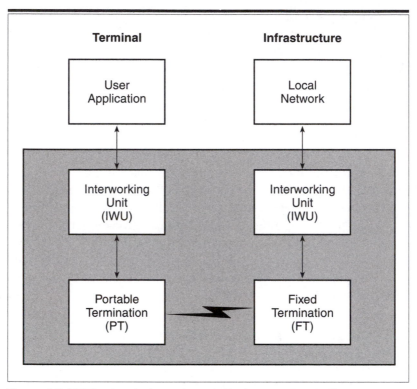

Figure 8.5 DECT network elements are in the large rectangle. The user application and local network are external to DECT.

to connect information devices to other information networks. One interworking unit shown in Figure 8.5 performs translations between DECT information formats and the formats of user devices. The other interworking unit performs translations between DECT formats and the formats of other networks.

Figure 8.6 is the formal specification of the DECT architecture. Here, the interworking functions of Figure 8.5 appear as reference points D2 and D4. At the heart of the standard is a *common interface*. The common interface is the compatibility specification that defines the portable termination, the fixed termination, and the information that flows between them. In addition to its compatibility specification, the DECT standard contains a *system description document* with detailed examples of DECT systems connected to several specific local and global networks, including:

- Public Switched Telephone Networks (PSTN),
- integrated services digital networks (ISDN),

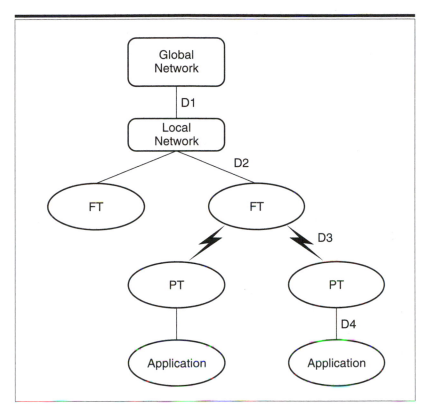

Figure 8.6 DECT network architecture.

- packet switched public data networks (PSPDN),
- GSM cellular networks,
- local area networks (LAN) conforming to IEEE 802 protocols,
- private branch exchanges (PBX), and
- integrated services private branch exchanges (ISPBX).

The first four of these are global networks, and when DECT is connected to one of them, the "local network" in Figure 8.5 is essentially transparent. On the other hand, a LAN, PBX, or ISPBX can contain a rich set of features that go beyond those of a global network, in which case DECT has to deliver these features to wireless terminals. Although the interworking functions of the system description are examples, rather than mandatory parts of a DECT system, they motivate many of the details of the communication protocols in the common interface.

The DECT-GSM interconnection in the above list of networks links two systems described in this book. It anticipates applications such as telepoints on trains or buses. Within the vehicle, terminals will communicate with the telepoint by means of DECT operations. Then, to move information between the vehicle and the outside world, an interworking unit will translate information between DECT and GSM formats.

8.2.2 Radio Transmission

DECT is a hybrid frequency-division/time-division multiple access system. It contains ten frequency carriers, with a spacing of 1.728 MHz, in the band 1.88–1.90 GHz. On each carrier, timing is organized in frames of duration 10 ms, with each frame divided into 24 time slots. In common with CT2, DECT employs time division duplex. Normally 12 time slots are dedicated to transmission from the base station and the other 12 time slots are dedicated to transmissions from terminals. However, it is also possible for a system to assign unequal numbers of time slots per frame to the two directions of transmission. For telephone communications, a two-way physical channel in DECT consists of a carrier and a pair of time slots, as shown in Figure 8.7. Figure 8.8 indicates that the DECT time slot duration is 480-bit intervals. Thus the bit rate is

$$\left(\frac{24 \text{ slots/frame}}{0.01 \text{ s/frame}}\right) \times 480 \text{ bits/slot} = 1.152 \text{ Mb/s} \tag{8.2}$$

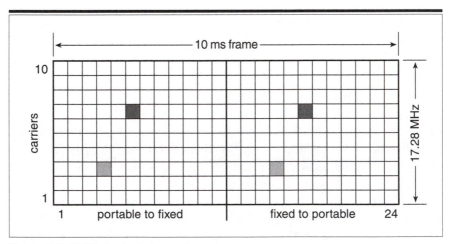

Figure 8.7 DECT physical channel.

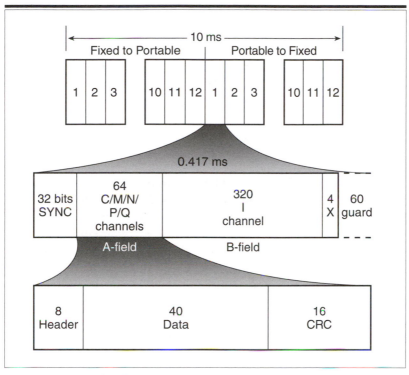

Figure 8.8 DECT time slot structure.

During each slot a base station or terminal transmits 420 bits, with the remaining 60 bit intervals (52 µs) left as a guard time. The beginning of each slot contains a 32-bit synchronization pattern, while the remaining 388 bits comprise a D-field, which carries user information and system control information in several logical channels. The D-field is further sub-divided into an A-field (64 bits), a B-field (320 bits), and an X-field (4 bits). The A-field carries network control information and the B-field carries user information. The X-field, consisting of 4 parity check bits, provides a means for terminals and base stations to monitor the quality of signal transmission. In common with GSM, the modulation technique is Gaussian minimum shift keying. However, the modulation efficiency is

$$\frac{1.152 \text{ Mb/s}}{1.728 \text{ MHz}} = \frac{2}{3} \text{ b/s/Hz,} \tag{8.3}$$

which is comparable to modulation efficiency of CT2 and about half that of GSM.

8.2.3 Logical Channels

DECT specifies two logical channels for the transfer of user information and five logical channels for network control information listed in Table 8.1. A DECT telephone signal travels in an unprotected information channel, I_N. This channel occupies one time slot in each frame with 320 bits per time slot. Therefore, the transmission rate of each telephone signal is

$$\frac{320 \text{ bits/frame}}{0.01 \text{ s/frame}} = 32 \text{ kb/s}. \tag{8.4}$$

In common with the other low-tier systems, the speech-coding technique is adaptive differential pulse code modulation. The only error control for the I_N channel is provided by the 4 parity check bits in the X-field (Figure 8.8), which give an overall indication of link quality.

Other types of user information travel in protected information channels (I_P) in which the B-field of each time slot is divided into four segments, each containing 80 bits. A segment contains 64 bits of user information and a 16-bit frame check sequence for error detection. With $4 \times 64 = 256$ information bits per time slot, the transmission rate of an I_P channel is 25.6 kb/s when the channel occupies one slot per frame. The DECT specification allows for considerable flexibility in assigning time slots to logical channels. It is possible to assign different numbers of slots per frame to different I-channels, with unequal transmission rates in the two directions.

Table 8.1 DECT Logical Channels

Channel	Rate (b/s)		Purpose
	Base Transmit	Terminal Transmit	
I_N	32,000	32,000	unprotected user information (telephone)
I_P	variable	variable	protected user information
C	0–3,200	0–3,200	call management special features
M	0–3,200	0–3,200	physical layer control
N	400–6,000	3,200–6,400	handshaking
P	0–2,400	0	paging
Q	400	0	system information

The control field (A) in Figure 8.8 statistically multiplexes five logical control channels, denoted C, M, N, P, and Q. The 64 bits in the A-field are divided into three groups, as shown in Figure 8.8. A header consisting of 8 bits indicates the logical channel (C, M, N, P, or Q) carried in the current time slot, and the nature of the message. The main part of the A-field, consisting of 40 bits, is the payload of the control channel. Finally, there is a 16-bit cyclic redundancy check (CRC) for error control.

The Q-channel is a broadcast control channel that carries general system information from each base station to all terminals in the service area of the base station. Each base station transmits Q-channel information in one frame in every group of 16 frames. Therefore, the bit rate of the Q-channel is 64 bits per 160 ms (the duration of 16 frames) or 400 b/s as shown in Table 8.1. To establish contact with DECT, a terminal locks onto one physical channel and decodes the information in the Q-channel to prepare for communications with the system.

In common with paging channels in cellular systems, the DECT paging channel broadcasts identities of terminals that are the targets of incoming calls. When a terminal receives a paging message, it uses the two-way M-channel to exchange media access control messages with the system. These messages coordinate channel allocation and handoff operations. Like the GSM access grant channel (AGCH, see Section 7.4.1), the DECT M-channel is active when the terminal and base station initially gain access to a physical channel, either at the beginning of a communication or during a handoff. The M-channel can use up to eight frames in each multiframe, or half of the DECT signaling system capacity. The M-channel has precedence over the C-channel in gaining access to transmission resources.

The N-channel is an identities channel used for handshaking. When a connection is established, the base station transmits an identification code on the fixed-to-portable channel and the terminal echoes this code on the portable-to-fixed half of the N-channel. This channel serves the same purpose as the supervisory audio tone (SAT) in AMPS and the digital verification color code (DVCC) in NA-TDMA. It allows the terminal and the base station to detect the presence of strong signals arriving from interfering sources that use the same physical channel. The N-channel occupies from 1 to 15 frames in each 16-frame fixed-to-portable multiframe, depending on the needs of the M-channel, C-channel, and P-channel. In the portable-to-fixed direction, the N-channel occupies at least 8 of the 16 frames per multiframe. The other frames are available to M-channel and C-channel messages as required. If these channels are idle, the N-channel makes use of the entire bandwidth of the B-field (6,400 b/s).

The other logical channel is the two-way C-channel dedicated to a specific connection. Its primary purpose is call management and delivery of supplementary services. The C-channel can use up to half of the control frames, provided they are not required by the M-channel.

8.2.4 Network Operations

In common with CT2, DECT radio resources management is part of the radio transmission system, rather than a network management operation controlled by a flow of messages between network elements. In DECT, terminals control the process of channel allocation and handoff. A terminal that requires a traffic channel selects an idle physical channel and initiates communications with a base station. DECT refers to two types of handoff. *Internal* handoff procedures transfer a call to a new physical channel at the serving fixed termination. *External* handoff moves the call to a new fixed termination. Like the cellular CDMA system (see Chapter 6), DECT terminals are capable of soft handoff. They can maintain communications on the original physical channel until they have a reliable connection on the new physical channel.

In addition to user information transfer and radio resources management, DECT specifies the following five network operations, shown in Figure 8.9:

- call control,

- supplementary services,

- connectionless message transfer,

- connection-oriented message transfer, and

- mobility management.

Many call control and supplementary services operations are performed by means of message transfers between the user application and the local network in Figure 8.5. In these operations, the DECT system simply relays messages between the external network elements. Call control includes establishing, maintaining, and releasing circuits for communication of user information between the application and the local network in Figure 8.5. Most of the DECT call control messages correspond to those defined in ISDN. Additional messages, exclusive to DECT, carry authentication data obtained from the mobility management service. In addition to setting up circuits at the beginning of a call and releasing circuits at the

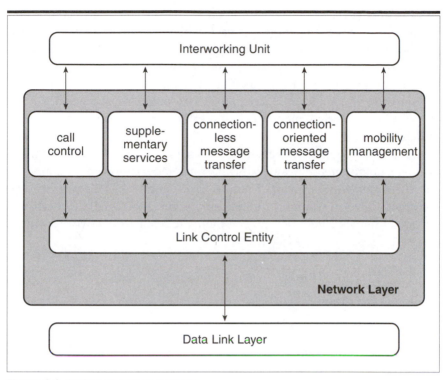

Figure 8.9 DECT network operations.

end of a call, the call management service establishes a connection with a new base station, and then releases the connection at the old base station when DECT performs an external handoff. Supplementary services give users additional control over their communications beyond the essentials of call setup and release.

The connectionless message service is similar to a cellular short message service. It allows terminals and base stations without calls in progress to exchange data in short bursts. Messages are of variable length with a maximum packet size of 63 bytes, including overheads and addresses. The transmissions may be point-to-point, as in conventional paging, or point-to-multipoint, as in group paging or broadcasting. The connectionless message service does not include acknowledged data delivery. If applications require acknowledgments, they must incorporate them in user protocols. The connection-oriented message service is a powerful DECT operation. It provides point-to-point packet data communication of user information. This service contains its own call establishment and release procedures, which are faster and simpler than those of the call management service for circuit switched communications. In establishing a connection for this service, DECT assigns to each call a transaction identifier that remains in effect for the duration of the call. This is essentially a virtual circuit identifier, which logically links the two ends of a connection, in contrast to the physical channels that physically link the endpoints of a circuit-switched connection. The connection-oriented message service delivers positive acknowledgments of all user information transmitted across the radio interface.

The DECT mobility management service incorporates mobility management operations and authentication and encryption operations discussed in general terms in Sections 2.4.2 and 2.4.3. To perform its responsibilities, the mobility management service contains procedures grouped in seven categories. Five of them are related to network security. They are

- identity procedures,

- authentication procedures,

- key allocation procedures,

- access rights procedures, and

- ciphering procedures.

Together they cause the portable termination to identify itself to the network, establish that it is a valid network user, and establish that it has the right to gain access to a specific type of local or global network. These network security procedures also distribute encryption keys for protecting network control information and user information.

The other two categories of mobility management procedures are location procedures and parameter retrieval procedures. Location procedures are necessary for setting up calls from the local or global network to DECT terminals. Although DECT location procedures provide mobility management information to interworking units, they do not specify how the information is used to manage databases. The databases and associated procedures for registration and location updating are external to the DECT standard. They are responsibilities of the local or global network. Parameter retrieval procedures are concerned with external handoff to a new base station. The terminal, which in DECT controls handoffs, can request channel quality information from the fixed termination. The fixed termination accepts the request by returning the parameters requested by the terminal. The fixed termination can also send a message that suggests to the portable termination that a handoff is necessary.

8.3 Personal Handyphone System (PHS)

The Ministry of Posts and Telecommunications in Japan initiated the creation of PHS technology in 1989. The Research and Development Center for Radio Communications, another Japanese government unit, developed the radio technology for PHS in the early 1990s, while the Telecommunications Technical Committee, a voluntary standards-setting organization, developed the network infrastructure specification. The standard was published at the end of 1993 [Research and Development, 1993] and subsequently the Japanese government issued operating licenses for PHS systems in 11 regions of the country. Following system trials, commercial service was launched in 1995. With the price of PHS about one half of that of cellular service, and attractive subscriber equipment available, the service was a great commercial success, attracting nearly four million subscribers in its first year of commercial operation. Attracted by the success of PHS in Japan, other Asian countries, including Thailand and Hong Kong, announced plans to establish PHS networks.

The aims of PHS span those of cordless and cellular systems. Like cordless systems, PHS provides wireless access to private communications systems such as business telephone systems and residential base stations. It also forms the basis of public networks accessible by subscribers moving at pedestrian speeds. In addition to telephone services, the PHS air interface supports voiceband data and facsimile communications at rates up to 9.6 kb/s. There is also a provision for direct digital access to information systems at rates of 32 kb/s and 64 kb/s. A version of PHS has been specified for operation in the band 1,910–1,930 MHz [Steer, 1994], allocated in the United States for unlicensed personal communications systems [ANSI, 1996a].

8.3.1 Network Architecture

Like CT2, PHS defines only two network elements: terminals and base stations. The nomenclature for a PHS terminal is *personal station* (PS) and a base station is referred to as a *cell station* (CS). PHS anticipates cell stations connected to public networks and also to private communications systems. Unique among the systems examined in this book, PHS also specifies direct radio communications between a pair of terminals. Figure 8.10 shows the PHS network elements and three standard interfaces. The six personal stations in the diagram fall in three categories. PS0, PS4, and PS5 are complete terminals containing all the equipment necessary to use a communications service. PS1, PS2, and PS3 are like the DECT portable termination. PS1, PS2, and PS3 correspond to the radio part of a terminal,

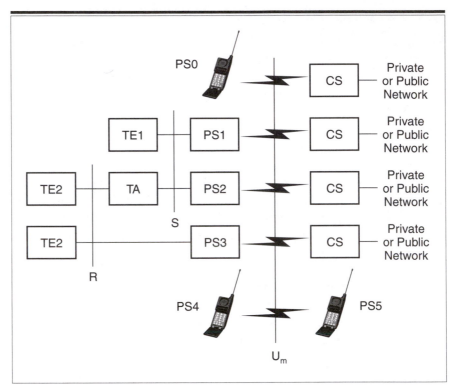

Figure 8.10 PHS architecture.

and Figure 8.10 shows them connected to other terminal equipment, designed to operate in conventional communications systems. This equipment is designated TE1 and TE2 according to ISDN conventions (see Section 9.11, Figure 9.15). PS1 and PS2 communicate with standard ISDN terminals or terminal adaptors by means of the ISDN S-interface. PS3 incorporates the R-interface that links ISDN equipment to a non-ISDN terminal. As in GSM, the air interface in PHS is labeled U_m. In a Personal Communications System, the most common arrangement is the one that includes PS0 in Figure 8.10, with a subscriber using a self-contained terminal to access a public telephone network.

8.3.2 Radio Transmission

The PHS band spans 23.1 MHz in the range of 1,895–1,918.1 MHz. In common with DECT, the access method is hybrid time-division/frequency-division multiple access with time-division duplex. There are 77 carriers in the PHS band with 300 kHz between adjacent carriers. The entire band is available for public operation. Private systems are confined to the

37 carriers at the low end of the band, 1,895–1,906.1 MHz. Direct communication between two personal stations (PS4 and PS5 in Figure 8.10) is permitted in the ten carriers at the low end of the band (1,895–1,898 MHz).

The PHS frame duration is 5 ms and each carrier contains four physical channels. Figure 8.11 indicates that a PHS physical channel occupies two time slots per frame, one for transmission from the personal station to a cell station, the other for transmission from the cell station to the personal station. (When two personal stations communicate directly, each station transmits in one time slot and receives in another time slot.)

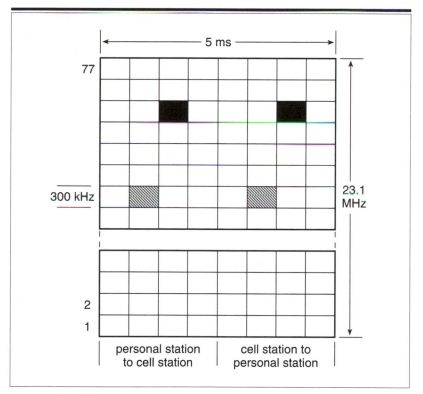

Figure 8.11 PHS physical channels.

Each time slot carries 240 bits. As in NA-TDMA, the modulation technique is differential quaternary phase shift keying (see Figure 5.4). The bit rate is

$$\frac{240 \text{ bits/slot} \times 8 \text{ slots/frame}}{0.005 \text{ s/frame}} = 384 \text{ kb/s.} \qquad (8.5)$$

The modulation efficiency is

$$\frac{384 \text{ kb/s}}{300 \text{ kHz}} = 1.28 \text{ b/s/Hz}, \tag{8.6}$$

intermediate between the modulation efficiencies of the cordless systems and the time-division cellular systems.

PHS specifies several time-slot formats corresponding to different logical channels. The formats fall in two categories. Common control channels use *control physical slots* and traffic channels use *communication physical slots*. PHS assigns each carrier to either common-control-channel operation or to traffic-channel operation. Therefore, all the time slots on a carrier are in a single category, either control physical slots or communication physical slots. Figure 8.12 shows the format of a communication physical slot and a control physical slot. All slot formats have a guard time of 41.7 μs, corresponding to 16 bits. This is followed by a 4-bit ramp time in which the personal station or cell station turns on its transmitter and a 2-bit start symbol (10) for establishing the phase of the remote demodulator. The other data field common to all time slot formats is a 16-bit cyclic redundancy check at the end of the slot. Every time slot also carries a *unique word*, known in advance by the receiver. Together, the start symbol, preamble, and unique word help synchronize the receiver to

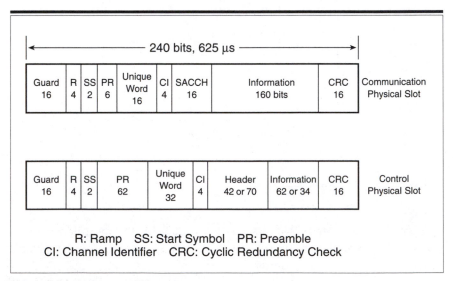

Figure 8.12 PHS time-slot formats.

the incoming signal. In all time slots, the start symbol and preamble together are repetitions of the pattern 1001. Personal stations and cell stations transmit different unique words. In a communication physical channel, these three fields contain a total of 24 bits. In a control physical channel, they contain 96 bits.

In each communication physical slot, the 4-bit channel identifier is similar to the flag bits in every GSM time slot (see Figure 7.8). In PHS, the channel identifier sequence 0000 indicates that the time slot carries user information, and 0001 indicates that the time slot carries a fast associated control channel (FACCH). As in NA-TDMA, the slow associated control channel (SACCH) field conveys out-of-band control signals between a cell station and a personal station with a call in progress. The most important data field in Figure 8.12 is the information field. For telephone conversations this field contains 160 bits from an adaptive differential pulse code modulation encoder. Therefore, in common with the other low-tier systems, the bit rate is

$$\frac{160 \text{ bits/frame}}{0.005 \text{ s/frame}} = 32 \text{ kb/s}. \tag{8.7}$$

Control physical slots have headers containing addresses. For some logical channels the header consists of a 42-bit cell station identification code and the information field contains 62 bits. In other logical channels the header contains a 42-bit cell station identification code and a 28-bit personal station identification code. In these channels, the length of the information field is 34 bits. This implies that the information rate for logical control channels is either

$$\frac{62 \text{ bits/frame}}{0.005 \text{ s/frame}} = 12,400 \text{ b/s or}$$
$$\frac{34 \text{ bits/frame}}{0.005 \text{ s/frame}} = 6,800 \text{ b/s}. \tag{8.8}$$

8.3.3 Logical Channels

PHS defines logical channels in two categories as indicated in Figure 8.13. Logical control channels are shared among all the terminals in a cell. They operate in control physical slots. Service channels are assigned to individual terminals. They operate in communication physical slots. Figure 8.13 indicates that there are four logical control channels. The broadcast con-

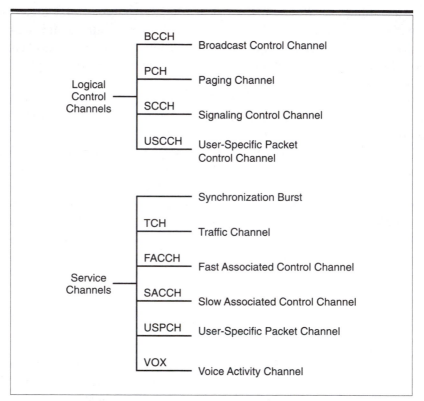

Figure 8.13 PHS logical channels.

trol channel (BCCH) and paging channel (PCH) transmit information from cell stations to personal stations. The BCCH carries general system information to all terminals in a cell. The PCH informs individual terminals of arriving calls. To set up a call, personal stations and cell stations exchange information on two-way signaling control channels (SCCH). For cell station transmissions, SCCH time slots are multiplexed with BCCH time slots and PCH time slots in logical control channel superframes. Cell stations establish the superframe structure and inform personal stations of the structure by means of messages transmitted on the BCCH. Corresponding to the sleep mode of digital cellular systems, PHS defines a *battery saving* mode of operation for personal stations that do not have calls in progress. In this mode, cell stations schedule the transmission of paging messages in a manner that enables terminals to turn off their receivers for a high proportion of the time. Personal stations contend for access to the reverse direction SCCH by means of the slotted ALOHA protocol [Tannenbaum, 1988, Section 3.2]. The fourth logical control channel, a user-specific packet control channel (USCCH), is optional in PHS.

This is a two-way channel for packet data communication of user information over control physical slots. This channel can be used for transmitting short messages to and from personal stations that do not have calls in progress.

Among the six information formats in the service channel category, the slow associated control channel (SACCH) and fast associated control channel (FACCH) are similar to their counterparts in NA-TDMA. Each time slot of a physical channel assigned to a terminal carries 16 bits in a SACCH format. Normally the information field of a service channel time slot carries user information in a traffic channel (TCH) format. However, when it has to transmit a message rapidly, a personal station or cell station interrupts the flow of user information to transmit a message in the FACCH format. Each message fits into the 160-bit information field of a single time slot.

Prior to exchanging user information, personal stations and cell stations transmit synchronization bursts containing a preamble field (62 bits) and a unique word field (32 bits) that are longer than those shown in Figure 8.12. The information field of a synchronization burst contains 108 bits that carry personal station and cell station identifiers. There is no SACCH field in a synchronization burst. The other two logical channels in the service channel category are optional in PHS. The USPCH is a user-specific packet data channel that uses a service channel assigned to a specific terminal. The time-slot structure of the USPCH resembles the communication physical slot in Figure 8.12. The difference is that the USPCH does not contain a SACCH field. Instead, the length of the information field is 176 bits. The other optional logical channel category is a VOX channel. The VOX channel operates in connection with a speech-activity detector that conserves battery power by turning off the personal station transmitter when it detects a pause in the speech of the subscriber using the personal station. When this occurs, the personal station transmits a special message in VOX format in one out of every four frames. This message describes the background noise at the personal station.

8.3.4 Network Operations

In a manner similar to DECT and GSM, PHS formally defines three network control protocols:

- radio frequency transmission management (RT),
- mobility management (MM), and
- call control (CC).

RT incorporates radio resources management functions and also signal encryption. In addition to the mobility management functions described in Section 2.4.2, the PHS MM protocol also performs authentication procedures.

PHS specifies three phases of network control operations performed during the course of each call: link channel establishment, service channel establishment, and communications. The link channel establishment phase consists of RT operations. In this phase, personal stations and base stations exchange messages on common control channels for the purpose of assigning a dedicated channel for communications. The personal station begins the process by sending a LINK CHANNEL ESTABLISHMENT REQUEST message on a signaling control channel (SCCH) to the cell station. The stimulus for this request is either a paging message received from the cell station, a call originated by the mobile subscriber, or a stimulus to register the location of the personal station. In contrast to the other low-tier systems, service channel allocation in PHS is the responsibility of the cell station. On receiving a request from the personal station, the cell station selects a service channel and sends a LINK CHANNEL ASSIGNMENT message to the personal station on the SCCH. This message identifies the physical channel (carrier and time slot) to be used to set up the call. The personal station tunes to the assigned channel and the two stations transmit synchronization bursts to establish communications on the new channel.

If the purpose of the channel allocation is to set up a call, the system enters the service channel establishment phase. In this phase, the personal station and cell station exchange mobility management (authentication) messages and call control messages, which closely resemble the messages used to set up a GSM call. When both parties are connected, the call enters the communications phase in which the principal activity is the exchange of user information. During a communication, either network element can command the other one to adjust its transmitted power by transmitting a TRANSMISSION POWER CONTROL message on either the slow associated control channel or the fast associated control channel. This message requests an adjustment between −32 dB and +32 dB relative to the current power.

In PHS, the terminology for handoff is *channel switching*. Both the cell station and personal station monitor channel quality during a call. The two quality measures are *received signal strength* and *frame error rate* as indicated by detected errors in the cyclic redundancy check in every time slot. In response to deteriorating quality, either station can initiate a handoff. PHS is capable of several types of channel switching. The type of channel switching that occurs depends on:

- whether the cell station or the personal station initiates the procedure,

- whether or not the call moves to a new cell station,

- whether the old cell station forwards the call to the new cell station after the handoff, or the network routes the call directly to the new cell station. Figure 8.14 illustrates these two alternatives.

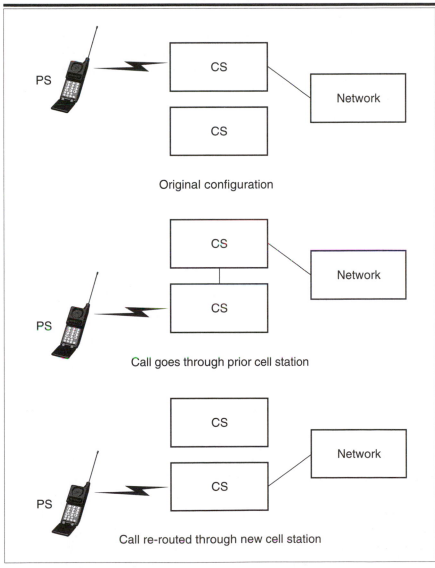

Figure 8.14 Alternative PHS channel switching procedures.

Consider a handoff initiated by the personal station. If the frame error rate is high and the received signal strength indicates that the terminal is still in the vicinity of the serving cell station, the terminal transmits a *TCH SWITCHING REQUEST* message to the cell station on the fast associated control channel. The cell station responds with a *TCH SWITCHING INDICATION* message, directing the personal station to a new physical channel. The two stations exchange synchronization bursts on the new channel and resume communications. If, on the other hand, the personal station determines that the received signal strength is too weak to maintain communications with the serving cell station, it selects a new cell station on the basis of signal strength measurements on common control channel signals received from surrounding cells. To move to a new cell station, the personal station transmits a *LINK CHANNEL ESTABLISHMENT REQUEST* message to the new cell station on a signaling control channel (SCCH). The new cell station assigns a channel to the call and informs the personal station by means of a *LINK CHANNEL ASSIGNMENT* message. The personal station and the cell station then synchronize their operations. If the call is to be routed through the previous cell station, the stations then resume communication of user information through the new cell station. If, however, the call is re-routed by the network to the new cell station, the personal station and new cell station exchange call control messages and authentication messages prior to transmitting user information through the new cell station.

8.4 Personal Access Communications System (PACS)

PACS [Noerpel, 1996; ANSI, 1996c] is the product of research conducted in the 1980s and early 1990s at Bell Communications Research (Bellcore), which was then owned by a consortium of United States local telephone companies. The original purpose of the research was to create a means of wireless access to conventional telephone services. The idea was to install, in a residential area, an array of radio base stations, separated from one another by about 400 meters. The base stations would communicate with radio transceivers in subscribers' homes and thereby eliminate the need for wires connecting telephone company equipment to customer premises. During the time PACS was developed, changes in telecommunications regulations in the United States made it possible for local telephone companies to offer mobile communications services. In response to these changes, the PACS radio transmission system and infrastructure were enhanced to support communications with mobile terminals in addition to the original application of wireless subscriber lines.

Originally, the name of the PACS system was WACS (Wide Area Communications System [Bell Communications Research, 1994]). When the United States telecommunications industry adopted standards for Personal Communications Services in the 1,900 MHz band [Cook, 1994], they decided to modify the WACS radio transmission system to make it similar to PHS. The modified system is now an American standard [ANSI, 1996c]. In 1997, the principal application of PACS is to provide telephone services to people carrying inexpensive, lightweight terminals while moving at pedestrian speeds or moderate vehicle speeds in urban and suburban environments. A modified version of PACS has been developed for operation in the band 1,910–1,930 MHz [Steer, 1994], allocated in the United States for unlicensed personal communications systems. Referred to as PACS-UB, this unlicensed version of PACS uses time-division duplex for two-way communications [ANSI, 1996b].

8.4.1 Architecture

The PACS network architecture, in Figure 8.15, reflects its original purpose: to provide access to existing local telephone systems. The PACS nomenclature for a terminal is *subscriber unit* (SU). Base station functions are performed by two network elements that correspond to base transceiver stations and base station controllers in GSM (see Figure 7.2). The PACS counterparts are radio ports (RP) and radio port control units

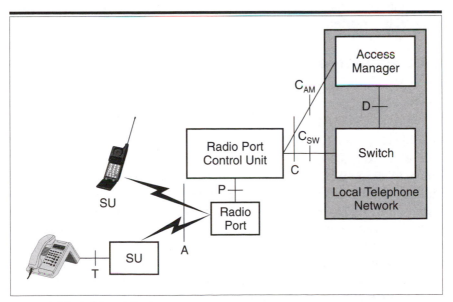

Figure 8.15 PACS architecture.

(RPCU). The major PACS departure from cellular systems is in the switch. Every cellular system has its own network element (referred to as a *mobile telephone switching office* or a *mobile switching center*) that performs switching functions and controls other network operations. Rather than specify a separate network element for switching, PACS relies on switching equipment in local telephone systems to perform switching functions as well as other network operations. To meet the additional requirements of a mobile wireless communications system, PACS specifies an *access manager* located in the local telephone system. Access managers and telephone company switches in PACS together perform the functions performed by mobile switching centers in cellular systems. The D-interface specifies communications between an access manager and telephone company switches. The C_{AM}-interface specifies communications between an access manager and a radio port control unit. The C_{SW}-interface specifies communications between a switch and a radio port control unit. Together, the C_{AM}-interface and the C_{SW}-interface comprise the PACS C-interface, which connects a radio port control unit to a local telephone network. The other PACS interfaces are:

- the P-interface, which links a radio port control unit to radio ports;

- the A-interface, which is the air interface between a subscriber unit and a radio port; and

- the T-interface, which links a conventional telephone to a fixed subscriber unit.

8.4.2 Radio Transmission

PACS is a hybrid time-division/frequency-division multiple access system with the slot and frame configuration of Figure 8.16. The frame duration is 2.5 ms, with each frame divided into eight time slots. The duration of a time slot is 2.5 ms/8 = 312.5 µs. The carrier spacing is 300 kHz. Initial applications of PACS are in the 1,900 MHz PCS bands in the United States. As shown in Figure 8.16, in this application PACS employs frequency-division duplex with a carrier separation of 80 MHz. The timing offset of 375 µs, just over one time slot, makes it unnecessary for terminals to transmit and receive simultaneously. Figure 8.17 shows a physical channel in a system operating with 30 carriers in one of the North American PCS bands. A full-rate physical channel occupies one time slot in each frame. PACS also specifies subrate channels that occupy one time slot in every two frames or one time slot in every four frames.

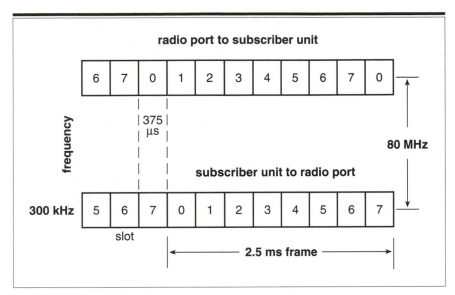

Figure 8.16 Time slots and frames.

As in NA-TDMA and PHS, the modulation technique is differential quaternary phase shift keying (DQPSK). Each time slot carries 120 bits, so that the transmission bit rate is

$$\frac{120 \text{ bits/slot} \times 8 \text{ slots/frame}}{0.025 \text{ s/frame}} = 384 \text{ kb/s} \tag{8.9}$$

identical to that of PHS.

DQPSK provides higher modulation efficiency than the modulation techniques of CT2 and DECT. In common with PHS, each PACS physical channel carries 1.28 b/s/Hz (Equation 8.6), compared with modulation efficiencies of approximately 0.7 b/s/Hz in CT2 and DECT. However, the higher PACS modulation efficiency comes at the expense of lower power efficiency in subscriber units and higher vulnerability to transmission impairments. To provide robust signal transmission, PACS prescribes antenna diversity at subscriber units as well as at base stations.

In common with NA-TDMA, PACS base stations transmit continuously on carriers that contain active time slots, even when some of the time slots are not assigned to physical channels. Therefore, there are no guard times in forward direction transmissions and the time slots carrying information from radio ports to subscriber units have different structures from the time slots carrying information from subscriber units to

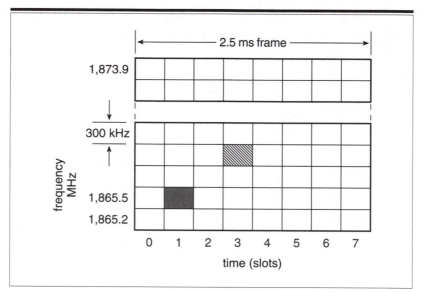

Figure 8.17 PACS physical channels.

radio ports. Figure 8.18 shows the two time-slot configurations. In the reverse direction there is a guard time of 12 bit intervals (31.25 µs) and a 2-bit training sequence to assist the demodulator at the radio port. In the forward direction, these fields are replaced by a 14-bit synchronization field. In both directions, there are two fields for information transfer, referred to in PACS as a *slow channel* field consisting of 10 bits per time

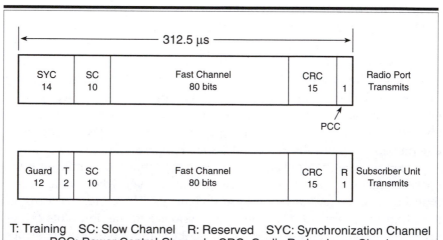

Figure 8.18 PACS time slots.

slot, and a *fast channel* field with 80 bits. In telephone applications, the fast channel field carries speech at a rate of

$$\frac{80 \text{ bits/time slot}}{0.025 \text{ s}} = 32 \text{ kb/s} \qquad (8.10)$$

in an adaptive differential pulse code modulation (ADPCM) format. The slow channel field carries network control information. A 15-bit cyclic redundancy check error-detecting code protects the 90 bits in the slow channel and fast channel fields. In the forward direction the remaining 1 bit per slot is used to send power control information to subscriber units. The corresponding bit in the reverse direction is unused in the initial PACS implementation.

8.4.3 Logical Channels

The PACS logical channels shown in Figure 8.19 fall in two categories. A system broadcast channel (SBC) is a common control channel, shared by all subscriber units using a radio port. A traffic channel (TC) is a point-to-point channel for communications between a radio port and an individual subscriber unit.

There is always a system broadcast channel active at each radio port. An SBC occupies time slot 5 of one carrier. The SBC statistically multiplexes five logical types of information:

- a synchronization channel (SYC) carried in the 14-bit synchronization field,

- channel rate information carried in the 10-bit slow-channel field,

- a system information channel (SIC) carried in the 80-bit fast channel field,

- an alerting channel (AC) carried in the 80-bit fast channel field, and

- a priority request channel (PRC) carried in the 80-bit fast channel field.

The SBC primarily carries information from a radio port to subscriber units. The synchronization information establishes the position of the current time slot in the PACS frame. Because an SBC always occupies time slot number 5, an SU initializes its communications with a radio port by searching for the 14-bit SYC sequence that characterizes time slot 5. A 4-bit sequence in the slow channel field distinguishes the system broadcast channel from a traffic channel that occupies time slot 5 on another carrier.

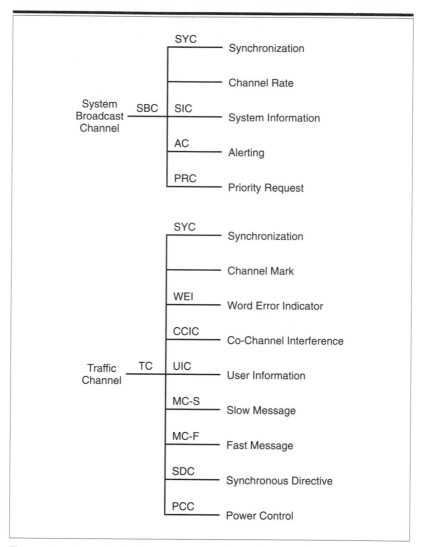

Figure 8.19 PACS logical channels and information formats.

If the subscriber unit finds time slot 5 on one carrier occupied by a traffic channel, it tunes to another carrier and examines time slot 5 to determine if it carries a system broadcast channel. A system can operate a SBC at one of four rates: 32 kb/s, 16 kb/s, 8 kb/s, or 4 kb/s. At 32 kb/s, the SBC occupies time slot 5 in every frame. At each of the lower rates, it occupies time slot 5 in a fraction of the frames. The other frames can then carry traffic channel information in time slot 5. Three bits of the slow channel field of the SBC indicate the SBC transmission rate.

The system information channel serves a function similar to broadcast control channels in cellular systems, and the alerting channel corresponds to cellular paging channels. The priority request channel is the only two-way logical channel in the SBC. It allows subscriber units to request special treatment in setting up calls, such as emergency calls.

Except for the physical channel that carries the SBC, all physical channels at a radio port operate as traffic channels (TC). A traffic channel carries the following logical channels and information categories:

- synchronization information (SYC) transmitted by the radio port as in an SBC,

- channel marking information transmitted in the 10-bit slow channel field,

- word error indicator (WEI), transmitted in the slow channel field,

- co-channel interference control (CCIC) transmitted in the slow channel field,

- a user information channel (UIC) transmitted in the 80-bit fast channel field of an active traffic channel time slot,

- a slow message channel (MC-S) transmitted in the slow channel field,

- a fast message channel (MC-F) transmitted in the fast channel field,

- a synchronous directive channel (SDC) transmitted in the slow channel field, and

- a power control channel (PCC) transmitted by the radio port.

As in the SBC, synchronization information transmitted by the radio port on a traffic channel indicates the position of the current time slot in the frame. It also indicates whether the traffic channel is busy or available. The channel-marking information transmitted in the slow channel field of each traffic channel enables a PACS system to perform channel assignment and handoff functions. At any particular time, each traffic channel is classified as busy or available depending on whether it carries user information. A channel carrying user information transmits a busy indication. An idle channel transmits channel-marking information that indicates whether the channel is available for any call, or only for priority access. It also indicates the available bit rate on the channel, which can be 32 kb/s, 16 kb/s, 8 kb/s, or 4 kb/s.

A word error indicator (WEI) occupies 1 bit of every slow channel field. The WEI transmitted by the radio port indicates whether it has successfully decoded the information most recently transmitted by the subscriber unit, and conversely the WEI transmitted by the subscriber unit indicates whether the information most recently transmitted by the radio port has been successfully decoded. Successful decoding refers to no errors in the cyclic redundancy check performed in each time slot. The co-channel interference control (CCIC) is a 4-bit code selected at random by the subscriber unit at the beginning of a call. During the call the subscriber unit and radio port transmit the CCIC from time to time as a handshaking procedure to verify that the incoming signal comes from the correct network element and not from an interfering element in another cell. The CCIC thus serves the same purpose as the supervisory audio tone (SAT) in AMPS and the DVCC (digital verification color code) in NA-TDMA.

In initial applications of PACS, the user information channel (UIC) occupies the fast channel field of one time slot of every frame. In this configuration, the UIC transmission rate is 32 kb/s. As in the other low-tier systems, the speech-coding format is adaptive differential pulse code modulation. PACS also specifies UIC transmission rates at 16 kb/s, 8 kb/s, and 4 kb/s. At these rates a UIC occupies one time slot in a fraction of the frames on a carrier. The slow and fast message channels (MC-S) and (MC-F) correspond to the slow and fast associated control channels in NA-TDMA, GSM, and PHS. They carry network control messages. MC-S is an out-of-band control channel carried in the slow channel fields of time slots occupied by active traffic channels. MC-F is a blank-and-burst channel that interrupts the flow of user information to send control messages in the fast channel fields of traffic channel time slots. The synchronous directive channel transmitted in slow channel fields carries brief radio resources management messages. The power control channel (PCC) occupies 1 bit of every traffic channel time slot transmitted by the radio port. Like the CDMA closed-loop power control signal (see Section 6.6.1), the radio port uses the PCC to command the subscriber unit to raise or lower its transmitter power by 1 dB. With a frame duration of 2.5 ms, PACS radio ports transmit power control commands at a rate of 400 b/s.

The fast and slow message channels (MC-F and MC-S) carry network control messages organized in 80-bit segments. Many messages fit into a single segment. The largest messages occupy four segments. The MC-F transmits a segment in the fast channel field of a single time slot. The MC-S occupies 8 bits of the slow channel field. (The other 2 bits are a word error indicator and a "busy" channel-marking bit.) Therefore a message segment of an MC-S is distributed over ten time slots, which implies a transmission time of 25 ms on a 32 kb/s traffic channel.

8.4.4 Network Operations

Like CT2 and DECT, PACS assigns much of the responsibility for radio resources management to subscriber units. Radio ports assist subscriber units by transmitting channel-marking information in the slow channel field of each time slot of a traffic channel. When it is necessary to establish communications over a traffic channel, a subscriber unit selects a carrier with a sufficiently high received signal strength. It then searches for a time slot with a 4-bit channel-marking indication stating that the time slot is available for communications at the bit rate required by the subscriber unit. On detecting a suitable time slot, the subscriber unit transmits an *INITIAL ACCESS REQUEST* message in the time slot it has selected. Depending on the quality of the received signal, the radio port control unit can accept or reject the request. When the request is accepted, the subscriber unit and the radio port exchange MC-F messages over the traffic channel to complete the call setup.

The PACS terminology for handoff is *automatic link transfer* (ALT). In a similar manner to initial channel allocation, the subscriber unit controls handoff in PACS. With a call in progress, the subscriber unit monitors the quality of communications on the user information channel by measuring received signal strength and also by monitoring the word error indicator bit in every received time slot. As occurs in the cellular systems with mobile-assisted handoff, the subscriber unit also measures the signal strength of other carriers to find candidate physical channels. The automatic link transfer procedure begins when the subscriber unit determines that the quality of the present physical channel is below an acceptable threshold and that another physical channel with sufficiently high quality is available. It then sends an *AUTOMATIC LINK TRANSFER REQUEST* message to the new radio port over the physical channel it has selected. Meanwhile, it continues to transfer user information over the old physical channel. If the PACS network accepts the link transfer request, it initiates communication on the new physical channel and sends an *AUTOMATIC LINK TRANSFER EXECUTE* message to the subscriber unit on the old channel. On receiving this message, the subscriber unit switches to the new physical channel and the network makes the old channel available for other communications. PACS identifies four categories of automatic link transfer, which depend on whether the new physical channel is associated with the same radio port, the same radio port control unit, or the same switch as the old physical channel. In addition to link transfer initiated by the subscriber unit, the radio port control unit can initiate a link transfer when it detects problems on the existing physical channel. To do so it sends a *PERFORM AUTOMATIC LINK TRANSFER* message to the subscriber unit.

The other novel radio resources management procedure in PACS is closed-loop power control facilitated by the power control channel that occupies 1 bit per time slot. Although the method of determining power adjustment commands is optional, the PACS standard gives an example of a procedure based on received signal strength measurements, a quality indication derived from the demodulator, and word error indication bits detected at the radio port.

In common with the North American cellular systems, PACS specifies cryptographic authentication and privacy procedures based on the CAVE algorithm (see Section 5.6.1).

Review Exercises

1. With respect to the design goals of Section 2.2, describe how low-tier systems differ from high-tier systems.

2. Why do low-tier systems have less protection against transmission errors than high-tier systems have?

3. All four low-tier systems have traffic channels with a bit rate of 32 kb/s. Comment on the capability of each system to support higher rate communications, for example 64 kb/s (ISDN B-channel) and 144 kb/s (ISDN basic rate).

4. To receive a telephone call, a terminal has to detect a paging message transmitted from a base station. How do terminals in the low-tier systems determine the channels that transmit paging messages?

5. How do radio receivers in the low-tier systems verify that an arriving signal comes from the correct transmitter rather than from an interfering transmitter?

6. How do network elements in the low-tier systems monitor the quality of communications?

7. What are some advantages and disadvantages of the link establishment and link re-establishment techniques used in CT2 compared to the approaches taken in cellular systems?

8. What are some advantages and disadvantages of the DECT handoff procedure compared to the approach taken in GSM?

9. Describe the similarities and differences between the radio transmission systems of PACS and PHS.

10. Describe the similarities and differences between the channel assignment and handoff procedures in PACS and PHS.

CHAPTER NINE

Science and Technology Tutorials

To confront the triple challenge of mobility, ether, and energy described in Section 1.5, personal communications systems harness a diverse set of technologies, many of which appear in other information systems. This book examines the way in which specific systems assemble these technologies in order to meet the priorities of the systems' designers. Each technology is the product of decades of research and development documented in journal publications, patents, textbooks, and university courses. People reading this book will already be familiar with some of these technical subjects, and perhaps have professional experience in one or more of them. On the other hand, many readers will not be acquainted with all of the diverse technical subjects relevant to personal communications. To assist those readers, this chapter introduces twelve technical subjects in a general way by presenting some of their most important properties and describing their roles in personal communications systems. These brief introductions will not, of course, substitute for the detailed literature on each of the topics. Instead, the tutorials are intended to give readers the information they need to understand the descriptions of personal communications systems in Chapters 1 through 8. Depending on their needs, readers can later plunge into details with the help of technical literature and classroom courses.

The tutorials begin with a discussion of multiple access (Section 9.1). Each of the systems that we study has its own method for establishing physical channels that enable a receiver to extract a desired signal from all of the other signals that arrive at its antenna. Many people regard the access technology as the property that most distinguishes a system from other systems. Indeed, in this book, we refer to two systems by their multiple access methods: NA-TDMA (North American time division multiple access, Chapter 5) and CDMA (code division multiple access, Chapter 6). Section 9.2 covers mobile radio signal propagation. Many of the techniques presented in subsequent sections are present in personal communications systems as a consequence of the properties of the signals arriving at radio receivers in personal communications systems. Section 9.3 addresses the important issue of spectrum efficiency, which depends on two properties of the signals transmitted in a personal communications system: compression efficiency and tolerance to interference. In order to achieve high spectrum efficiency, systems employ a diverse array of signal-processing operations, including the four techniques described in Sections 9.4 through 9.7. Channel coding (Section 9.4), interleaving (Section 9.5), and adaptive equalization (Section 9.6) all promote high tolerance to interference. Linear prediction coding of speech (Section 9.7) contributes to compression efficiency. Sections 9.8 and 9.9 address topics relevant to the CDMA system described in Chapter 6: soft capacity and Walsh Hadamard matrices.

The final three tutorials describe communications networks that have influenced the design of many personal communications systems. Section 9.10 introduces the Open Systems Interconnection model, which provides an architecture for the design of digital networks. Integrated Services Digital Networks (ISDN) (Section 9.11) and Signaling System Number 7 (Section 9.12) are collections of communication protocols widely used in public telephone networks. Many of these protocols also appear in personal communications systems.

9.1 Multiple Access

In a personal communications system, the access technology makes it possible for a receiver to separate the desired signal from interfering signals. As in broadcast radio and television, the original approach to media access in cellular and cordless systems is frequency division multiple access (FDMA), in which each signal occupies its own band of frequencies. This situation can be illustrated by a frequency-time diagram, such as Figure 9.1. Two different signals, indicated with different shading, simultaneously occupy the frequency-time plane. In an FDMA system, a physical channel corresponds to a band of frequencies. The frequency at the center of the band is the *carrier*. Thus, in Figure 9.1, each horizontal stripe represents one carrier capable of transmitting one signal. The total number of signals that can be transmitted is equal to the number of carriers. The modulation technique determines the required carrier spacing. Figure 9.1 represents a system with 12 physical channels. This book

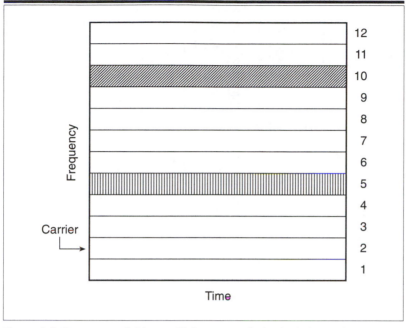

Figure 9.1 Frequency division multiple access. A physical channel corresponds to a carrier. Each signal occupies one carrier for the duration of a communication.

describes two FDMA systems. The analog cellular system, AMPS (see Chapter 3), contains 832 carriers, with adjacent carriers separated by 30 kHz. The digital cordless telephone system, CT2 (see Section 8.1), contains 40 carriers, with adjacent carriers separated by 100 kHz.

The complement of FDMA is time division multiple access (TDMA), in which each signal uses the entire frequency band of the system for a fraction of the time, as shown in Figure 9.2. To perform TDMA, a transmitter stores the source information that arrives in a time interval referred to as a TDMA *frame*. A TDMA system transmits each signal for a fraction of the frame. The time interval occupied by one signal is a *slot*. In TDMA, a physical channel corresponds to one time slot. When its time slot begins, a transmitter sends the stored information at an accelerated speed, so that all the information recorded in one frame is transmitted in a slot. (Like the "Donald Duck" sounds produced when an audio tape is played back at a higher speed than the recording speed, the transmitted signal occupies a wider band of frequencies than the source signal.) The receiver records the signal arriving in its slot and plays it back at the original slower rate. The playback interval is one frame and the received information, restored

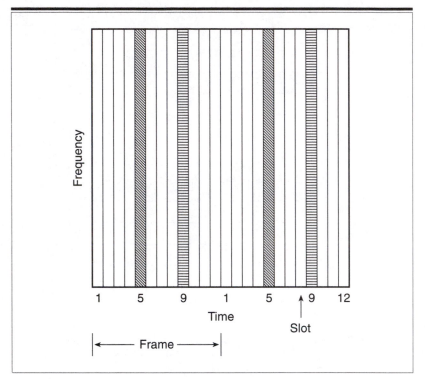

Figure 9.2 Time division multiple access. Each signal occupies the entire frequency band for a fraction of the time. The signal appears in one time slot per frame. The time slot corresponds to a physical channel.

to the lower rate, emerges in a continuous stream with no gaps. In principle, time division techniques could be used for analog communications. In practice, however, they appear only in digital systems, where it is a simple matter to store binary signals arriving at the source rate, and then release them at the faster channel rate.

The number of physical channels in a time division system is the number of slots per frame, which is the ratio of the frame duration to the slot duration. In Figure 9.2, each frame has 12 slots corresponding to the 12 carriers in Figure 9.1. Two shaded vertical stripes in Figure 9.2 together occupy the same area as the correspondingly shaded horizontal stripe in Figure 9.1.

It is a basic principle of communications that two regions in the time-frequency plane with equal areas can carry the same amount of information regardless of its shape. Thus, if the signals in Figures 9.1 and 9.2 occupy the same area, they are equivalent in their information capacity. There are myriad ways of dividing up the time-frequency plane. In

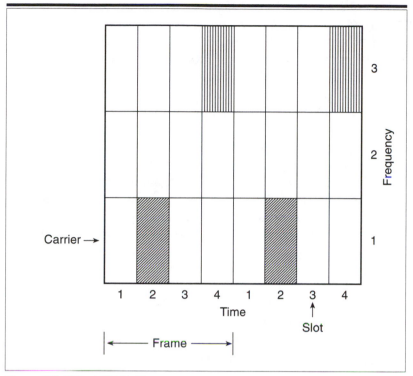

Figure 9.3 Hybrid FDMA/TDMA. A physical channel is a time slot within one carrier.

practical personal communications systems, the multiple access techniques referred to as TDMA take the form of Figure 9.3, which is a combination of frequency division and time division. In Figure 9.3, there are three carriers, each capable of carrying four signals. These four signals use time division to share the carrier, and there are four slots per frame. In this hybrid approach, a physical channel corresponds to a time slot and a carrier. The shaded parts of Figure 9.3 have the same total area as each of the shaded parts of Figures 9.1 and 9.2, and once again there are 12 physical channels in the multiple access technique. The majority of the systems covered in this book employ hybrid time-division/frequency-division multiple access.

The third type of multiple access encountered in this book is CDMA (code division multiple access). In CDMA, many signals simultaneously occupy the same wide band of frequencies. The cellular CDMA system (see Chapter 6) employs direct sequence spread spectrum modulation. With this technique, a physical channel corresponds to a binary code. Each physical channel has its own binary code, which is a sequence of binary symbols referred to as *chips*. To transmit one information bit, a CDMA system transmits its entire code if the bit is a 1. For a 0, it transmits the complement of its code. CDMA differs from TDMA and FDMA in the mutual effects that different signals produce when they arrive at the same receiver. TDMA and FDMA are similar in that a receiver can completely separate the signals arriving on different physical channels. This is not the case in CDMA. Owing to the nature of the physical channels, the output of a receiver contains small components of all of the input signals. Provided that the number of simultaneous transmissions is not too high, each receiver, with high probability, accurately detects the transmitted information signal. As the number of simultaneous transmissions increases, the interference increases, causing receivers to make an increasing number of digital errors. The channel capacity definition for CDMA is different from the definition for TDMA and FDMA. In TDMA and FDMA, capacity corresponds to the number of physical channels. CDMA capacity is the highest number of simultaneous transmissions consistent with an error-rate objective established for a system. For voice transmission, a typical error rate limit is 0.001 errors per bit. Section 9.8 describes CDMA capacity in more detail.

9.2 Mobile Radio Signals

The nature of the radio signals arriving at a receiver in a personal communications system motivates a high proportion of the technology of personal communications. Signals undergo a variety of alterations as they travel from transmitters to receivers. Some of the changes are due to the distance between transmitter and receiver. Others are due to the physical environment of the propagation path, including terrain, buildings, and vehicles in outdoor systems. Indoors, the radio signals are influenced by building materials and furniture. The signals are also influenced by the motion of the terminal. Together these physical conditions produce four main effects:

- attenuation that increases with distance,
- random variations due to environmental features,
- signal fluctuations due to the motion of a terminal, and
- distortions due to the fact that components of the signal travel along different paths from a transmitter to a receiver.

The following paragraphs describe these effects individually.

9.2.1 Attenuation Due to Distance

On average, the signal strength decreases with distance according to the relationship:

$$P_{\text{receive}} = P_{\text{transmit}} \frac{\text{const}}{x^{\alpha}} \tag{9.1}$$

where P_{receive} watts and P_{transmit} watts are the power of the received and transmitted signals, respectively. The distance between the terminal and the base station is x meters and α is an exponent that characterizes the steepness of the decrease. In physics, we learn that radiated power conforms to an inverse square relationship, corresponding to $\alpha = 2$. Although this describes radio waves propagating in free space, there is usually additional attenuation in wireless networks, which brings $\alpha > 3$ with the precise value dependent on terrain and other environmental factors such as buildings. In metropolitan areas, $\alpha = 4$ is often used for planning purposes. However, it is also possible to find $\alpha \le 2$ as a result of wave-guide effects that appear in indoor corridors and in cities when antenna heights are substantially lower than surrounding buildings.

9.2.2 Slow Fading Due to Random Environmental Effects

Although Equation 9.1 represents the average signal strength, the power measured at a specific location is likely to be above or below this average. For example, if there are significant obstacles between the transmitter and receiver, the signal strength will be lower than P_{receive}. In other locations the signal strength will be higher. As a terminal moves, the signal strength gradually rises and falls with significant changes occurring over tens of meters. This phenomenon is referred to as *shadow fading* or *slow fading*. To describe shadow fading, we characterize the received power P watts as a random variable with a log-normal probability model with mean value P_{receive}. In personal communications systems, it is customary to express signal strength in units of dBm, decibels relative to 1 milliwatt. Thus,

$$S = 10 \log_{10}(1000P) \text{ dBm.} \tag{9.2}$$

The log-normal distribution of P implies that S has a normal distribution with mean value

$$S_{\text{receive}} = 10 \log_{10}(1000P_{\text{receive}}) \text{ dBm.} \tag{9.3}$$

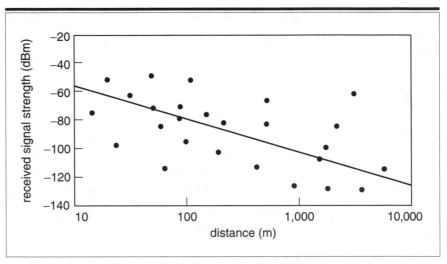

Figure 9.4 Relationship of received signal strength to distance between transmitter and receiver. The line corresponds to $\alpha = 3.5$.

The standard deviation of S is a quantity σ dB, that also depends on the physical environment. Typical values are 4 dB $\leq \sigma \leq$ 10 dB. It is common to observe low values of σ in rural areas with uniform terrain and high values of σ in cities with complex features including hills, water, and a variety of building types. Together Equations 9.1 and 9.3 imply a linear relationship between average received signal strength measured in decibel units and the logarithm of the distance between transmitter and receiver:

$$S_{\text{receive}} = S_{\text{transmit}} + \text{const} - 10\alpha \log_{10}(x) \text{ dBm}. \qquad (9.4)$$

Figure 9.4 is typical of signal strength measurements. The vertical axis represents power averaged over several milliseconds (millions of cycles of the carrier frequency). The line represents Equation 9.4 with $\alpha = 3.5$. The solid circles are individual measurements.

9.2.3 Rayleigh Fading Due to Motion of Terminals

This section examines the fine detail of a received signal. The signal arrives at a receiver as a collection of rays coming from different directions. As the terminal moves, each ray undergoes a Doppler shift, causing the apparent wavelength of the signal to either increase (if the terminal is moving in the same direction as the ray) or decrease (if the terminal is moving in a direction opposite to the direction of a ray). Rays arriving at right angles to the direction of motion experience no shift in wavelength. The different

Doppler shifts in the many rays arriving at the receiver cause the rays to arrive with different relative phase shifts that depend on the exact position of the terminal. At some locations, the rays reinforce each other and the received signal is relatively strong. At other locations, the rays cancel each other out and the received signal is weak.

These fluctuations occur much faster than the changes due to environmental effects described in Section 9.2.2. Significant changes occur over a distance equal to the wavelength of the carrier frequency, which is on the order of 30 cm in cellular radio bands and 15 cm in PCS frequency bands. It follows that the speed of the changes is related to the carrier frequency of the signal and to the velocity of the terminal. In a cellular system operating at 850 MHz with a terminal moving at 50 km/hr, the signal exhibits approximately 40 peaks and nulls per second. The phenomenon is referred to as *Rayleigh fading* or *fast fading*. This fading presents a severe challenge to the quality of received signals. Even when the average signal strength (see Figure 9.4) is high, there are instants when the signal is very weak and difficult to demodulate accurately.

9.2.4 Intersymbol Interference Due to Different Signal Paths

In a personal communications system, there are many ways for a signal to travel from a transmitter to a receiver. Sometimes there is a direct, unobstructed path. The signal can also be reflected from buildings, hills, vehicles, and other objects. Figure 9.5 shows five different transmission paths between a transmitter and a receiver. Components of the transmitted signal travel on each of the paths. The receiver assembles all of them. However, each path has its own propagation time that is proportional to path length. The path differences can be on the order of microseconds or tens of microseconds depending on location. The effect of this multipath propagation is to distort the signal, with the degree of distortion dependent on the relative path lengths. At home, we observe this distortion from time to time as a "ghost" appearing in a television picture when an airplane passes. In a digital system, distortion due to multipath propagation is referred to as *intersymbol interference*. Figure 9.6 shows the effect of intersymbol interference on a digital signal for two different situations. The graph on the left shows the original binary signal. The center graph shows the received signal when the maximum delay difference between all the paths is $0.3T$ seconds, where T seconds is the bit period. The intersymbol interference is modest. The graph on the right shows the received signal when the path difference exceeds 1 bit period. In this case the maximum path delay difference is $1.7T$ seconds. The distortion is severe.

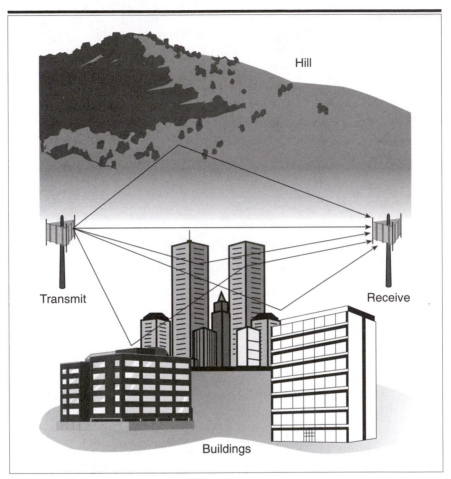

Figure 9.5 Multipath propagation.

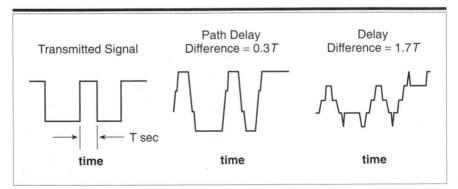

Figure 9.6 Intersymbol interference caused by multipath propagation.

9.2.5 Technology Implications

The four signal transmission conditions described in the preceding sections all influence the design and operation of personal communications systems. Section 9.3 describes how the signal attenuation determines the assignment of channels to cells. Systems employ power control to overcome the effects of slow fading. Fast fading and multipath propagation introduce severe distortions to signals and together pose a major challenge to the design of radio transmitters and receivers, particularly in the high-tier systems that cater to terminals moving at vehicular speeds. Systems use a large array of techniques to respond to this challenge, including

- channel coding (see Section 9.4),
- interleaving (see Section 9.5),
- equalization (see Section 9.6),
- RAKE receivers (see Section 6.3),
- slow frequency hopping (see Section 7.3.3), and
- antenna diversity.

All personal communications systems employ antenna diversity at their base stations. The PACS system (Section 8.4) prescribes antenna diversity in terminals as well. With antenna diversity, there are at least two receiving antennas, installed in a manner that makes it likely that when a signal is very weak at one antenna, the signal at another antenna will be adequately strong. This occurs with high probability because the spatial patterns of signal reinforcement and cancellation (Section 9.2.3) are statistically independent at locations separated by at least one wavelength.

9.3 Spectrum Efficiency

As discussed in Section 2.3.19, the spectrum efficiency of a personal communication system has a strong influence on the cost of providing service. In addition to personal communications, there are many information systems with economic incentives to reduce the bandwidth (or bit rate) necessary to carry each signal. These incentives have stimulated great advances in compression technologies, including low-bit-rate source coding and bandwidth-efficient modulation. These technologies pack signals into narrow frequency bands (or, in the case of computer storage, into small disk areas).

In personal communications, signal compression is only half of the story. The other half is tolerance to interference. This phenomenon is conveyed by the term *reuse*, which refers to the fact that within a limited geographical area, several different signals occupy the same frequency band. Because signals travel through the air, they are vulnerable to interference from one another. If signals have a high tolerance of interference, the same frequency band can be reused within a relatively small geographical area, and thus serve a high density of subscribers. Signals with low tolerance of interference require a longer distance between interfering transmitters and a lower density of subscribers using the same physical channel. Compression efficiency and reuse together determine spectrum efficiency. They are both properties of a system's signal processing and transmission technologies, including source coding, channel coding, and modulation.

In studying personal communications it is essential to be aware of both properties of a system because it is usually necessary to strike a balance between them. This is because techniques with high compression efficiency tend to be vulnerable to interference and thus have a low reuse density. These issues are discussed quantitatively in the following paragraphs. Section 2.3.19 describes the motivation for adopting conversations per base station per MHz as the efficiency measure for a telephone-oriented system. For other information services there are different efficiency measures, for example, bits per second per base station. However, the same general principles apply.

9.3.1 Compression Efficiency and Reuse Factor

To describe compression efficiency we adopt the notation C *conversations per MHz* to denote the compression efficiency, defined as the number of conversations, per MHz of assigned bandwidth, that would be possible in a single cell with no interference from surrounding cells. In a digital system, the compression efficiency is the ratio of modulation efficiency to source coding rate. The modulation efficiency, expressed in b/s/Hz, is the ratio of the transmission rate to the channel bandwidth. The source coding rate, in b/s, is the total bit rate of one speech signal including channel coding.

The other influence on spectrum efficiency is the *channel reuse factor, N*. This is a dimensionless quantity that takes into account the effects at each cell of interference from signals generated in other cells. Specifically, the channel reuse factor indicates the capacity reduction due to interference from signals transmitted in other cells. With a reuse factor N, the number of conversations per base station per MHz goes from C, for a single-cell system, to $E = C/N$ conversations per base station per MHz for a personal communications system. The reuse factor depends on each signal's tolerance of interference from other signals. A measure of this tolerance is the signal-to-interference ratio, $(S/I)_{req}$, required to meet transmission quality goals. In this ratio, the numerator is the received power of the signal directed to a base station or terminal. The denominator is the combined power of all interfering signals at the base station or terminal receiver. A high tolerance to interference promotes cellular efficiency. It is reflected in a low $(S/I)_{req}$, and a low reuse factor, N. Figure 9.7 is a summary of the relationships between the two phenomena, compression efficiency and interference tolerance, that determine spectrum efficiency.

In practical systems, the reuse factor takes different forms depending on the method of radio resources management. Cellular telephone systems based on TDMA and FDMA employ fixed channel allocation. With reuse factor N, a cellular system divides the total number of channels into N groups. Each base station operates with one group of channels (see Section 9.3.2). In cordless systems, with dynamic channel allocation, the reuse factor varies randomly with the locations and calling patterns of terminals. In CDMA, the reuse factor is also a random variable that reflects the average interference received from surrounding cells. In CDMA systems and systems with dynamic channel allocation, the reuse factor in any cell depends on activity in surrounding cells. When surrounding cells have a small number of calls in progress, the reference cell

---

Spectrum Efficiency

$E = C/N$ conversations/cell/MHz

C = compression efficiency conversations/MHz (one-cell system)

N = reuse factor

N increases with required signal-to-interference ratio $(S/I)_{req}$

low $(S/I)_{req}$ implies high tolerance to interference
and high efficiency

Figure 9.7 Spectrum efficiency depends on compression efficiency and tolerance of interference.

can operate with a high fraction of the C conversations/MHz that would be possible in a single-cell system. This implies a low reuse factor N. Conversely, with surrounding cells active, the reference cell operates with a relatively high reuse factor.

9.3.2 Reuse Planning

A book on personal communications would not be complete without pictures of hexagons. These pictures arise from channel plans, which are important tools for radio resource management in cellular FDMA and TDMA systems. A *channel plan* is a method of assigning channels to cells in a way that guarantees a minimum reuse distance between cells using the same channel. The *reuse distance*, a dimensionless quantity, is the ratio D/R, of two distances. The denominator (R meters) is the radius of a cell. It is the distance from the center of the cell to the furthest point in the cell from the center. The numerator (D meters) is the distance between a base station and the nearest base station that uses the same channels. This is illustrated in Figure 9.8, which shows a set of hexagonal cells. Cells using the same channels have the same shading. A signal's tolerance of interference is expressed by the required signal-to-interference ratio $(S/I)_{req}$. This requirement implies that $D/R > (D/R)_{req}$, which is the minimum reuse distance for which $S/I > (S/I)_{req}$. In order to guarantee that all cells with the same shading have $D/R > (D/R)_{req}$, it is necessary to divide the available channels into N separate groups, where N is the reuse factor described in Figure 9.7. The geometry of hexagons [MacDonald, 1979] implies that

$$N \geq \frac{1}{3}\left(\frac{D}{R}\right)^2_{req}. \tag{9.5}$$

A radio propagation model and the ratio D/R together determine the signal-to-interference ratio, S/I, encountered at terminals and base stations. S/I is an increasing function of the reuse factor, N, in Figure 9.7. The purpose of the frequency plan is to ensure that with high probability the actual signal-to-interference ratio equals or exceeds $(S/I)_{req}$. This is achieved by selecting a sufficiently high value of reuse factor N and dividing the available channels into N groups. Thus each cell has access to the fraction $1/N$ of the available traffic channels.

Cellular systems based on FDMA and TDMA employ a wide range of reuse factors, depending on radio transmission techniques, radio propagation conditions, and antenna configurations. Practical values of N range

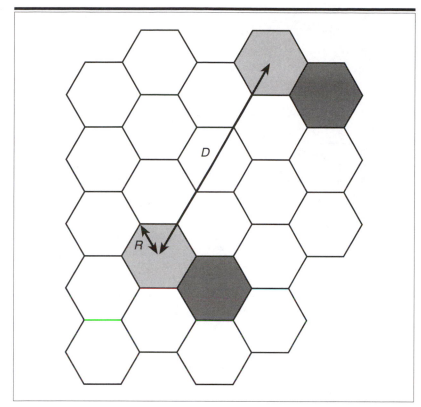

Figure 9.8 Definitions of D and R in a hexagonal cell frequency plan.

from 3 ($D/R = 3$ cell radii) to 21 ($D/R = 7.9$). Figure 9.9 shows a channel plan with $N = 7$ reuse ($D/R = 4.6$), the factor most frequently used in analog cellular systems. The total number of channels is divided into seven groups, numbered 1 through 7. The number in each hexagonal cell indicates the group of channels assigned to that cell.

An important property of a frequency plan is the fact that it represents a compromise between the design goals of high spectrum efficiency, obtained with a low reuse factor, N, and high transmission quality, obtained with a high value of N. The frequency plan, therefore, has a critical influence on system quality and economics. The discussion here is the most elementary possible introduction to the subject. Readers with an interest in more details are encouraged to study the extensive technical literature on the subject [Lee, 1989; Rappaport, 1996].

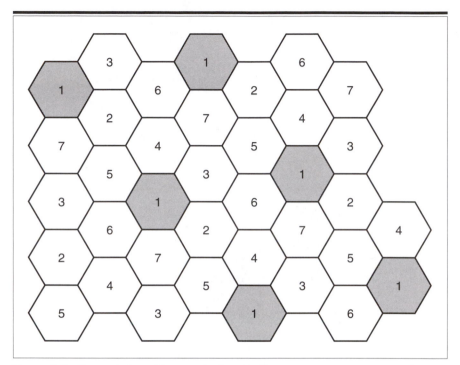

Figure 9.9 Channel plan with reuse factor, $N = 7$.

9.4 Channel Coding

Channel codes [Wicker, 1995] protect information signals against the effects of interference and fading. In doing so, they decrease the required signal-to-interference ratio $(S/I)_{req}$ and they decrease the reuse factor N (Figure 9.7). However, they also increase the bandwidth occupied by a signal and decrease the compression efficiency C. Usually, the net effect is to increase the overall spectrum efficiency $E = C/N$ of a system [Goodman and Sundberg, 1985]. An important property of a channel code is the code rate

$$r = \frac{\text{information rate}}{\text{transmission rate}} \leq 1. \tag{9.6}$$

A low value of r introduces strong protection against transmission impairments. However, it also implies a high channel bandwidth. Each channel coder performs a set of logical operations on the input bit sequence in order to produce an output sequence with a higher bit rate. Channel codes in personal communications systems come in two categories: block codes, described in Section 9.4.1, and convolutional codes, described in Section 9.4.2.

Channel codes can serve two purposes: error detection and forward error correction. When a system uses a code for error detection, the decoder examines a set of bits coming from the demodulator and decides whether all of the bits in the set have been received correctly. When the decoder determines that there are errors in the set of received bits, it discards them. Depending on system design, this can produce a variety of effects. A common effect is to stimulate the transmitter to send the affected data again, thereby slowing down the average rate of information transfer while promoting high accuracy of received information (see Table 2.1). This is a common procedure in the transfer of numerical data, such as telephone numbers and network control messages.

When a system uses a code for forward error correction, the decoder processes the received signal in order to estimate the transmitted bit stream, even when there are errors in the received signal. With forward error correction, the output rate is constant, but signal quality can be poor when transmission impairments are high. Personal communications systems use forward error correction on speech signals that require prompt delivery. For numerical data, many systems apply two codes: an error-correcting code to overcome a moderate level of channel impairment, followed by an error-detecting code to prompt corrective action when the impairments are too strong for the error-correcting code to correct all errors.

9.4.1 Block Codes

A binary block coder divides an information stream into blocks of k bits, referred to as information code words. For each information code word, the encoder computes, from the k information bits, a channel code word of length $n > k$ bits. The ability of the code to protect the information against transmission errors is indicated by d_{min}, the minimum distance between all pairs of code words. The distance between two code words is the number of bits that must be inverted (0s changed to 1s or vice versa) in order to change one code word to the other. The code rate $r = k/n$ indicates the proportion of the channel bandwidth that carries user information. The remaining fraction, $1 - r$, protects user information against the effects of transmission errors. In general, a high degree of protection, reflected in a high d_{min}, is achieved at the cost of a low r.

With k information bits in a code, there are 2^k possible code words. A block code transmits each code word as a sequence of n bits. Allowing for the possibility of transmission errors, the number of possible received code words is $2^n > 2^k$. When a decoder receives a code word that is not among the set of valid transmitted code words, it has two alternatives. It can either generate the nearest valid code word, thus attempting to correct the transmission errors in the block, or it can discard the received signal and let the system take further action. For example, the system can request retransmission of the missing information. With a minimum distance d_{min}, a decoder can attempt to correct all patterns with up to $(d_{min} - 1)/2$ errors in the block. If it tries to correct c errors, it is capable of detecting all combinations of up to $d_{min} - c - 1$ errors. For example, when $d_{min} = 5$, there are three possible decoder actions:

- The decoder can correct no errors ($c = 0$) and detect all combinations of up to four errors. In this case, code word errors occur only when 5 or more bits in a block are in error. However, any combination of up to four errors causes the block to be discarded.

- It can correct one error ($c = 1$) and detect all combinations of two or three errors. In this case, four or five binary errors in a block cause a code word error, and two or three errors cause the block to be discarded.

- It can correct two errors ($c = 2$), in which case three or more binary errors in a block produce a code word error.

The performance of a block code is measured by the relationship of three probabilities related to the probability of binary transmission errors. The three performance indicators (which add up to one) are the probability of receiving a correct code word, the probability of a detected error, and the probability of an undetected error. The systems studied in this book employ a wide variety of error-detecting and error-correcting block codes to protect information transmitted across the air interface.

One type of error-detecting block code that appears in all of the digital systems covered in this book (see Chapters 5 through 8) is a cyclic redundancy check (CRC). Because they are effective (low probability of detecting an incorrect block), efficient (high code rate, r), and economical to implement, CRC codes are used repeatedly in digital communications systems of all types. To perform CRC encoding with k information bits and n transmitted bits per block, a CRC encoder computes $n - k$ parity check bits per block. The encoding technique is similar to treating each k-bit information block as a binary integer and multiplying that number by a constant binary integer with $n - k$ bits. This constant is referred to as the code generator. Representing the information as X, and the code generator as G, the product is $Y = GX$, which is the n-bit transmitted signal block.

The receiver obtains Y'. In the absence of transmission errors $Y = Y'$. At the receiver, the CRC decoder divides the received signal by the code generator integer to obtain $X' = Y'/G$. If X' is an integer, the decoder assumes that $X = X'$. When transmission errors occur, $Y' \neq Y$ and, for most patterns of errors, X' is not an integer. In this case the decoder reports that an error has occurred and the receiver takes corrective action. Although we describe the coding and decoding operations as analogous to multiplication and division, the operations are performed in practice by simple logic circuits containing binary shift registers.

9.4.2 Convolutional Codes

In contrast to block codes, which segregate input bits into blocks and produce longer blocks of output bits, convolutional codes treat the input as a continuous stream of bits. Each time a new input bit arrives at the coder, the coder produces m new output bits. To do so, it examines the j most recent input bits and performs m different binary logic operations referred to as *convolutions*. Each convolution produces one binary output. When a new information bit arrives, it enters a binary shift register of length j and the encoder obtains m output bits by performing m convolutions on the j bits in the shift register. The code rate is $r = 1/m$ input bits per output bit and the number of bits in each convolution, j, is referred to as the *constraint length* of the convolutional code. The decoder performs forward error correction. To recover the transmitted information, a convolutional decoder performs a complex signal processing procedure referred to as the *Viterbi algorithm*.

The resistance of a convolutional code to channel impairments is an increasing function of m, the number of convolutions per bit. It also increases with the constraint length, j, the number of prior inputs used in each convolution. However, the complexity of the decoder increases rapidly as j increases. The digital cellular systems studied in Chapters 5 through 7 employ convolutional codes with $m = 2$, 3, and 4 (code rates $r = 1/2, 1/3, 1/4$). The codes have constraint lengths $j = 5$ (GSM), $j = 6$ (NA-TDMA), and $j = 9$ (CDMA).

In all of the systems, the convolutional codes work on data sequences of fixed length (typically between 100 and 200 bits). In order to perform convolutions on the final bits in the sequence, a coder adds $j - 1$ extra bits that do not carry any information. They are referred to as *tail bits*.

9.5 Interleaving

Most error-correcting codes are effective only when transmission errors occur randomly in time. Decoding accuracy suffers when errors occur in clusters. Fades in cellular radio signals last for several bit intervals, with the average duration dependent on the velocity of the terminal. To prevent errors from clustering in decoder bit streams, cellular systems introduce interleaving. They systematically scramble (permute) the order of bits generated by a channel coder. Receivers perform the inverse permutation in order to return the bits to the sequence in which they leave the encoder. With the correct sequence restored, the signal goes to the decoder. The effect of the interleaver is to take transmission errors that occur in clusters in the channel and spread them over longer intervals in the decoder bit stream. This process can be illustrated by transmitting the letters:

WHAT_I_TELL_YOU_THREE_TIMES_IS_TRUE   (A)

through a noisy channel. Including spaces, the sequence contains 35 symbols. If they are transmitted through a channel that produces four consecutive errors in the middle (symbols 13–16) of the sequence. The result is

WHAT_I_TELL_YBVOXHREE_TIMES_IS_TRUE   (B)

Alternatively, it is possible to interleave the symbols using a 5 × 7 interleaving matrix. An interleaver writes the original message column-wise into the matrix, to obtain

W	I	L	_	E	E	_	
H	_	_	T	_	S	T	
A	T	Y	H	T	_	R	(C)
T	E	O	R	I	I	U	
_	L	U	E	M	S	E	

The system then transmits the symbols row by row in the sequence

WIL_EE_H__T_STATYHT_RTEORIIU_LUEMSE   (D)

Errors in symbol positions 13–16 result in

WIL_EE_H__T_BVOXYHT_RTEORIIU_LUEMSE   (E)

The errors transform matrix (C) to

W	I	L	_	E	E	_
H	_	_	T	_	B	V
O	X	Y	H	T	_	R
T	E	O	R	I	I	U
_	L	U	E	M	S	E

(F)

Finally the sequence presented to the decoder is obtained by reading matrix (F) column-wise:

WHOT_I_XELL_YOU_THREE_TIMEB_IS_VRUE   (G)

The original message is (A). Without interleaving, errors transform it to (B). Even though there is redundancy in printed English, the clustering of the four errors makes it hard to recover the original message. Due to interleaving, the same errors are dispersed in message (G), making it easier to decode.

In contrast to the built-in redundancy of English text, block and convolutional channels coders introduce redundancy systematically, in a manner that guarantees that isolated errors can be corrected. Interleaving transforms error clusters into isolated errors.

9.6 Adaptive Equalization

Adaptive equalizers are important components of data transmission equipment such as telephone modems. An adaptive equalizer operates in two modes: a *training mode,* followed by a *tracking mode.* In the training mode, the modem transmits a signal, referred to as a *training sequence,* that is known to the receiver. The receiving modem processes the distorted version of the training sequence to obtain a channel estimate. Then, in the tracking mode, the equalizer uses this channel estimate to compensate for distortions in the unknown information sequence. It also tracks changes in the channel by making small changes to the channel estimate as it processes the unknown information signal. In a telephone modem

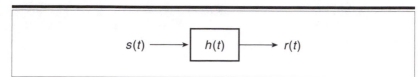

Figure 9.10 Equalizer time functions.

session, the channel changes slowly over the course of a call and the modem does all of the training at the beginning of the call.

In personal communications systems based on TDMA, equalizers suppress intersymbol interference due to multipath propagation (see Section 9.2.4). With terminals moving at high speed, propagation characteristics can undergo rapid changes in a short period of time. Therefore, TDMA systems perform a new training operation to receive the information in each time slot. The SYNC fields in Figure 5.6 and the TRAIN field in Figure 7.8 contain training sequences for the NA-TDMA and GSM signals, respectively.

An adaptive equalizer works with three time functions shown in Figure 9.10: a transmitted signal, $s(t)$; a received signal, $r(t)$; and a channel impulse response, $h(t)$. The equalizer operates on the principle that given any two of these functions, it is possible to calculate the third one. Owing to the presence of noise and interference, the result of the calculation is an estimate rather than a precise representation. Table 9.1 summarizes equalizer operations in the training and tracking modes. During training, the input signal is the training sequence $s_T(t)$, which the equalizer knows in advance. It observes the received signal $r_T(t)$ and computes an estimate, $h'(t)$, of the unknown channel impulse response $h(t)$. During the tracking interval, the transmitted information signal $s_I(t)$ is unknown at the receiver. The equalizer uses the received signal $r_I(t)$ and the estimated channel impulse response, $h'(t)$, to calculate $s_I'(t)$, an estimate of $s_I(t)$. It then demodulates $s_I'(t)$.

Table 9.1 Summary of Equalizer Operations

	known	measured	unknown	estimated
training	s_T	r_T	h	$h' = fcn(s_T, r_T)$
tracking	h'	r_I	s_I	$s_I' = fcn\,(h', r_I)$

9.7 Linear Prediction Coding

A linear prediction vocoder analyzes a sequence of speech samples in order to derive a representation that corresponds to a model of human speech production. This model has three components, as indicated in Figure 9.11. An *excitation waveform* and a *long-term predictor* represent the action of the lungs and vocal chords. A linear filter, sometimes referred to as a *short-term predictor,* represents the shape of the vocal tract. The linear filter is characterized by a set of linear prediction coefficients and by a gain factor, corresponding to the energy of the speech segment.

The process of deriving the filter coefficients and the gain factor is similar in all linear prediction coders. Coders differ in the number of coefficients and their digital representations. Coders typically derive a new set of linear prediction coefficients every 20 ms. The long-term predictor is represented by a time delay estimate, corresponding to the pitch period of the speech in the analysis block. Typically, a coder derives a separate time delay estimate to represent each 5 ms segment of speech. The process that most distinguishes linear prediction speech coders from one another is the derivation of the excitation waveform. The coders in NA-TDMA and CDMA select excitation waveforms from a code book consisting of a large number of candidate waveforms. The GSM coder uses linear filtering techniques to derive the excitation waveform from the input speech. Vocoders generate excitation waveforms corresponding to 5 ms blocks of speech samples.

To summarize, the task of the encoder is to examine a sequence of speech samples, and derive digital representations of a short-term predictor, a long-term predictor, and an excitation waveform. A transmitter sends this representation to a receiver, where a decoder performs the operations indicated in Figure 9.11.

There are three principal figures of merit of a speech-coding technique: the bit rate, the quality of the reconstructed speech, and the amount of computation necessary to encode the speech. Each technique represents a

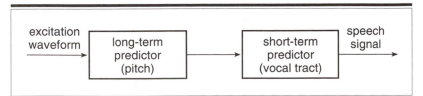

Figure 9.11 Speech production model for linear prediction coder.

tradeoff among these three figures of merit. A long-term trend in speech-processing technology is a rapid increase in available computational power in digital signal processing microprocessors. Each new generation of microprocessors makes it possible to improve speech coders by either decreasing the bit rate or increasing the speech quality, or a combination of the two. This progress in speech coding is apparent in the evolution of digital cellular systems.

To take advantage of this progress, standard organizations periodically publish specifications of new speech coders for specific systems. During a call setup procedure, terminals and base stations exchange information about the speech-coding techniques they are capable of performing. They select the best coder available to both the terminal and base station for use on digital traffic channels.

9.8 CDMA Soft Capacity

This tutorial describes the simplest possible configuration of a direct sequence spread spectrum system in order to explain the basic principles of operation and the nature of the soft capacity of a spread spectrum base station [Viterbi, 1995]. We begin by referring to the digital modulator in Figure 6.3. This modulator combines two binary signals: an information signal with a low binary signaling rate, R b/s, and a digital carrier with a high signaling rate, W ch/s. The signal elements of the information signal are referred to as *bits*. The signal elements of the carrier signal are referred to as *chips*. If we assign values +1 and −1 to the two possible signal levels, the spread spectrum signal in Figure 6.3 is the product of the information signal and the carrier signal. The number of chips per bit is the processing gain

$$G = \frac{W}{R} \text{ ch/b.} \tag{9.7}$$

The heart of the digital demodulator is a correlator (Figure 6.4) that multiplies the received signal by the digital carrier signal and adds the products derived from all of the chips in one bit interval. With +1 transmitted, the sum of the G products is +G. With −1 transmitted, the sum is −G. This is the situation with one signal transmitted. In a CDMA system the demodulator processes the combined signals from many transmitters. Consider a system with K transmitters, labeled $k = 0, 1, 2, \ldots, K - 1$. We examine the output of the correlator when transmitter k sends 1 bit, $b_k = \pm 1$. To do so, it transmits the sequence of G chips, $c_{nk}, n = 0, 1, 2, \ldots, G - 1$, if $b_k = 1$. Otherwise it sends $-c_{nk}, n = 0, 1, 2, \ldots, G - 1$.

The output of the demodulator for signal $k = 0$ contains the number $\pm G$ due to the signal arriving from transmitter 0. It also contains interference components due to the signals from all of the other transmitters. For transmitter k, this component is $\pm r_{0k}$, the cross-correlation of the chip sequences for transmitter 0 and transmitter k. Mathematically, we have at the output of the digital demodulator for transmitter 0,

$$R_0 = Gb_0 + \sum_{k=1}^{K-1} b_k r_{0k} = \pm G + I_0 \qquad (9.8)$$

where the cross-correlation of the digital carriers for transmitter k and transmitter 0 is

$$r_{0k} = \sum_{n=1}^{G} c_{0n} c_{kn} \qquad (9.9)$$

and I_0 is the aggregate interference in receiver 0.

The demodulator for transmitter 0 decides that $b_0 = -1$ if $R_0 < 0$. Otherwise, it decides $b_0 = +1$. Equation (9.8) indicates that for $b_0 = +1$ transmitted, an error occurs ($R_0 < 0$) when $I_0 < -G$. When $b = -1$ is transmitted, an error occurs ($R_0 \geq 0$) when $I_0 \geq G$. Therefore, a necessary condition for an error is $|I_0| \geq G$. The probability of error therefore depends on I_0, which in this simplified example is a binomial random variable with mean value 0 and variance $G(K-1)$. It follows that the probability of error is an increasing function of K, the number of transmitters in the system. Figure 9.12 is a graph of the error probability as a function of K when $G = 128$. As K increases, the probability of error approaches 0.5. In CDMA, this increasing probability of error limits the number of simultaneous transmissions and thereby determines spectrum efficiency. Thus, in the graph shown in Figure 9.12, if the probability of error objective is $P_e < 0.001$, the capacity is $K = 14$ transmitters per cell.

This is referred to as a *soft capacity* because it is possible for a system to admit more than 14 transmitters to a cell. For example, if $K = 15$, the binary error probability is 0.0013, slightly higher than the objective. This is in contrast to the situation in an FDMA system, such as AMPS or a TDMA system, where each cell has a fixed number of physical channels that determines its capacity. In CDMA, the number of possible physical channels is virtually unlimited (2^{127} when $G = 128$), and capacity is determined by performance considerations.

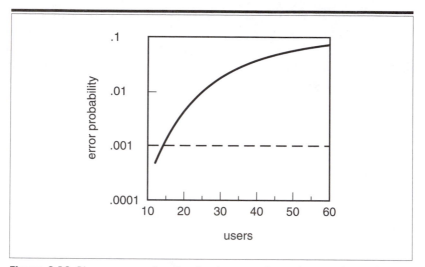

Figure 9.12 Binary error rate of a simple spread spectrum system with 128 chips/bit.

The purpose of this tutorial is to explain, in the simplest possible terms, spread spectrum interference and soft capacity. To analyze a practical system, it is necessary to take into account a large number of phenomena including additive noise, the relative time delays of the signals arriving at a base station, imperfections in power control, and error-correcting codes. As the system model becomes more realistic, the complexity of the analysis increases rapidly. One analysis [Rappaport, 1996: Appendix C] that considers additive noise and relative time delays between arriving signals uses 98 equations to derive graphs corresponding to more accurate versions of Figure 9.12.

When the purpose of the analysis is to estimate spectrum efficiency, the analysis proceeds in three steps:

1. Determine the relationship between the demodulator error rate and E_b/N_0. E_b is the signal energy per bit. In the simple analysis presented here E_b is proportional to G. N_0 is the power spectral density of the noise and interference in the receiver. In the simple example of this section, N_0 is proportional to I_0.

2. Determine that the minimum value of E_b/N_0 meets system error objectives.

3. The capacity estimate for a single cell and perfect power control is

$$K \approx \frac{G}{E_b/N_0} = \frac{W/R}{E_b/N_0} \text{ conversations/base station,} \qquad (9.10)$$

in which the value of E_b/N_0 is the requirement derived in step 2. Equation (9.10) is identical to Equation (6.9), the starting point for the discussion of spectrum efficiency of the cellular CDMA system (Section 6.3.3).

9.9 Walsh Hadamard Matrix

The CDMA system uses a 64×64 Walsh Hadamard matrix in two ways. In reverse direction transmissions, it uses this matrix to perform orthogonal Walsh modulation, which is equivalent to an error-correcting block code with $(n,k;d_{min}) = (6,64;32)$ (see Section 9.4). In the forward direction, each row of the matrix serves as a digital carrier. This brief tutorial introduces the Walsh Hadamard matrix and states the properties that make it suitable for these two roles.

We use the notation W_{64} for the matrix in the CDMA system. It is a member of a general class of Walsh Hadamard matrices W_L of dimension $L \times L$, where L is a power of two. The matrix entries are binary, with values either 0,1 or −1,+1, depending on the properties of the matrix that are being exploited. In this tutorial we will begin with the 0,1 notation, which is convenient for indicating how the matrix is used to produce a block code.

The simplest way to specify W_L is to begin with

$$W_1 = [0] \qquad (9.11)$$

and produce a sequence of matrices each with a dimension two times the previous one. The rule is

$$W_L = \begin{bmatrix} W_{L/2} & W_{L/2} \\ W_{L/2} & W'_{L/2} \end{bmatrix}. \qquad (9.12)$$

In this notation, each element w'_{ij} of $W'_{L/2}$ is the complement of the corresponding element w_{ij} of $W_{L/2}$. That is, $w'_{ij} = 1 - w_{ij}$. Following this rule we go from W_1 to

$$W_2 = \begin{bmatrix} 0 & 0 \\ 0 & 1 \end{bmatrix}, \qquad (9.13)$$

and from W_2 to

$$W_4 = \begin{bmatrix} 0 & 0 & 0 & 0 \\ 0 & 1 & 0 & 1 \\ 0 & 0 & 1 & 1 \\ 0 & 1 & 1 & 0 \end{bmatrix}. \tag{9.14}$$

In each case the first row of the matrix consists entirely of 0s and each of the other rows contains $L/2$ 0s and $L/2$ 1s. Moreover, the distance between any pair of rows is exactly $L/2$. This property indicates how W_{64} is used as a block code. Each sequence of $k = 6$ information bits identifies one row of the matrix. (There are $2^6 = 64$ possible sequences.) The 64 entries in that row are the $n = 64$ code bits of the block. The minimum distance between code words is $64/2 = 32$ bits. In fact all distances are 32 bits.

The other interesting property of W_{64} is that all rows are mutually orthogonal. This is relevant to the use of each row as the digital carrier of a physical channel in the forward direction of a CDMA link. To describe this property, it is most helpful to assign values $-1,+1$ instead of $0,1$ to the matrix entries. Then we find that

$$\sum_{k=0}^{63} w_{ik} w_{jk} = 0 \tag{9.15}$$

for all rows i and j. This demonstrates that multiplying any pair of rows bit by bit and adding the products always results in 0. By specifying the rows of W_{64} to be the physical channels of forward direction transmissions, the CDMA system ensures that there is no interference among signals transmitted by the same base station. With respect to the notation in Section 9.8, the code for user k is

$$c_{kn} = w_{kn}, k = 0, 1, 2, \ldots K - 1. \tag{9.16}$$

Then Equations 9.9 and 9.15 imply that $r_{ij} = 0$ for all $i \neq j$. In this situation, Equation 9.8 reduces to $R_0 = \pm G$, independent of K, the number of transmissions from the base station, which suggests that all the co-channel interference in forward direction CDMA transmissions comes from other cells.

9.10 Open Systems Interconnection

Figure 9.13 is the starting point for discussing many communications networks. It shows the seven-layer model formulated in the early 1980s by the International Standards Organization [Halsall, 1992]. The model simplifies the design of networks by identifying network functions that can be implemented independently of one another. This reduces an extremely complex procedure to a set of simpler, linked procedures. A potential disadvantage of a layered approach is reduced efficiency. Each interface between network layers introduces overheads with the total cost of the result higher than that of the best unified design. However, this "best"

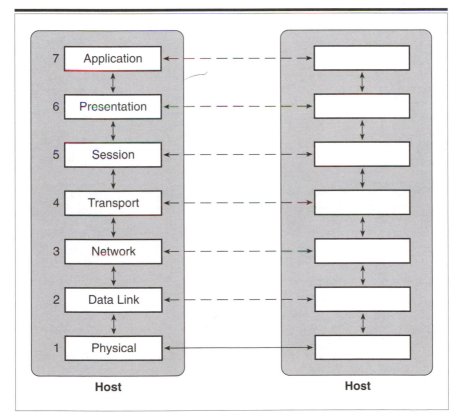

Figure 9.13 Open Systems Interconnection reference model.

overall design is impossible to achieve, and even good approximations to it are expensive to modify. On balance, the advantages of a layered network usually compensate for inefficiencies.

The OSI model originally applied to computer communications, and hence much of the terminology derives from computing. The model has proved so useful that it has been applied to many other networks. Figure 9.13 depicts communication between two *hosts*, machines that serve as the original sources and final destinations of information. Figure 9.14 illustrates the movement of information through networks connecting the hosts. In OSI terminology, the communications path consists of a sequence of *nodes*, with *links* connecting them. The links and nodes function only at

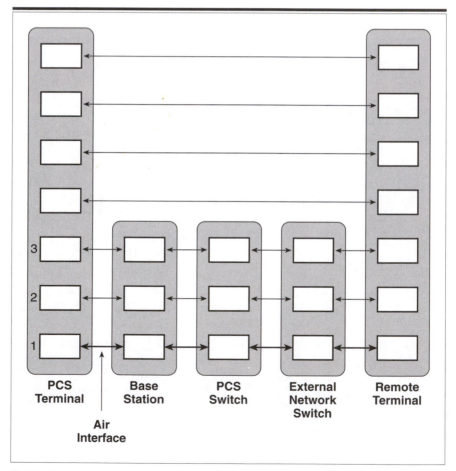

Figure 9.14 Reference model showing intermediate nodes.

the lowest three layers of the model. In a conventional telephone network, the nodes are switching machines. In a computer network, a node can be a special-purpose router or a computer that stores arriving information and forwards it to another computer. Figure 9.14 indicates that in personal communications, the communications system nodes include base stations and switches.

To examine the OSI model, we refer to the 14 boxes in Figure 9.13, seven in each host. Solid lines indicate direct communication between two boxes. Broken lines are virtual communications. The two boxes at the end of a broken line function as if they are interacting directly, even though the information that passes from one to another goes through a sequence of boxes connected by solid lines. In Figure 9.13, a solid line at the physical layer directly connects the two hosts. The physical layer is concerned with the energy transfer necessary to cause digital symbols that exist at one place in the network to appear someplace else. Layer 1 specifies all of the details of the transfer of bits from one place in a network to another.

At the other extreme of the protocol stack is the application layer, which is related to the purpose of the communication, such as electronic mail, home banking, or file transfer. The application is a remote abstraction from the energy transfer at the bottom of the model, and the intermediate layers move step by step from the tangible flow of physical energy to the user's purpose in performing the information exchange.

The four lowest layers have the job of transferring information between hosts. The upper three layers are closer to the user's interaction with a particular information service. In OSI terminology, each layer *provides a service* to the layers above it and *uses* the layers below it. The significance of this terminology is evident at the transport layer. The lowest end-to-end protocol, layer 4 has the job of establishing, maintaining, and finally discontinuing communications between two hosts. This layer also takes responsibility for end-to-end service quality. In the transmitting host, the transport layer accepts and interprets user information from the session layer. It then performs various information processing tasks, such as translating the name of the receiving host to a network address. Next, it adds its own header information including the network address, and submits the aggregate to the network layer for transmission to the receiving host. In the receiving host, the transport layer obtains data from the lower layers, verifies its accuracy, and delivers it to the session layer.

Layers 1–3, as shown in Figure 9.14, function across individual links. As information makes its way from source to destination, layers 1–3 deliver it from one node to the next in a complex network. The principal focus in this book is on the air interface between a terminal and a base station. In studying each system, this book describes three properties of the air interface: radio transmission, logical channels, and messages. They correspond to operations at layers 1, 2, and 3, respectively, of the OSI model.

9.11 Integrated Services Digital Networks

ISDN protocols reflect the efforts of the worldwide telephone industry to transform hundreds of coupled voice communication networks, (referred to, in aggregate, as the PSTN—public switched telephone network) into one global network, that carries many kinds of information [Ramteke, 1994]. At the moment the PSTN carries mostly voice traffic augmented by an increasing amount of other information, such as facsimile and computer data. Non-voice digital information is generally disguised, by modems, to resemble voice signals. ISDN is a set of standards published by the International Telecommunications Union (ITU) for moving digital information of all kinds through the PSTN over transmission links that are designed at the outset to carry digital (rather than analog voice) signals.

The principal purpose of an ISDN is to move a wide variety of user information among information terminals. To do so, ISDN contains connection management protocols for setting up and releasing connections between terminals. Among personal communications systems, GSM (see Chapter 7) and PHS (see Section 8.3) make most extensive use of these protocols. Other important objectives of ISDN are to provide special services that give users control over the network. Examples of services and network control features are calling number identification, alarms, telemetry, call forwarding, conference calls, virtual private networks, teletext, and directory access.

ISDN is also relevant to personal communications systems because standard personal communications protocols are in many respects similar to ISDN protocols. ISDN specifications that have close counterparts in wireless networks are devices, reference points, and logical channels.

Figure 9.15 shows one configuration of ISDN devices and reference points. The subscriber owns or leases the terminal equipment (TE) and network termination equipment (NT2 and NT1), while the switch is the property of the network operator. The ISDN U-reference point is the interface between user and operating company. GSM and PHS both use the terminology U_m (where m denotes mobile) for the air interface between subscriber terminals and base stations.

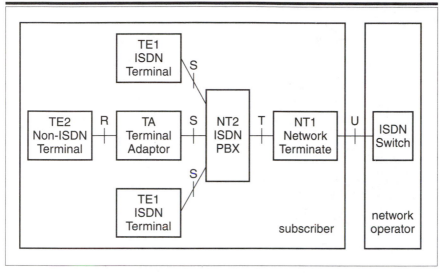

Figure 9.15 ISDN devices and reference points.

The information moving through the reference points is carried on ISDN *logical channels*, including B-(bearer) and D-(data) channels. B-channels carry user information at a rate of 64 kb/s. D-channels, operating at 16 kb/s, primarily carry network control information but are also capable of carrying a low volume of user information. ISDN also specifies higher rate, H, channels, which carry multiples of 64 kb/s. As in personal communications systems, ISDN logical channels are multiplexed onto physical channels.

A standard ISDN terminal (TE1) communicates with other devices by means of an ISDN basic rate multiplex, referred to as 2B + D, consisting of two B-channels and one D-channel. Typically, one B-channel carries telephone conversations and the other B-channel delivers a digital data service such as a connection to a remote computer. The D-channel helps ISDN networks deliver advanced calling features. After the basic rate, the next ISDN multiplex level is the primary rate that unfortunately differs in different parts of the world. In some places (North America), the primary rate multiplex carries 23 B-channels and 1 D-channel, while elsewhere (Europe), the primary rate is 30B + D. As shown in Figure 9.16, primary rate multiplexes connect customer switching equipment to the facilities of an operating company.

Like ISDN, personal communications systems also organize information in logical channels. Owing to the complexity of the network control

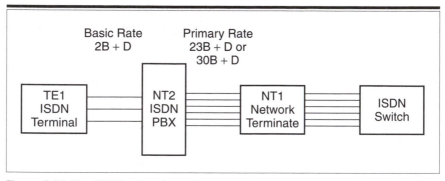

Figure 9.16 The ISDN basic rate multiplex links terminal equipment to customer switching equipment. The primary rate multiplex carries information between customer switching equipment and the public network.

operations of personal communications systems, terminals have access to a large set of logical channels. These channels are in turn multiplexed onto the air interface that connects terminals to base stations.

9.12 Signaling System Number 7

A signaling system moves network control information between network elements [Ramteke, 1994]. When we use a telephone, we have direct experience of some aspects of the signaling system, such as dial tones, audible busy signals, dial pulses in the form of clicks, and multifrequency signaling tones (DTMF) generated in pushbutton phones. For many years, signals such as these, usually audible to callers, were also used to control many internal functions of the telephone network. Since 1980, there has been a big move toward replacing these beeps and clicks and musical tones with a message-oriented signaling system based on digital communication over packet switched signaling channels that are physically distinct from the circuits that carry user information. The culmination of this trend is the adoption of SS7, an international standard with various regional accents, as the worldwide means of public telecommunications network control.

The purpose of SS7 is to move control messages between elements of the network it is controlling. In its specifications, SS7 is a complete communications system, conforming to a layered model, shown in Figure 9.17, that approximately parallels the OSI model. The SS7 layers are called *parts,* and include message transfer parts at the lower layers and user parts at the higher layers. The user parts contain the network control messages that are moved from point to point by the message transfer parts.

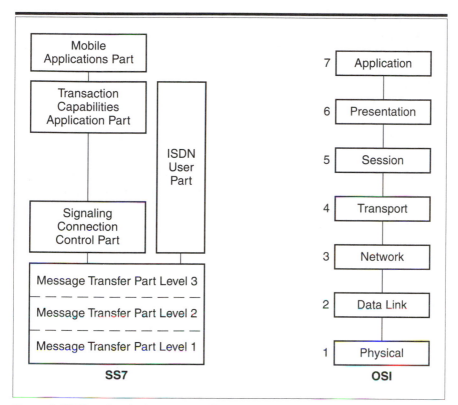

Figure 9.17 Signaling System Number 7 parts and corresponding OSI layers.

There is also a signaling connection control part, corresponding to the OSI network layer (layer 3).

The ISDN user part of SS7 consists of messages carried to and from ISDN standard devices (Section 9.11). Prominent among these are call management messages defined in the ISDN standard Q.931. Other applications, including IS-41 and GSM, use the transactions capabilities application part to perform network control.

SS7 is important to the study of personal communications systems because it controls the fixed networks to which the personal communications systems provide access, and because it participates in the control of the personal communications systems themselves. To perform network control operations unique to personal communications, GSM adds a mobile applications part (MAP) to SS7. IS-41 (see Chapter 4) incorporates a similar, but not identical, MAP. As indicated in Figure 9.17, the MAP makes use of the transactions capabilities applications part.

Physically, SS7 presents itself to a personal communications system in two ways. Some networks use SS7 protocols over dedicated facilities that connect network elements. The communication links that connect GSM base station controllers and mobile switching centers by means of the A-interface is a prominent example. In many other applications, including IS-41 roaming services, SS7 protocols carry information over a separate network that provides signaling services to separate communications systems. Figure 9.18 shows elements of a typical SS7 network. The signal transfer points are SS7 packet switches. A service switching point is an interface between a telecommunications switch, such as a mobile switching center, and an SS7 network. A service control point is a database system. The network shown in Figure 9.18 is designed to maintain operations even when there are failures in network devices or in communications links between devices.

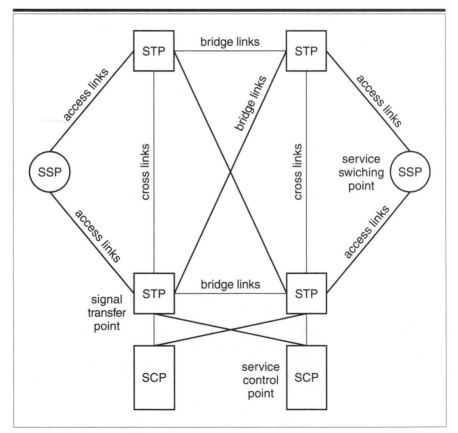

Figure 9.18 Typical configuration of an SS7 network.

Review Exercises

1. Describe some of the factors that influence the shape of the rectangles (carrier spacing × time-slot duration) in a hybrid FDMA/TDMA system (Figure 9.3). What are the advantages and disadvantages of:

   • a large carrier spacing?

   • a large time-slot duration?

2. How does the value of the propagation exponent α in Section 9.2.1 influence the figures of merit of personal communications systems? Under what circumstances does a system benefit from a high value of α? When is a low value beneficial?

3. Section 9.2.5 lists several signal processing techniques for overcoming the effects of interference and the signal distortions described in Sections 9.2.1 through 9.2.4. Why are these techniques more prominent in high-tier systems than in low-tier systems?

4. Explain how cellular systems use radio resources management techniques to combat the signal distortions described in Sections 9.2.1 through 9.2.4?

5. Explain how personal communications systems use error-detecting codes and error-correcting codes to improve performance. In what circumstances are error-detecting codes more effective than error-correcting codes? In what circumstances are error-correcting codes more effective than error-detecting codes?

6. How does the code rate r of a channel code influence compression efficiency C and tolerance of interference $(S/I)_{req}$ in personal communications systems?

7. How do the dimensions (width and height) of an interleaving matrix influence its ability to combat channel impairments? What limits the total size of the matrix (width × height)?

8. Section 9.7 describes three figures of merit of a speech coding technique: bit rate, speech quality, and complexity. How do these figures of merit influence the figures of merit of a personal communications system described in Section 2.3?

9. How can soft capacity benefit a personal communications system? Is it possible for a TDMA or FDMA system to operate with soft capacity? Explain your answer.

10. Why is it impractical for a CDMA system to use orthogonal carriers, such as the Walsh Hadamard codes described in Section 9.9, in the reverse direction?

APPENDIX A

Personal Communications Symbols and Abbreviations

AC	alerting channel
AC	authentication center
ACCH	associated control channel
ACELP	algebraic code excited linear prediction
ADPCM	adaptive differential pulse code modulation
AGCH	access grant channel
ALT	automatic link transfer
AMPS	Advanced Mobile Phone System
B-ISDN	broadband integrated services digital network
b/s	bits per second
BCCH	broadcast control channel
BCH code	Bose-Chaudhuri-Hocquenghem block code
BER	binary error rate
BMI	base station, mobile switching center, and interworking function
BSC	base site controller
BSMC	base station manufacturer code
BSS	base station system
BTS	base transceiver station
CAVE	cellular authentication and voice encryption
CC	call control
CCH	control channel
CCIC	co-channel interference control
CDMA	code division multiple access
CELP	code excited linear prediction
CEPT	Conference of European Postal and Telecommunications administrations

CFP	cordless fixed part
ch	chip
CPP	cordless portable part
CRC	cyclic redundancy check
CS	cell station
CSS	cellular subscriber station
CT2	cordless telephone, second generation
CT2Plus	CT2, enhanced version
dB	decibels
dBm	decibels relative to 1 mW
DCC	digital color code
DCCH	digital control channel
DCS 1800	digital cellular system at 1,800 MHz
DCS 1900	digital cellular system at 1,900 MHz
DECT	Digital European Cordless Telecommunications
DL	digital control channel locator
DTCH	digital traffic channel
DTMF	dual-tone multiple frequency
DTX	discontinuous transmission
DVCC	digital verification color code
E-BCCH	extended broadcast control channel
EIR	equipment identity register
ESN	electronic serial number
ETSI	European Telecommunications Standards Institute
F-BCCH	fast broadcast control channel
FACCH	fast associated control channel
FCC	Federal Communications Commission
FCCH	frequency correction control channel
FDMA	frequency division muliple access
FM	frequency modulation
FOCC	forward control channel
FSK	frequency shift keying
FT	fixed termination
FVC	forward voice channel
GEO	geostationary orbit satellite system
GMSK	Gaussian minimum shift keying
GPS	global positioning system
GSM	Global System for Mobile Communications
HLR	home location register
IMEI	international mobile equipment identity
IMSI	international mobile subscriber identity

IS	Interim Standard
ISDN	Integrated Services Digital Network
ISPBX	integrated services private branch exchange
ITU	International Telecommunications Union
IWU	interworking unit
kb/s	kilobits per second
kHz	kilohertz
Ki	secret encryption key in a GSM system
LAN	local area network
LEO	low Earth orbit satellite system
LPC - RPE	linear prediction coding with regular pulse excitation
MACA	mobile-assisted channel allocation
MAHO	mobile-assisted handoff
MAP	mobile applications part
Mb/s	megabits per second
MC-F	fast message channel
MC-S	slow message channel
MEO	medium Earth orbit satellite system
MHz	megahertz
MIN	mobile identification number
MM	mobility management
MSC	mobile switching center
MTSO	mobile telephone switching office
mW	milliwatt
NA-TDMA	North American Time Division Multiple Access
NAMPS	Narrowband Advanced Mobile Phone Service
NMT	Nordic Mobile Telephone
NT	network termination
OA&M	operations administration and maintenance
OACSU	off air call setup
OQPSK	offset quadrature phase shift keying
OSI	Open Systems Interconnection
PACS	Personal Access Communications System
PBX	private branch exchange
PCC	power control channel
PCS	personal communications services
PDC	Pacific digital cellular
PHS	Personal Handyphone System
PIM	personal information machine
PRC	priority request channel
PS	personal station

PSPDN	packet switched public data network
PSTN	Public Switched Telephone Network
PT	portable termination
QCELP	Qualcomm code excited linear prediction
RACH	random access channel
RECC	reverse control channel
RP	radio port
RPCU	radio port control unit
RRM	radio resources management
RSSI	received signal strength indication
RT	radio frequency transmission management
RVC	reverse voice channel
S-BCCH	short messsage service broadcast control channel
SACCH	slow associated control channel
SAPI	service access point identifier
SAT	supervisory audio tone
SBC	system broadcast channel
SCCH	signaling control channel
SCF	shared channel feedback
SCH	synchronization control channel
SCM	station class mark
SDC	synchronous directive channel
SDCCH	stand-alone dedicated control channel
SFP	superframe phase
SIC	system information channel
SID	system identifier
SIM	subscriber identity module
SOC	system operator code
SPACH	short message service, paging, and access response channel
SS7	signaling system number 7
SSD	shared secret data
ST	supervisory tone
SU	subscriber unit
TC	traffic channel
TCAP	transactions capabilities applications part
TCH	traffic channel
TCH/F	full-rate traffic channel
TCH/H	half-rate traffic channel
TDMA	time division multiple access
TE	terminal equipment

TETRA	trans-European trunked radio system
TIA	Telecommunications Industry Association
TMSI	temporary mobile subscriber identity
UIC	user information channel
U_m	interface between terminal and base station
USCCH	user-specific packet control channel
USPCH	user-specific packet channel
VLR	visitor location register
VMAC	mobile attenuation code for voice signals
VOX	voice operated transmission
VSELP	vector sum excited linear prediction
W	watt
WACS	wide area communications system
WEI	word error indicator
μs	microsecond

APPENDIX B

High-Tier Systems

	AMPS	NA-TDMA	CDMA	GSM
Frequency bands (MHz)	824–849 869–894	824–849 869–894	824–849 869–894	890–915 935–960
		1,850–1,910 1,930–1,990	1,850–1,910 1,930–1,990	1,710–1,785 1,805–1,885
				1,850–1,910 1,930–1,990
Carrier spacing (kHz)	30	30	1,230	200
Physical channels per carrier	1	3	soft capacity	8
Access	FDMA	FDMA/ TDMA	CDMA	FDMA/ TDMA
Duplex	frequency	frequency	frequency	frequency
Frame duration (ms)	n/a	40	20	4.6
Modulation method	FM	DQPSK	PSK	GMSK
Channel rate (kb/s)	10	48.6	1,228,800	270.833
Modulation efficiency (b/s/Hz)	0.33	1.6	1.0	1.4
Maximum terminal transmitter power (W)	4	4	6.3	8

(continued)

	AMPS	NA-TDMA	CDMA	GSM
Speech coding	analog FM	VSELP	QCELP	RPE
Speech rate, including channel coding (kb/s)	n/a	13	variable	22.8
Cellular efficiency (conversations/ cell/MHz)	2.3	7.0	12.1–45.1	5.0–6.6
Associated control channels	blank-and-burst	in each slot	dim-and-burst*	separate slots in each frame
Out-of-band ACCH rate (b/s)	n/a	600	4,400–7,600*	950

* Speech quality reduced when control channel is active.

APPENDIX C

Low-Tier Systems

	CT2	DECT	PHS	PACS
Frequency bands (MHz)	864–868	1,880–1,900	1,895–1,918	1,850–1,910 1,930–1,990
Carrier spacing (kHz)	100	1,728	300	300
Physical channels per carrier	1	12	4	8
Access	FDMA	FDMA/ TDMA	FDMA/ TDMA	FDMA/ TDMA
Duplex	time	time	time	frequency
Frame duration (ms)	2	10	5	2.5
Modulation method	GFSK	GMSK	DQPSK	DQPSK
Channel rate (kb/s)	72	1,152	384	384
Modulation efficiency (b/s/Hz)	0.72	0.67	1.3	1.3
Maximum terminal transmitter power (mW)	10	240	10	200
Speech coding	ADPCM	ADPCM	ADPCM	ADPCM
Speech rate, including channel coding (kb/s)	32	32	32	32
Associated control channels	in each slot	in each slot	in each slot	in each slot
Out-of-band ACCH rate (b/s)	2,000	0–3,200	3,200	4,000

Glossary of Personal Communications Terms

A-key encryption key used in North American cellular systems

adaptive differential pulse code modulation a 32 kb/s speech-coding technique used in low-tier systems

adjacent channel interference impairment to a signal transmitted at one carrier frequency due to another signal transmitted at a nearby carrier frequency

Advanced Mobile Phone System U.S. standard for analog cellular communications

air interface set of techniques for transferring information between base stations and terminals

alerting channel a common control channel that carries paging messages in the PACS system

algebraic code excited linear prediction advanced speech-coding technique used in NA-TDMA and GSM

associated control channel a control channel that uses the same physical channel as a traffic channel

authentication center a secure database for storing encryption keys

automatic link transfer handoff procedure in the PACS system

blank-and-burst interruption of user information in order to transmit system control information

burst the signal transmitted in one time slot in a TDMA system

cell station base station in a PHS system

cellular authentication and voice encryption a computational procedure used to obtain encryption keys

chip transmitted binary signal element in a spread spectrum system

co-channel interference impairment to a signal caused by another signal transmitted at the same carrier frequency

common control channel a control channel shared by all of the terminals in a cell (see dedicated control channel)

control channel a logical channel that carries network control information

cordless fixed part base station in a CT2 cordless telephone system

cordless portable part wireless terminal in a CT2 cordless telephone system

cyclic redundancy check a type of error-detecting block code used in many data transmission systems

dedicated control channel a control channel dedicated to one communication session (see common control channel)

digital cellular system at 1,800 MHz GSM system operating in European 1,800 MHz personal communications band

digital cellular system at 1,900 MHz GSM system operating in North American 1,900 MHz personal communications band

digital color code a 2-bit identifier assigned to each base station in the AMPS system

discontinuous transmission an operating mode in which a terminal reduces its transmitter power when no input speech is detected

downlink from a base station to a terminal (*forward direction* is a synonym)

fixed termination the radio part of a DECT base station

forward direction from a base station to a terminal (*downlink* is a synonym)

geostationary orbit satellite system a communications system with satellites in stationary orbits 35,800 km above the Earth

handoff procedure for transferring a communication from one base station to another base station

handover synonym for handoff

international mobile subscriber identification a telephone number that conforms to an international standard format (E.212)

intersystem handoff handoff procedure in which the two base stations are connected to different switches

intracell handoff procedure for transferring a communication from one channel to another channel at the same base station

linear prediction coding with regular pulse excitation the speech-coding technique used in GSM

local loop the communication link between customer premises and telephone company equipment

low Earth orbit satellite system a communications system with satellites in orbits 500–2,000 km above the Earth

medium Earth orbit satellite system a communications system with satellites in orbits approximately 10,000 km above the Earth

mobile applications part a set of Signaling System Number 7 protocols used to control mobile communications networks

mobile identification number telephone number of a cellular subscriber

mobile station control order AMPS control message sent to a specific terminal

modulation efficiency number of transmitted bits per second for each Hz of bandwidth

multipath propagation a radio transmission phenomenon that causes replicas of the same signal to arrive at a receiver at different times

North American time division multiple access dual-mode cellular system specified in TIA Interim Standard 136

off-air call setup a call setup procedure that assigns a traffic channel after the called party accepts a call

overhead message a control message broadcast to all AMPS terminals in a cell

personal information machine a communication device carried by a person

personal station wireless terminal in a PHS system

portable termination the radio part of a DECT terminal

private branch exchange a business telephone system owned by a private company and connected to a public telephone system

processing gain the ratio of transmitted chip rate to source bit rate in a spread spectrum system

Qualcomm code excited linear prediction speech coding technique used in CDMA

radio port the radio part of a base station in the PACS system

radio port control unit the controller of a base station in the PACS system

RAKE receiver spread spectrum demodulator with several correlators, each synchronized to a different multipath component of the received signal

received signal strength indication received power measured at a terminal or a base station

reverse direction from a terminal to a base station (*uplink* is a synonym)

roaming   using a personal communications service outside one's home service area

sleep mode   an operating condition in which a terminal turns off its receiver for a substantial fraction of the time

soft handoff   handoff procedure in which a communication uses two base stations simultaneously

station class mark   a code that identifies the capabilities of a wireless terminal

subscriber identity module   an electronic card that can be connected to a GSM terminal. The SIM contains information about a telephone subscription

subscriber unit   a wireless terminal in the PACS system

supervisory audio tone   a sine wave transmitted on AMPS traffic channels at 5,070 Hz, 6,000 Hz, or 6,030 Hz

supervisory tone   a 10 kHz sine wave transmitted on AMPS traffic channels

system broadcast channel   a physical channel that carries network control information in the PACS system

system identifier   a code assigned by the U.S. government to identify a cellular service area

system information channel   a common control channel in the PACS system

telepoint   wireless public telephone system

traffic channel   a logical channel that carries user information

trans-European trunked radio system   European standard for a wireless data communications system in frequency band 380–393 MHz

uplink   from a terminal to a base station (*reverse direction* is a synonym)

vector sum excited linear prediction   speech coding technique used in NA-TDMA

voice-operated transmission   a technique for reducing transmitter power when there is no speech input to a terminal

REFERENCES AND BIBLIOGRAPHY

Abrishamkar, F., and Siveski, Z. (1996). "PCS Global Mobile Satellites," *IEEE Communications* **34** (9), 132–136.

"Advanced Mobile Phone Service," *Bell System Technical Journal* **58** (1), 1–269.

American National Standards Institute. (1996a). "Personal Access Communications Systems Unlicensed (Version A) Air Interface Standard Supplement," T1/TIA Joint Standard ANSI J-STD-014a-1997.

American National Standards Institute. (1996b). "Personal Access Communications Systems Unlicensed (Version B) Air Interface Standard Supplement," T1/TIA Joint Standard ANSI J-STD-014b-1996.

American National Standards Institute. (1996c). "Personal Access Communications Systems Air Interface Standard," T1/TIA Joint Standard, ANSI J-STD-014-1996.

Badrinath, B., and Imielinski, T. (1994). "Wireless Mobile Computing: Challenges in Data Management," *Communications of the ACM,* October: 19–27.

Bell Communications Research. (1994). *Generic Criteria for Version 0.1 Wireless Access Communications Systems (WACS), TR-INS-001313 Revision 1.*

Buckingham, C., Wolterink, G. K., and Åkerberg, D. (1991). "A Business Cordless PABX Telephone System on 800 MHz Based on the DECT Technology," *IEEE Communications* **29** (1), 105–110.

CDPD Forum. (1995). *Cellular Digital Packet Data System, Release 1.1.*

Clark, G. C., and Cain, J. B. (1981). *Error Correction Coding for Digital Communications,* Plenum Press, New York.

Cook, C. I. (1994). "Development of Air Interface Standards for PCS," *IEEE Personal Communications* **1** (4), 30–34.

Cox, D. C. (1995). "Wireless Personal Communications: What Is It?" *IEEE Personal Communications* **2** (2), 20–35.

Department of Trade and Industry. (1987). *Performance Specification: Radio Equipment for Use at Fixed and Portable Stations in the Cordless Telephone Service,* MPT 1334, April.

Department of Trade and Industry. (1989a). "Phones on the Move," January.

Department of Trade and Industry. (1989b). "Common Air Interface Specification," MPT 1375, May.

Electronic Industries Association. (1989). *Mobile Station Land Station Compatibility Specification, ANSI/EIA/TIA Standard 553.*

European Telecommunications Standards Institute. (1991a). *Common Air Interface Standard to be Used for the Interworking Between Cordless Telephone Apparatus in the Frequency Band 864.1 MHz, ETSI.*

European Telecommunications Standards Institute. (1991b). *Digital European Cordless Telephone Common Air Interface, ETSI.*

Federal Communications Commission. (1990a). "Special Provisions for Alternative Cellular Technologies and Alternative Services," *Memorandum Opinion and Order Docket 87-390,* February: 3.

Federal Communications Commission. (1990b). "In the Matter of Amendment of the Commission's Rules to Establish New Personal Communications Services," Docket 90-314, June.

Feher, K. (1991). "Modems for Emerging Digital Cellular Mobile Radio Systems," *IEEE Trans. on Vehicular Technology* **40** (2), 355–365.

Fukasawa, A., Sato, T., Takizawa, Y., Kato, T., Kawabe, M., and Fisher, R. (1997). "W-CDMA System for Personal Radio Communications," *IEEE Communications* **34** (11), 116–123.

Garg, V. K., Smolik, K., and Wilkes, J. E. (1997). *Applications of CDMA in Wideband/ Personal Communications,* Prentice Hall, Upper Saddle River, New Jersey.

Goodman, D. J., and Saleh, A. A. M. (1987). "The Near/Far Effect in Local ALOHA Radio Communications," *IEEE Trans. on Vehicular Technology* **36** (1), 19–27.

Goodman, D. J., and Sundberg, C. E. (1995). "The Effect of Channel Coding on the Efficiency of Cellular Mobile Radio Systems," *IEEE Trans. on Communications* **33** (3), 288–291.

Halsall, F. (1992). *Data Communications, Computer Networks and Open Systems,* 3rd edition, Addison-Wesley, Reading, Massachusetts.

Holtzman, J. M. (1991). "CDMA Power Control for Wireless Networks," *Third-Generation Wireless Information Networks,* S. Nanda and D. J. Goodman, eds., Kluwer Academic Publishers, Boston: 299–311.

Institute of Electrical and Electronic Engineers. (1996). "Wireless LAN Media Access Control and Physical Layer Specification," *IEEE 802.11.*

Jakes, W. C., ed. (1974). *Microwave Mobile Communications,* Wiley, New York.

Khan, M., and Kilpatrick, J. (1995). "Mobitex and Mobile Data Standards," *IEEE Communications* **33** (3), 96–101.

Kinoshita, K., Kuramoto, M., and Nakajima, N. (1991). "Development of a TDMA Digital Cellular System Based on Japanese Standard," Proceedings of the 41st IEEE Vehicular Technology Conference, 642–645.

LaMaire, R. O., Krishna, A., Bhagwat, P., and Panian, J. (1996). "Wireless LANs and Mobile Networking: Standards and Future Directions," *IEEE Communications* **34** (8), 86–94.

Landolsi, M. A., Veeravalli, V. V., and Jain, N. (1996). "New Results on the Reverse Link Capacity of CDMA Cellular Networks," Proceedings of the 46th IEEE Vehicular Technology Conference, 1462–1466.

Lee, W. C. Y. (1989). *Mobile Cellular Telecommunications Systems.* McGraw-Hill, New York.

MacDonald, V. H. (1979). "The Cellular Concept," *Bell Sys. Tech. J.* **58** (1), 15–43.

Madhavapeddy, S., Basu, K., and Roberts, A. (1995). "Adaptive Paging Algorithms for Cellular Systems," Proceedings of the Fifth WINLAB Workshop, 347–361.

Mehrotra, A. (1994). *Cellular Radio: Analog and Digital Systems.* Artech House, Norwood, Massachusetts.

Mouly, M., and Pautet, M.-B. (1992). *The GSM System for Mobile Communications,* Palaiseau, France.

Mouly, M., and Pautet, M.-B. (1995). "Current Evolution of the GSM Systems," *IEEE Personal Communications* **2** (5), 9–19.

Noerpel, A. R. (1996). "PACS: Personal Access Communications System: An Alternative Technology for PCS," *IEEE Communications* **34** (10), 138–150.

Pahlavan, K., and Levesque, A. H. (1995). *Wireless Information Networks,* Wiley, New York.

Radio Advisory Board of Canada. (1990). "CT2Plus," Issue 2.0, December.

Ramteke, T. (1994). *Networks,* Prentice Hall, Englewood Cliffs, New Jersey.

Rappaport, T. S. (1996). *Wireless Communications Principles and Practice.* Prentice Hall, Upper Saddle River, New Jersey.

Raychaudhuri, D. (1996). "Wireless ATM Networks: Architecture, System Design and Prototyping," *IEEE Personal Communications* **3** (4), 42–49.

Research and Development Center for Radio Systems. (1993). *Personal Handy Phone System,* RCR STD-28, December.

Sheng, S., Chandrakasan, A., and Brodersen, R. W. (1992). "A Portable Multimedia Terminal," *IEEE Communications* **30** (12), 64–75.

Steele, R. ed. (1992). *Mobile Radio Communications,* Pentech Press, London.

Steer, D. G. (1994). "Coexistence and Access Etiquette in the United States Unlicensed PCS Band," *IEEE Personal Communications* **2** (3), 36–43.

Tannenbaum, A. S. (1988). *Computer Networks,* 2nd edition, Prentice Hall, Englewood Cliffs, New Jersey.

Telecommunications Industries Association. (1991). "Cellular Radio-Telecommunications Intersystem Operations," *Interim Standard 41B,* December.

Telecommunications Industries Association. (1992) "Cellular System Dual-Mode Mobile Station-Base Station Compatibility Standard," *Interim Standard 54B,* December.

Telecommunications Industries Association. (1993a). "Mobile Station Land Station Compatibility Specification for Dual-Mode Narrowband Analog Cellular Technology," *Interim Standard 88,* January.

Telecommunications Industries Association. (1993b). "Mobile-Station Base Station Compatibility Standard for Dual-Mode Wideband Spread Spectrum Cellular System," *Interim Standard 95,* July.

Telecommunications Industries Association. (1995a). "Speech Service Option Standard for Wideband Spread Spectrum Digital Cellular System," *Interim Standard 96A,* April.

Telecommunications Industries Association. (1995b). "Data Service Option Standard for Wideband Spread Spectrum Digital Cellular System," *Interim Standard 99,* July.

Telecommunications Industries Association. (1995c). "800 MHz Cellular Systems TDMA Radio Interface Radio Link Protocol 1," *Interim Standard 130,* April.

Telecommunications Industries Association. (1995d). "800 MHz Cellular Systems TDMA Services Async Data and Fax," *Interim Standard 135,* April.

Telecommunications Industries Association. (1996a). "TDMA Cellular/PCS Radio Interface Enhanced Full Rate Speech Codec," *Interim Standard 641,* May.

Telecommunications Industries Association. (1996b). "Short Message Services for Wideband Spread Spectrum Cellular Services," *Interim Standard 637,* August.

Telecommunications Industries Association. (1996c). "Enhanced Variable Rate Code, Speech Service Option 3 for Wideband Spread Spectrum Digital Systems," *Interim Standard 127,* September.

Telecommunications Industries Association. (1996d). "800 MHz TDMA Cellular Radio Interface Mobile Station Base Station Compatibility," *Interim Standard 136A,* October.

Tuttlebee, W. H. W., ed. (1990). *Cordless Telecommunications in Europe.* Springer Verlag, London.

Viterbi, A. J. (1995). *CDMA Principles of Spread Spectrum Communication.* Addison-Wesley, Reading, Massachusetts.

Wickelgren, I. J. (1996). "Local-Area Networks Go Wireless," *IEEE Spectrum* **33** (9), 34–40.

Wicker, S. B. (1995). *Error Control Systems for Digital Communication and Storage.* Prentice Hall, Englewood Cliffs, New Jersey.

Wilkinson, T., Phipps, T. G. C., and Barton, S. K. (1995). "A Report on HIPERLAN Standardization," *International J. of Wireless Information Networks* **2** (2), 99–119.

Yu, C. C., Morton, D., White, R. G., Wilkes, J. E., and Ulema, M. (1997). "Low-Tier Wireless Local Loop Radio Systems," *IEEE Communications* **35** (3), 84–98.

Yue, O. (1983). "Spread Spectrum Mobile Radio, 1977–1982," *IEEE Trans. on Vehicular Technology* **32** (1), 98–105.

INDEX

The Addison-Wesley Wireless Communications Series

Dr. Andrew J. Viterbi, Consulting Editor
http://www.awl.com/cseng/wirelessseries/

0-201-63374-4
272 pages
Hardcover

CDMA
Principles of Spread Spectrum Communication
by Andrew J. Viterbi

CDMA, a wireless standard, is a leading technology for relieving spectrum congestion caused by the explosion in popularity of wireless communications. This book presents the fundamentals of CDMA so that you can develop systems, products, and services for this demanding market.

Andrew J. Viterbi is a pioneer of wireless digital communications technology. He is best known as the creator of the digital decoding technique used in direct broadcast satellite television receivers and in wireless cellular telephones, as well as numerous other applications. He is cofounder, Chief Technical Officer, and Vice Chairman of QUALCOMM Incorporated, a developer of mobile satellite and wireless land communication systems employing CDMA technology.

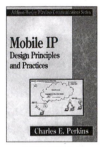

0-201-63469-4
288 pages
Hardcover

Mobile IP
Design Principles and Practices
by Charles E. Perkins

This book introduces the TCP/IP savvy reader to the design and implementation of Internet protocols that are useful for maintaining network connections while moving from place to place. Written by Charles E. Perkins, a leader in the networking field, this book addresses Mobile IP; route optimization; IP version 6; dynamic Host Configuration Protocol (DHCP); encapsulation; and source routing.

Charles Perkins is currently a senior staff engineer at Sun Microsystems. Previously he was associated with the IBM T. J. Watson Research Center. Mr. Perkins is a recognized leader and innovator in the field of mobile computing. He is the document editor for the Mobile IP working group of the Internet Engineering Task Force (IETF). Charles Perkins is also is a co-author of DHCPv6 (version 6) and is an Internet Architecture Board (IAB) member. He is a three-time recipient of IBM's Invention Achievement Award.

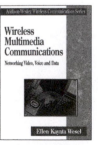

0-201-63394-9
544 pages
Hardcover

Wireless Multimedia Communications
Networking Video, Voice, and Data
by Ellen Kayata Wesel

This book is a comprehensive guide to understanding the design of wireless multimedia communication systems. *Wireless* is synonymous with *mobile*, enabling the computer user to remain connected while moving from one place to another. *Multimedia* denotes a mix of video, voice, and data information—each of which has different transfer requirements. The author has made significant contributions to the theory, design, standardization, and policies of several wireless multimedia systems. The book is comprehensive in scope, and addresses in detail each of the key elements of a complete wireless multimedia system, including propagation characteristics, modulation, intersymbol interface mitigation, coding, medium access protocols, and spectrum and standards networking, while defining their relationship to one another.

Dr. Ellen Kayata Wesel is a Senior Scientist at Hughes Communications, Inc. (HCI), where she designs the next generation of wireless multimedia communication systems. Prior to joining HCI, she designed the physical and data link layers for high data rate wireless LANs at Apple Computer.